ISNM
International Series of
Numerical Mathematics
Vol. 117

Edited by
K.-H. Hoffmann, München
H. D. Mittelmann, Tempe

Mathematical Modelling and Simulation of Electrical Circuits and Semiconductor Devices

**Proceedings of a Conference held at the
Mathematisches Forschungsinstitut, Oberwolfach, July 5-11, 1992**

Edited by

R.E. Bank
R. Bulirsch
H. Gajewski
K. Merten

Springer Basel AG

Editors

R.E. Bank
Dept. of Mathematics
University of California
San Diego
La Jolla, CA 92093
USA

R. Bulirsch
Mathematisches Institut
TH München
Postfach 20 24 20
D-80290 München
Germany

H. Gajewski
Institut für Angewandte
Analysis und Stochastik
Mohrenstr. 39
D-10117 Berlin
Germany

K. Merten
ZTI DES2
Siemens AG
Otto-Hahn-Ring 6
D-81739 München
Germany

A CIP catalogue record for this book is available from the Library of Congress, Washington D.C., USA

Deutsche Bibliothek Cataloging-in-Publication Data
**Mathematical modelling and simulation of electrical circuits
and semiconductor devices** : proceedings of a conference held
at the Mathematisches Forschungsinstitut, Oberwolfach, July 5
– 11, 1992 / ed. by R. E. Bank ... – Basel ; Boston ; Berlin :
Birkhäuser, 1994
   (International series of numerical mathematics ; Vol. 117)
   ISBN 978-3-0348-9665-8   ISBN 978-3-0348-8528-7 (eBook)
   DOI 10.1007/978-3-0348-8528-7
NE: Bank, Randolph E. [Hrsg.]; Mathematisches Forschungsinstitut
   <Oberwolfach>; GT

© 1994 Springer Basel AG
Originally published by Birkhäuser Verlag in 1994
Camera-ready copy prepared by the editors
Printed on acid-free paper produced from chlorine-free pulp
Cover design: Heinz Hiltbrunner, Basel

ISBN 978-3-0348-9665-8

9 8 7 6 5 4 3 2 1

Preface

Progress in today's high-technology industries is strongly associated with the development of new mathematical tools. A typical illustration of this partnership is the mathematical modelling and numerical simulation of electric circuits and semiconductor devices.

At the second Oberwolfach conference devoted to this important and timely field, 35 scientists from around the world, mainly applied mathematicians and electrical engineers from industry and universities, presented their new results.

The contributions to this conference are presented in this proceedings. They cover electric circuit simulation, device simulation and process simulation, including discussions on experiences with standard software packages and improvements of such packages.

In electric circuit simulation three different types of problems can be distinguished depending on the size of a circuit: small circuits with less than 20 basic elements and an oscillating behaviour; middle-sized circuits up to 500 elements; very large circuits. Today the simulation of middle-sized circuits is well understood. Current focal points for oscillating circuits include new discretization schemes, limit cycle computation and the transient phase. Parallel methods and multirate strategies are suggested and tested for the very large circuits.

In the semiconductor area special lectures were given on new modelling approaches, numerical techniques and existence and uniqueness results. Among these, for example, mention is made of mixed finite element methods, an extension of the Baliga-Patankar technique for a three dimensional simulation, and the connection between semiconductor equations and the Boltzmann equations.

The editors are grateful to Georg Denk and Peter Rentrop for their effective and efficient help in organizing the conference.

The editors

Table of Contents

Circuit Simulation

Device Simulation

List of Participants

Dr. G. Albinus
Institut für Angewandte Analysis
und Stochastik
Mohrenstr. 39

10117 Berlin

Dr. R. P. Brinkmann
ZFE BT ACM 31
Siemens AG
Otto-Hahn-Ring 6

81739 München

J.-F. Chabilan
Fachbereich Mathematik
AG 8 Numerische Mathematik
Technische Hochschule
Darmstadt
Schloßgartenstr. 7

64289 Darmstadt

Prof. Dr. P. Deuflhard
Konrad-Zuse-Zentrum für
Informationstechnik Berlin
Heilbronner Str. 10

10711 Berlin

Prof. Dr. R. E. Bank
Dept. of Mathematics, C-012
University of California, San Diego

La Jolla, CA 92093
USA

Prof. Dr. R. Bulirsch
Mathematisches Institut
der TU München

80290 München

Dr. G. Denk
Mathematisches Institut
der TU München

80290 München

Dr. U. Feldmann
ZFE BT SE 43
SIEMENS AG
Otto-Hahn-Ring 6

81739 München

Prof. Dr. H. Gajewski
Institut für Angewandte Analysis
und Stochastik
Mohrenstr. 39

10117 Berlin

Dr. J. Gawriljuk
Warschauer Str. 14/1602

99089 Erfurt

Prof. Dr. F. Grund
Institut für Angewandte Analysis
und Stochastik
Mohrenstr. 39

10117 Berlin

Prof. Dr. Karl-Heinz Hoffmann
Institut für angewandte Mathematik
und Statistik
der TU München

80290 München

Prof. Dr. T. Kerkhoven
Department of Computer Science
University of Illinois at
Urbana-Champaign
1304 W. Springfield Av.

Urbana, IL 61801
USA

Dr. K. Gärtner
IPS
ETH-Zentrum

CH-8092 Zürich

Dr. Albert Gilg
ZFE SPT 3
SIEMENS AG
Otto-Hahn-Ring 6

81739 München

R. Hiptmair
Mathematisches Institut
der TU München

80290 München

Prof. Dr. R. Hoppe
Mathematisches Institut
der TU München

80290 München

Dr. M. Kiehl
Mathematisches Institut
der TU München

80290 München

Prof. Dr. R. März
Institut für Angewandte Mathematik
Fachbereich Mathematik
Humboldt-Universität Berlin
Unter den Linden 6

10099 Berlin

Prof. Dr.-Ing. W. Mathis
Bergische Universität
Gesamthochschule Wuppertal
FB 13 Elektrotechnik
Fuhlrottstr. 10

42119 Wuppertal

Dr. B. Meinerzhagen
Institut für Theoretische
Elektrotechnik
RWTH Aachen
Kopernikusstr. 16

52074 Aachen

Dr. K. Merten
Siemens Corporate Research, Inc.
755 College Road East

Princeton, N. J. 08540
USA

Prof. Dr. H. D. Mittelmann
Department of Mathematics
Arizona State University

Tempe, AZ 85287-1804
USA

F. Montrone
Mathematisches Institut
der TU München

80290 München

Dr. M. Paffrath
ZFE BT SE 43
Siemens AG
Otto-Hahn-Ring 6

81739 München

Prof. Dr. E. Rank
Anwendung numerischer Methoden
im Bauwesen
Universität Dortmund
August-Schmidt-Str.

44227 Dortmund

Prof. Dr. P. Rentrop
Mathematisches Institut
der TU München

80290 München

Dr. W. Schmidt
Mathematisches Institut
der TU München

80290 München

Prof. Dr. K. Taubert
Institut für Angewandte Mathematik
Universität Hamburg
Bundesstr. 55

20146 Hamburg

Prof. Dr. J. Wick
FB Mathematik
Universität Kaiserslautern
Erwin-Schrödinger-Straße

67663 Kaiserslautern

H. Wriedt
Regionales Rechenzentrum der
Universität Hamburg
Schlüterstr. 70

20146 Hamburg

Dr. Q. Zheng
ZFE BT SE 43
SIEMENS AG
Otto-Hahn-Ring 6

81739 München

Prof. Dr. G. Wachutka
Physical Electronics Laboratory
ETH Höngerberg, HPT

CH-8093 Zürich

Dr. W. Wiedl
Regionales Rechenzentrum der
Universität Hamburg
Schlüterstr. 70

20146 Hamburg

Dr. D. Wrzosek
Institute of Applied Mathematics
and Mechanics
Warsaw University
ul. Banacha 2

02-097 Warsaw
Poland

Circuit Simulation

International Series of Numerical Mathematics, Vol. 117, © 1994 Birkhäuser Verlag Basel

A new efficient numerical integration scheme for highly oscillatory electric circuits

Georg Denk

Mathematisches Institut, TU München

Abstract. This paper presents a new numerical integration scheme for second order ordinary differential equations which can integrate highly oscillating electric circuits with high efficiency. This is shown by numerical results. The discretization scheme is based on the principle of coherence proposed by Hersch which will be described shortly. The analysis reveals important properties of the new method such as consistency. Some problems (e. g. cancellation) make an efficient implementation difficult, solutions are given.

AMS Subject Classification: 65L05, 65L06.

Key words: ordinary differential equations, oscillatory solutions, multistep method, consistency, convergence, circuit simulation.

1. Introduction

Circuit simulation is a standard task for the computer aided design of electronic circuits. However, there are two types of circuits which need special proceeding: These are very large circuits (e. g. memory chips) and highly oscillating circuits (e. g. quartz oscillators). The first type can be handled by the exploitation of latency [7, 8]. Oscillating circuits require a quite different approach, see e. g. [16]. In the paper presented here, a new discretization scheme will be described that is able to use some information about the ordinary differential equation (ODE) to compute highly oscillatory solutions with high efficiency.

Most integration schemes solve the general initial value problem

$$y'(x) = f(x, y(x)), \quad y(x_0) = y_0 \text{ with } y : \mathbf{R} \to \mathbf{R}^n.$$

As no special information about the ODE is given, they require a large amount of computing time for the integration of highly oscillatory ODEs: Every oscillation of y has to be followed which leads to very small integration steps and subsequently to rounding errors. This is the motivation for special integration schemes for special ODEs.

An ODE describing oscillatory behavior with possible damping can be written as

$$y''(x) + a\,y'(x) + b\,y(x) = f(x, y(x)), \quad y(x_0) = y_0, \; y'(x_0) = y'_0, \tag{1}$$

with $a, b \in \mathbb{R}$, $y : \mathbb{R} \to \mathbb{R}^n$. As this equation does not reflect the standard form $y'(x) = f(x, y(x))$ for an ODE, standard techniques for the construction of a numerical integration scheme cannot be used. A transformation of (1) into standard notation would yield to a loss of the special information about the ODE which is given by the parameters a and b.

One approach for this special ODE was used by Deuflhard [6] for the construction of an extrapolation scheme. His work is based on the *principle of coherence* [11, 12]. This principle will also be used in this paper for the construction of a multistep formula for (1) and will be described in the next section.

If the parameter a in (1) equals 0, the ODE is called *special second order ODE*. For this kind of ODE many approaches can be found in literature. One of the first investigators were Stiefel and Bettis [20] who modified the polynomial ansatz of Cowell's method in order to take oscillatory solutions into account. Here, some part of the polynomial basis is replaced by trigonometric functions. A generalization of this approach has been used by Skelboe [19] for the simulation of mildly nonlinear electric circuits. As the ODEs describing the circuit are given in standard notation $z'(x) = g(x, z(x))$, the additional information about the frequency has to be passed to the algorithm in an appropriate manner.

Another approach for the numerical computation of special second order ODEs is the adaptation of parameters of the discretization scheme. This has be done for the construction of multistep schemes [1, 3, 18] as well as for Runge-Kutta-Nyström type methods [14, 15].

2. The principle of coherence

The principle of coherence was formulated by Hersch [12] as "Successive approximations should not contradict each other". This yields some conditions on the coefficients of a numerical method which have to be fulfilled. The main idea will become clear by the following example: Consider the ODE $z''(x) + \lambda z(x) = 0$, $\lambda > 0$. A standard approach for this ODE is the difference equation

$$z(x - h) - A(h)z(x) + z(x + h) = 0. \tag{2}$$

Here, $A(h)$ is the coefficient of the method and it depends on the step size h. If the step size is multiplied with 2, the difference equation is

$$z(x - 2h) - A(2h)z(x) + z(x + 2h) = 0.$$

A formula for the step size $2h$ can be constructed also by a linear combination of three equations of type (2), centered at $x - h$, x, and $x + h$, resp. This leads to the expression

$$z(x - 2h) - (A(h)^2 - 2)z(x) + z(x + 2h) = 0. \tag{3}$$

Comparing (3) with (2) gives the coherence condition $A(2h) = A(h)^2 - 2$ which has to be satisfied for a coherent discretization scheme. This condition holds for $A(h) = 2\cos(\kappa h)$ with $\kappa \in \mathbf{R}$. For $h \to 0$ this leads to $\kappa = \sqrt{\lambda}$, as the difference equation should reflect the differential equation. The coherent approximation of $z''(x) + \lambda z(x) = 0$ is

$$z(x - h) - 2\cos(\sqrt{\lambda}h)z(x) + z(x + h) = 0$$

instead of the classical approach yielding

$$z(x - h) - (2 - \lambda h^2)z(x) + z(x + h) = 0.$$

The mathematical approach for the construction of coherent discretization schemes is the transformation of the homogeneous part of the ODE (in the example $z''(x) + \lambda z(x)$) into a difference equation. The solution of the latter one fulfills the ODE, too. The principle of coherence can be used for the construction of a number of integration schemes depending on the kind of the homogeneous part to be considered. In Table 1, some of the basic discretization formulas for equidistant step sizes are given. The well-known Adams-Bashforth/Moulton schemes can be found as well as the Cowell/Störmer methods, they belong to the homogeneous parts $y'(x)$, and $y''(x)$, resp.

homogeneous part	discretization scheme
$y'(x)$	$y(x + h) = y(x)$
$y'(x) + b\,y(x)$	$y(x + h) = \exp(-b\,h)y(x)$
$y''(x)$	$y(x + h) = 2y(x) - y(x - h)$
$y''(x) + b^2 y(x)$	$y(x + h) = 2\cos(b\,h)y(x) - y(x - h)$
$y^{(4)}(x)$	$y(x + 2h) = 4y(x + h) - 6y(x) + 4y(x - h) - y(x - 2h)$
$y^{(4)} - b^4 y(x)$	$y(x + 2h) = (2\cosh(b\,h) + 2\cos(b\,h))y(x + h)$ $- (2 + 4\cosh(b\,h)\cos(b\,h))y(x)$ $+ (2\cosh(b\,h) + 2\cos(b\,h))y(x - h) - y(x - 2h)$

Table 1: Coherent discretization schemes due to Hersch [11].

The construction of the coherent discretization scheme for the ODE (1) is complicated by the dependency on the values of a and b. As the transformation of the homogeneous part of the ODE into a difference equation is done via the calculus of distribution and the distribution depends on the fundamental solutions $g_1(x)$ and $g_2(x)$ of the homogeneous ODE, there exist five different cases. These cases are given in Table 2. The most important case is $b \neq 0$, $b > a^2/4$ giving oscillatory solutions and will be considered in the following. For this case, the coherent discretization scheme is

$$y(x + h) = 2 \exp(-\alpha h) \cos(\beta h) y(x) - \exp(-2 \alpha h) y(x - h).$$

Please note that there is no evaluation of $y'(x)$ necessary despite the term $ay'(x)$ in (1). A derivation of this formula for the non-equidistant case can be found in [4, 5].

		$g_1(x)$	$g_2(x)$
$b = 0$	$a = 0$	1	x
$b = 0$	$a \neq 0$	1	$\exp(a h)$
$b \neq 0$	$b > a^2/4$	$\exp(\frac{a}{2} x) \cos(\sqrt{b - a^2/4}\, x)$	$\exp(\frac{a}{2} x) \sin(\sqrt{b - a^2/4}\, x)$
$b \neq 0$	$b = a^2/4$	$\exp(\frac{a}{2} x)$	$x \exp(\frac{a}{2} x)$
$b \neq 0$	$b < a^2/4$	$\exp(\frac{a}{2} x) \cosh(\sqrt{a^2/4 - b}\, x)$	$\exp(\frac{a}{2} x) \sinh(\sqrt{a^2/4 - b}\, x)$

Table 2: Fundamental solutions for the homogeneous part of (1).

Up to now, the inhomogeneous part $f(x, y(x))$ of (1) has not taken into account. With an approach similar to classical multistep methods as the Cowell/Störmer family, the following ansatz is used for a discretization scheme with $k + 1$ function evaluations:

$$\sum_{i=k-2}^{k} \alpha_i y(x_{n+i}) = h^2 \sum_{i=0}^{k} \beta_i f(x_{n+i-s}, y(x_{n+i-s})). \tag{4}$$

α_i, $i = k - 2, k - 1, k$, denotes the coefficients derived from the transformation of the homogeneous part, β_i, $i = 0, \ldots, k$, describes the unknown coefficients for the right-hand side. For an explicit method the parameter s equals 1; an implicit scheme is characterized by $s = 0$. The unknowns β_i in this ansatz are computed in such a way that the formula (4) is exact for $f(x, y(x)) \in\, < x^0, x^1, \ldots, x^k >$. Due to the non-standard form of (1), the coherent discretization scheme is more complicated than standard methods. This is especially true for the important case of non-equidistant step sizes.

3.   Consistency of the new method

For the theoretical analysis of the coherent discretization scheme, standard theorems for numerical analysis found in textbook (see e. g. [21]) can not be used without modification. This is due to the form of the ODE (1) which is solved by the method. The analysis becomes even more complicated as the coefficients α_i, β_i of the method depend on the step size h even in the case of equidistant step sizes. This is in contrast to classical multistep methods.

To make notation simple, the following definitions are introduced: A multistep method which solves the general second order ODE $z(x)'' = g(x, z(x), z'(x))$ can be written as

$$\sum_{\imath=0}^{k} a_\imath z(x_{n+\imath}) = h \sum_{\imath=0}^{k} b_\imath z'(x_{n+\imath-s}) + h^2 \sum_{\imath=0}^{k} c_\imath g(x_{n+\imath-s}, z(x_{n+\imath-s}), z'(x_{n+\imath-s}))$$

with $a_\imath = a_\imath(h)$, $b_\imath = b_\imath(h)$, $c_\imath = c_\imath(h)$, and $x_{n+\imath} = x_n + \imath h$ for the equidistant case. This defines the operator

$$\mathcal{L}[z, x_{n+k}, h] := \sum_{\imath=0}^{k} a_\imath z(x_{n+\imath}) - h \sum_{\imath=0}^{k} b_\imath z'(x_{n+\imath-s}) - h^2 \sum_{\imath=0}^{k} c_\imath g(x_{n+\imath-s}, z(x_{n+\imath-s}), z'(x_{n+\imath-s})),$$

and the *characteristic polynomials*

$$\rho(x; h) := \sum_{\imath=0}^{k} a_\imath(h) x^{\imath+s}, \quad \sigma(x; h) := \sum_{\imath=0}^{k} b_\imath(h) x^\imath, \quad \tau(x; h) := \sum_{\imath=0}^{k} c_\imath(h) x^\imath.$$

The consistency of a method with coefficients depending on the step size h even in the equidistant case was computed by Lyche [17] for the special ODE $y^{(r)}(x) = g(x, y(x))$. This fits into the notation used above if $r = 2$ and $b_\imath = 0$ for $\imath = 0, \ldots k$. Following Lyche, $a_\imath(h)$ and $c_\imath(h)$ are expanded in Taylor's series around h yielding

$$a_\imath(h) = \sum_{\jmath=0}^{q} a_{\imath,\jmath} h^\jmath + \mathcal{O}(h^{q+1}), \quad c_\imath(h) = \sum_{\jmath=0}^{q} c_{\imath,\jmath} h^\jmath + \mathcal{O}(h^{q+1}).$$

Expanding $z(x_{n+\imath})$ in Taylor's series around x_n, we find after some rearranging

$$\mathcal{L}[z, x_{n+k}, h] = \sum_{\imath=0}^{q} \left(\sum_{\jmath=0}^{\imath} C_{\jmath, \imath-\jmath} z^{(\jmath)}(x_n) \right) h^\imath + \mathcal{O}(h^{q+1})$$

with

$$C_{\jmath, \imath} = \frac{1}{\jmath!} \left(0^\jmath a_{0,\imath} + 1^\jmath a_{1,\imath} + \ldots + k^\jmath a_{k,\imath} \right) \quad \text{for } \jmath = 0, 1,$$

$$C_{2,\imath} = \frac{1}{2!} \left(a_{1,\imath} + 2^2 a_{2,\imath} + \ldots + k^2 a_{k,\imath} \right) - \left(c_{0,\imath} + c_{1,\imath} + \ldots + c_{k,\imath} \right),$$

$$C_{\jmath,\imath} = \frac{1}{\jmath!} \left(0^\jmath a_{0,\imath} + 1^\jmath a_{1,\imath} + \ldots + k^\jmath a_{k,\imath} \right)$$
$$- \frac{1}{(\jmath - 2)!} \left(0^{\jmath-2} c_{0,\imath} + 1^{\jmath-2} c_{1,\imath} + \ldots + k^{\jmath-2} c_{k,\imath} \right)$$
$$\text{for } \jmath = 3, 4, \ldots$$

This leads to (see [17])

Definition 1: *The operator $\mathcal{L}[z, x_{n+k}, h]$ of a multistep method solving $z''(x) = g(x, z(x))$ is said to be of order of consistency p, if $C_{j,i} = 0$ for $0 \leq i+j \leq p+1$ and $C_{j,i} \neq 0$ for some i, $j \geq 0$ such that $i+j = p+2$.*

The main difference to the classical definition of order of consistency is the appearance of more terms $z^{(j)}(x_n)$ for one h-order. This makes the analysis complicated as well as tedious. The following theorem makes the computation of p simpler, especially if programs for computer algebra are used:

Theorem 2: *The operator $\mathcal{L}[z, x_{n+k}, h]$ of a multistep method solving $z''(x) = g(x, z(x))$ is said to be of order of consistency p if*

$$C_j(h) = C_{j,p+2-j}h^{p+2-j} + \mathcal{O}(h^{p+3-j})$$

holds for $j = 0, \ldots, p+1$. $C_j(h)$ is defined as

$$
\begin{aligned}
C_j(h) &= \frac{1}{j!}\left(0^j a_0(h) + 1^j a_1(h) + \ldots + k^j a_k(h)\right) \quad for j = 0, 1, \\
C_2(h) &= \frac{1}{2!}\left(a_1(h) + 2^2 a_2(h) + \ldots + k^2 a_k(h)\right) \\
&\quad - \left(c_0(h) + c_1(h) + \ldots + c_k(h)\right), \\
C_j(h) &= \frac{1}{j!}\left(0^j a_0(h) + 1^j a_1(h) + \ldots + k^j a_k(h)\right) \\
&\quad - \frac{1}{(j-2)!}\left(0^{j-2}c_0(h) + 1^{j-2}c_1(h) + \ldots + k^{j-2}c_k(h)\right) \\
&\qquad\qquad for\ j = 3, 4, \ldots
\end{aligned}
$$

Proof: The coefficients $C_j(h)$ are expanded in Taylor's series by expanding the coefficients $a_i(h)$ and $c_i(h)$ in Taylor's series around h. This leads to

$$C_j = C_{j,0} + hC_{j,1} + h^2 C_{j,2} + \ldots + h^{p+2-j}C_{j,p+2-j} + \mathcal{O}(h^{p+3-j}).$$

For a method of order p, Definition 1 yields

$$C_{j,n} = 0 \quad \text{für } 0 \leq n \leq p+1-j,$$

which reduces the Taylor's series of $C_j(h)$ to

$$C_j = h^{p+2-j}C_{j,p+2-j} + \mathcal{O}(h^{p+3-j}).$$

$\square$

With the help of the characteristic polynomials, the computation of the order of consistency can be simplified even more:

Theorem 3: *The operator $\mathcal{L}[z, x_{n+k}, h]$ of a multistep method solving $z''(x) = g(x, z(x))$ is said to be of order of consistency p if*

$$\rho(e^h; h) - h^2 \cdot \tau(e^h; h) = \tilde{C}_{p+2}h^{p+2} + \mathcal{O}(h^{p+3}), \quad \tilde{C}_{p+2} \neq 0.$$

$\tilde{C}_{p+2}$ is defined as

$$\tilde{C}_{p+2} = \sum_{j=0}^{p+2} C_{j,p+2-j}.$$

Proof: Consider the special case $z(x) = e^x$. For a discretization scheme of order p, the following expression holds:

$$\mathcal{L}[e^x, x_{n+k}, h] = h^{p+2} \sum_{j=0}^{p+2} C_{j,p+2-j} e^{x_n} + \mathcal{O}(h^{p+3}).$$

On the other hand,

$$\mathcal{L}[e^x, x_{n+k}, h] = e^{x_n} \left(\rho(e^h, h) - h^2 \cdot \tau(e^h, h) \right).$$

The Taylor's series of $\rho(e^h, h) - h^2 \cdot \tau(e^h, h)$ yields after some calculation

$$\rho(e^h, h) - h^2 \cdot \tau(e^h, h) = \sum_{i=0}^{q} \left(\sum_{j=0}^{i} C_{j,i-j} \right) h^i + \mathcal{O}(h^{q+1}).$$

If the integration scheme is of order p, then — according to Definition 1 —

$$C_{j,i-j} = 0 \quad \text{for } 0 \leq i \leq p+1.$$

This cancels the terms $h^0, \ldots, h^{p+1}$ and

$$\rho(e^h, h) - h^2 \cdot \tau(e^h, h) = \sum_{i=p+2}^{q} \left(\sum_{j=0}^{i} C_{j,i-j} \right) h^i + \mathcal{O}(h^{q+1})$$

is the final expression. This finishes the proof. □

To compute the order of consistency of the discretization scheme presented here, Theorem 3 has to be generalized to take the term $ay'(x)$ in (1) into account. Before this can be done, the ODE (1) has to be transformed into standard notation which reads as $z''(x) = g(x, z(x), z'(x))$. This requires the definitions

$$a_i(h) := \alpha_i(h) - \beta_i(h)h^2 b, \quad b_i(h) := \beta_i(h)ah, \quad c_i(h) := \beta_i(h).$$

Theorem 4: *The operator $\mathcal{L}[z, x_{n+k}, h]$ of a multistep method solving*

$$z''(x) = g(x, z(x), z'(x))$$

is said to be of order of consistency p if

$$\rho(e^h; h) - h \cdot \sigma(e^h; h) - h^2 \cdot \tau(e^h; h) = \tilde{C}_{p+2} h^{p+2} + \mathcal{O}(h^{p+3}), \quad \tilde{C}_{p+2} \neq 0.$$

The proof of this theorem is similar to that of Theorem 1–3 and therefore omitted.

It is not possible to compute the actual value of p in a straightforward manner as the exponential and trigonometric terms make the Taylor's expansion rather complicated. Therefore, the computer algebra program MAPLE [2] has been used. For both explicit and implicit methods, the order of consistency has been computed for $1,\ldots,6$ function evaluations. The computation results in the following theorem:

Theorem 5: *The coherent discretization scheme (4) exhibits an order of consistency of $k+1$ for $k = 0,\ldots,5$.*

Up to now, only equidistant step sizes have been considered. This restriction makes the numerical integration scheme inefficient for practical usage. The proceeding for the computation of p can be repeated for non-equidistant step sizes. We introduce the auxiliary values η_i as

$$\eta_i := \frac{x_{n+i} - x_{n-1}}{x_{n+k} - x_{n+k-1}}.$$

The characteristic polynomials change into functions which are no polynomials any longer as the exponents are differences of two η_i-terms which is in general not a natural number. The computation with MAPLE requires now a large amount of CPU time and confirms Theorem 5. As an example, the Taylor's series for an explicit method with four function evaluations is given.

$$
\begin{aligned}
\mathcal{L}[z, x_{n+k}, h] \;=\; & \frac{1}{1440} h^6 \left(z^{(6)}(x_n) + a\, z^{(5)}(x_n) + b\, z^{(4)}(x_n) \right) (\eta_7 - \eta_5)(\eta_7 - \eta_6) \\
& \Big(\eta_7\,\eta_6^3 + \eta_5^3\eta_6 + \eta_5^2\eta_6^2 + \eta_5\,\eta_6^3 + \eta_6\,\eta_7^3 + \eta_5\,\eta_7^3 + \eta_6^2\eta_7^2 \\
& + \eta_5^2\eta_7^2 + \eta_5^3\eta_7 + 2\,\eta_5^3\eta_4 + 2\,\eta_4\,\eta_6^3 + 2\,\eta_4\,\eta_7^3 - \eta_7^4 \\
& - \eta_5^4 - \eta_6^4 - 3\,\eta_4\,\eta_6\,\eta_7^2 - 2\,\eta_5\,\eta_6^2\eta_7 - 3\,\eta_4\,\eta_5^2\eta_6 \\
& - 3\,\eta_4\,\eta_5\,\eta_7^2 - 3\,\eta_4\,\eta_7\,\eta_6^2 - 2\,\eta_5^2\eta_6\,\eta_7 - 3\,\eta_4\,\eta_5^2\eta_7 \\
& - 3\,\eta_4\,\eta_5\,\eta_6^2 - 2\,\eta_5\,\eta_6\,\eta_7^2 + 12\,\eta_4\,\eta_5\,\eta_6\,\eta_7 \Big) \\
& + \mathcal{O}(h^7).
\end{aligned}
\tag{5}
$$

This gives an impression about the complexity which is necessary for the computation for the order of consistency. The usage of non-equidistant step sizes produces another problem: It is possible that, under certain relations for the values of η_i, the order p increases by 1. In the example above, this will be true if the expression $(\eta_7\,\eta_6^3 + \ldots + 12\,\eta_4\,\eta_5\,\eta_6\,\eta_7)$ equals 0. This can cause problems for the step size control.

The next step in the analysis of the discretization scheme is the investigation of stability. The results are very promising regarding the usefulness for highly oscillatory electric circuits and can be formulated as

Theorem 6: *The coherent discretization scheme (4) is*

- *0-stable,*
- *P-stable,*
- *has no phase-lag.*

It is beyond the scope of this paper to give the proof of Theorem 6, see e. g. [4, 5].

4.   Implementation

The coherent discretization scheme described above has been implemented as a Fortran subroutine HERSCH. It uses a predictor-corrector formulation with automatic step size and order selection. The interface of the subroutine is similar to standard integrators like LSODE [13] with additional parameters such as a and b of (1). Despite the fact that the estimation of the local truncation error *LTE* can be done with Milne's device (for a proof see [4]), an approximation of the *LTE* is computed by using divided differences. This is due to the possible anomalies in the order of consistency described at the end of Section 3 and due to the complexity of (5).

As the discretization scheme presented here solves a second order ODE the initial values $y(x_0)$ and $y'(x_0)$ have to be transformed into $y(x_0)$ and $y(x_0 + h_0)$ with h_0 denoting the step size of the first integration step. This transformation delivers the integration scheme with the appropriate history information needed for computing the next y-value. Expanding $y(x_0 + h_0)$ in a Taylor's series around x_0 yields

$$y(x_0 + h_0) = \left(1 - b\frac{h_0^2}{2}\right) y(x_0) + h_0 \left(1 - a\frac{h_0}{2}\right) y'(x_0) + \frac{h_0^2}{2} f(x_0, y(x_0)) + \mathcal{O}(h_0^3).$$

If the initial step size h_0 is small enough, the truncation error will be smaller than the prescribed error tolerances for the integration.

The computation of the coefficients α_i and β_i of (4) is rather tricky, a straightforward implementation would result in an integration scheme which is theoretically consistent but practically inconsistent. This is due to severe cancellation for small step sizes h making the straightforward implementation useless. During the construction of the linear equation system for β_i the right-hand side has to be computed. For $a = 1$, $b = 1$ and equidistant step sizes h, the fourth component will look like

$$
\begin{aligned}
RHS(4) \;=\; &+ \frac{125 \exp(-h) - 432 \exp(-h/2) \cos(\sqrt{3}\,h/2) + 343}{h^2} \\
&+ \frac{-75 \exp(-h) + 216 \exp(-h/2) \cos(\sqrt{3}\,h/2) - 147}{h^3} \\
&+ \frac{6 \exp(-h) - 12 \exp(-h/2) \cos(\sqrt{3}\,h/2) + 6}{h^5}.
\end{aligned}
$$

The Taylor's series of $RHS(4)$ yields

$$RHS(3) = 219 - \frac{1507\,h}{15} + \mathcal{O}(h^2)$$

indicating the cancellation of $h^{-5}, \ldots, h^{-1}$. For the modest step size $h = 10^{-3}$, following expression has to be computed:

$$
\begin{aligned}
RHS(4) \quad = \quad &+ \quad 36091170.40718087575790696250\ldots \\
&- 6033091451.50762527480417321882\ldots \\
&+ 5997000499.99999166666696423611\ldots \\
= \quad &\qquad\quad 218.89954726762069797979\ldots
\end{aligned}
$$

The cancellation of 7 digits certainly influences the accuracy of the computed coefficients of the coherent discretization scheme. As the step size control will indicate an error too large for the prescribed error tolerances, the step size will be reduced resulting in an increase of the cancellation. Thus, the integration will fail.

A remedy to this situation is the computation of Taylor's series for the right-hand side of the linear equation system instead of the direct approach. This is done only for those situations which could cause cancellation, i. e. small step sizes. With an implementation modified in such a manner, the integration scheme proofs to be consistent both theoretically and practically. Most authors of similar integration schemes (e. g. [3, 14, 15, 18]) do not encounter this problem: As they use only equidistant step sizes, the coefficients can be computed in multiple precision. Here, the computation of the coefficients has to be done for every integration step which would make the scheme inefficient.

5. Numerical results

The coherent discretization scheme implemented as Fortran subroutine HERSCH has been successfully used for problems in circuit simulation, mechanics, and celestial mechanics. The first example presented here is the test equation

$$
y''(x) + y(x) = x^2, \quad y(0) = 1, \ y'(0) = 0 \text{ for } x \in [0, 500]. \tag{6}
$$

The exact solution of (6) is $y(x) = 3\cos(x) - 2 + x^2$ and shows small oscillations around $z(x) = x^2$ for large values of x. The ODE (6) has been integrated with HERSCH using 10^{-8} as prescribed error tolerances both for the absolute and relative error. This computation is compared with the results of standard integration methods, namely LSODE [13] (multistep method), DOPRI5 [9] (explicit Runge-Kutta method), RADAU5 [10] (implicit A-stable Runge-Kutta method). For these methods, the error tolerances have been set in such a way that the gained relative error at $x = 500$ is approximately 10^{-8}. In order to integrate the second order ODE (6) with standard methods, it has to be transformed into a system of first order ODEs by introducing $z_1(x) = y(x)$, $z_2(x) = y'(x)$:

$$
\begin{aligned}
z_1'(x) &= z_2(x), \\
z_2'(x) &= -z_1(x) + x^2.
\end{aligned}
$$

In Table 3, the statistics of the integration are given. NST denotes the number of integration steps necessary for the integration of (6), $NFCN$ gives the number of function evaluations for it (one-dimensional for HERSCH, two-dimensional otherwise). The column "rel. error" shows the relative error at the end of the integration interval. Table 3 shows clearly the ability of HERSCH to solve this ODE with high efficiency. It needs only the minimum number of integration steps, which is determined by the heuristic parameters of the code. HERSCH needs only about 1% of the function evaluations $NFCN$ of the best standard method and gives a result which is about 4 digits more accurate.

method	NST	$NFCN$	rel. error
HERSCH	25	51	$1.7 \cdot 10^{-12}$
LSODE	4132	4578	$5.0 \cdot 10^{-8}$
DOPRI5	1394	8365	$3.5 \cdot 10^{-8}$
RADAU5	1437	6717	$1.5 \cdot 10^{-8}$

Table 3: Statistics for the integration of (6).

The second numerical experiment is the simulation of a small electric circuit which is given in Figure 1.

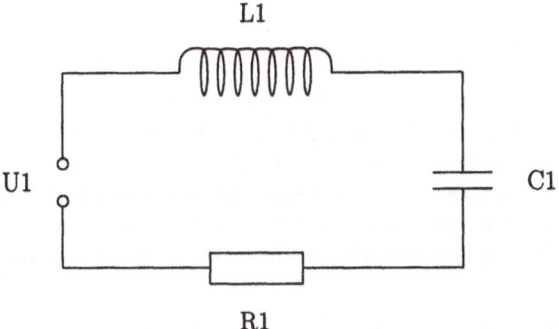

Figure 1: Electric circuit

The voltage U_R, U_C, and U_L, resp., at the elements R1, C1, and L1, resp., are modeled by the relations

$$
\begin{aligned}
U_R &= R I_R, \\
U_C &= \frac{1}{C} Q, \\
U_L &= L \frac{dI_L}{dx},
\end{aligned}
\tag{7}
$$

with $I_R = I_R(x)$ and $I_L = I_L(x)$ denoting the current through R1 and L1. $Q = Q(x)$ is the charge of the capacitor, R is the resistance of R1, C the capacity of C1, and L is the flux through L1.

Kirchhoff's laws are valid for every electric circuit, they can be stated as

- The algebraic sum of currents traversing each cutset of the network must be equal zero at every instant of time.

- The algebraic sum of voltages around each loop of the network must be equal zero at every instant of time.

These laws result in the following equations:

$$
\begin{aligned}
0 &= U(x) - U_R - U_C - U_L, \\
0 &= I_L(x) - \frac{dQ(x)}{dx}, \\
0 &= I_R(x) - \frac{dQ(x)}{dx},
\end{aligned}
$$

where $U(x)$ describes the voltage across U1. Substituting U_R and U_L with (7) and eliminating I_L, I_R, and $Q(x)$ yields the following second order ODE:

$$
LC\,U_C''(x) + RC\,U_C'(x) + U_C(x) = U(x). \tag{8}
$$

This equation will be integrated by HERSCH.

It is obvious that the influence of oscillations in the circuit will be reflected by the efficiency of standard methods for solving (8). Therefore, two parameter sets have been investigated, they are given in Table 4. The increase of R in Test 2 damps the high oscillatory part of the solution rather rapidly, the slow varying part becomes dominant. In Test 1, the oscillations can be noticed for the first half of the integration interval $x \in [0, 5 \cdot 10^{-3}]$. In Test 2, however, the oscillations were damped out already after about 10% of the interval.

	R	C	L	$U(x)$
Test 1	5	10^{-9}	10^{-3}	$5\sin(10^3 x)$
Test 2	50	10^{-9}	10^{-3}	$5\sin(10^3 x)$

Table 4: Parameters for the integration of (8).

The electric circuit has been simulated with HERSCH, LSODE, DOPRI5, and RADAU5 for various prescribed error tolerances. In Figure 2, the number of function evaluations *NFCN* is plotted versus the maximum of the absolute error δ during the integration process in a double-logarithmic scale.

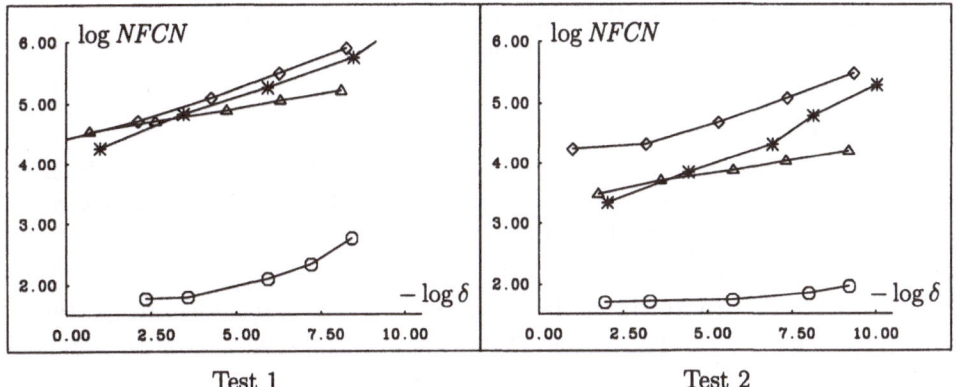

Test 1   Test 2

Figure 2: Number of function evaluations *NFCN* versus the maximum absolute error δ for the simulation of (8).
($\circ$ = HERSCH, $\triangle$ = LSODE, $\diamond$ = DOPRI5, $*$ = RADAU5)

The results for the coherent discretization scheme HERSCH are clearly separated from the curves for the standard schemes. This shows again the ability of the new method to solve highly oscillatory ODEs with efficiency. The standard methods behave quite similar for Test 1 indicating that A-stability is not important for this test equation. In Test 2, there are some differences between the standard methods with the explicit method DOPRI5 being the most inefficient one. In both tests, the slope of LSODE is the flattest one among the standard methods. This reflects the well-known differences between one-step and multistep methods.

The gap between the standard integration schemes and the coherent method becomes smaller for a large damping in the circuit. This is the expected behavior as the right-hand side for the standard schemes is getting smoother. The right-hand side for HERSCH does not change with the damping, only the parameters a and b differ between Test 1 and Test 2. Nevertheless, the difference between HERSCH and the best standard method is quite impressive: For $\delta \approx 10^{-8}$, HERSCH needs about 70 function evaluations for Test 2 compared with about 10000 required by LSODE. For Test 1, HERSCH needs about 200 function evaluations, LSODE as the most efficient standard scheme about 140000, and for DOPRI5 the value is 500000.

Conclusion

The new numerical integration method based on the principle of coherence has proved to solve highly oscillatory second order ODEs efficiently. Together with the excellent stability properties, this indicates the usefulness of the discretization scheme for the simulation of electric circuits. Depending on the relation between the homogeneous part and the right-hand side of the ODE, the number of function evaluations can be reduced up to 1 per mill compared with standard integration schemes.

Acknowledgment

The author acknowledges the support for this work given by Prof. Dr. Dr. h. c. R. Bulirsch and Prof. Dr. P. Rentrop. He is indebted to the members of the Numerical Analysis group at the TU München for stimulating discussions.

The author thanks the Bayerische Forschungsstiftung for the support in FORTWIHR: Bayerischer Forschungsverbund für Technisch-Wissenschaftliches Hochleistungsrechnen. This work is part of the project 4.4 "Numerical Simulation of Electric Circuits and Semiconductor Devices".

References

[1] G. vanden Berghe, H. De Meyer and J. Vanthournout, A modified Numerov integration method for second order periodic initial-value problems, *Internat. J. Comput. Math. 32 (1990), pp. 233–242.*

[2] B. W. Char et al., Maple V: Language Reference Manual, *Springer Verlag, New York, 1990.*

[3] M. M. Chawla and P. S. Rao, A Noumerov-type method with minimal phase-lag for the integration of second order periodic initial-value problems, *J. Comput. Appl. Math. 11 (1984), pp. 277–281.*

[4] G. Denk, Ein neues Diskretisierungsverfahren zur effizienten numerischen Lösung rasch oszillierender Differentialgleichungen, *Ph. D. thesis, Technische Universität München, Germany, 1992.*

[5] G. Denk, A new numerical method for the integration of highly oscillatory second order differential equations, Appl. Numer. Math. 13 (1993), pp. 57–67.

[6] P. Deuflhard, A study of extrapolation methods based on multistep schemes without parasitic solutions, *ZAMP 30 (1979), pp. 177–189.*

[7] U. Feldmann, U. Wever, Q. Zheng, R. Schultz and H. Wriedt, Algorithms for modern circuit simulation, *AEÜ 46 (1992), pp. 274–285.*

[8] M. Günther and P. Rentrop, Multirate Rosenbrock-Wanner methods and latency of electric circuits, *Appl. Numer. Math. 13 (1993), pp. 83–102.*

[9] E. Hairer, S. P. Nørsett and G. Wanner, Solving ordinary differential equations I, *Springer Verlag, Berlin, 1987.*

[10] E. Hairer and G. Wanner, Solving ordinary differential equations II, *Springer Verlag, Berlin, 1991.*

[11] J. Hersch, Contribution à la méthode des équations aux différences, *ZAMP IXa, no. 2 (1958), pp. 129–180.*

[12] J. Hersch, Eine Kohärenzforderung für Differenzengleichungen, *in J. Albrecht and L. Collatz, Numerische Methoden bei Differentialgleichungen und mit funktionalanalytischen Hilfsmitteln, ISNM 19 (1974), pp. 121–124.*

[13] A. C. Hindmarsh, ODEPACK, a systemized collection of ODE solvers, *Lawrence Livermore National Laboratory, Report UCRL-88007, 1982.*

[14] P. J. van der Houwen and B. P. Sommeijer, Explicit Runge-Kutta (-Nyström) methods with reduced phase errors for computing oscillating solutions, *SIAM J. Numer. Anal. 24 (1987), pp. 595–617.*

[15] P. J. van der Houwen and B. P. Sommeijer, Diagonally implicit Runge-Kutta-Nyström methods for oscillatory problems, *SIAM J. Numer. Anal. 26 (1989), pp. 414–429.*

[16] W. Kampowsky, P. Rentrop and W. Schmidt, Classification and numerical simulation of electric circuits, *Surv. Math. Ind. 2 (1992), pp. 23–65.*

[17] T. Lyche, Chebyshevian multistep methods for ordinary differential equations, *Numer. Math. 19 (1972), pp. 65–75.*

[18] T. E. Simos and A. D. Raptis, Numerov-type methods with minimal phase-lag for the numerical integration of the one-dimensional Schrödinger equation, *Computing 45 (1990), pp. 175–181.*

[19] S. Skelboe, Numerical methods for the analysis of mildly nonlinear electrical circuits, *Report 75-08, Institute for Numerical Analysis, Technical University of Denmark, 1975.*

[20] E. Stiefel and D. G. Bettis, Stabilization of Cowell's method, *Numer. Math. 13 (1969), pp. 154–175.*

[21] J. Stoer and R. Bulirsch, Introduction to numerical analysis, *Springer Verlag, New York, 1993.*

Dr. Georg Denk
Mathematisches Institut
Technische Universität München
D-80290 München
Germany

e-mail: denk@mathematik.tu-muenchen.de

International Series of Numerical Mathematics, Vol. 117, © 1994 Birkhäuser Verlag Basel

NUMERISCHE LÖSUNG VON HIERARCHISCH STRUKTURIERTEN SYSTEMEN VON ALGEBRO–DIFFERENTIALGLEICHUNGEN

FRIEDRICH GRUND

1. EINLEITUNG

Beim Entwurf mikroelektronischer Schaltungen ist u.a. das elektrische Verhalten des Schaltkreises zu simulieren. Mathematisch ist hierzu ein System von Algebro–Differentialgleichungen zu lösen. Da die Anzahl der Gleichungen und in engem Zusammenhang die der Transistoren groß und die Schaltungen strukturiert sind, bieten sich für die Simulation diakoptische Methoden an. Für die Illustration der Größenordnungen sei erwähnt, daß mit dem Netzwerksimulator MAGNUS [1] u.a. eine Teilschaltung eines 4M–Bit–Speichers mit über 30 000 Transistoren simuliert worden ist.

Nach der Formulierung der Aufgabenstellung werden im Abschnitt 3 die Methoden für die Lösung von hierarchisch strukturierten Systemen von Algebro–Differentialgleichungen dargelegt. Im Abschnitt 4 werden, da die sequentiellen Methoden bei sehr großen Problemen erhebliche Rechenzeiten erfordern, parallele Methoden auf dem Niveau

– der Algebro–Differentialgleichungen
– der nichtlinearen Gleichungssysteme
– der linearen Gleichungssysteme

entwickelt und über Ergebnisse bei einem linearen System wird berichtet.

2. AUFGABENSTELLUNG

Gegeben sei das folgende implizite System von Algebro–Differentialgleichungen

$$F_1(x(t), \dot{x}(t), t) = 0, \quad t \geq t_0 \tag{2.1}$$
$$F_1 : \mathbb{R}^n \times \mathbb{R}^n \times \mathbb{R} \to \mathbb{R}^n$$

mit der Anfangswertbedingung

$$x(t_0) = x^0 \tag{2.2}$$

und $x^0 \in \mathbb{R}^n$ und $t \in [t_0, t_e]$.

Gesucht ist die Funktion $x(t)$ mit $t \in [t_0, t_e]$, so daß das System von Algebro–Differentialgleichungen (2.1) und die Anfangswertbedingung (2.2) erfüllt sind.

1991 *Mathematics Subject Classification.* 65W05, 65L05, 65H10, 65F50; Secondary 68Q35.
Key words and phrases. Algebro–Differentialgleichungen, strukturierte nichtlineare Gleichungssysteme, Gleichungssysteme mit sparsam besetzten Matrizen, Parallelisierung numerischer Verfahren.

Es wird vorausgesetzt, daß eine Lösung von (2.1) und (2.2) existiert.

Für alle weiteren Betrachtungen wird angenommen, daß (2.1) die folgende Struktur hat:

$$F(u_1(t), \dot{u}_1(t), y_1(t), \dot{y}_1(t), \ldots, u_r(t), \dot{u}_r(t), y_r(t), \dot{y}_r(t), w(t), \dot{w}(t), t) = 0, \quad t \geq t_0 \ (2.3)$$

$$F : \mathbb{R}^{\sigma_1} \times \mathbb{R}^{\sigma_1} \times \mathbb{R}^{\sigma_1} \times \mathbb{R}^{\sigma_1} \times \cdots \times \mathbb{R}^{\sigma_r} \times \mathbb{R}^{\sigma_r} \times \mathbb{R}^{\sigma_r} \times \mathbb{R}^{\sigma_r} \times \mathbb{R}^{\rho} \times \mathbb{R}^{\rho} \times \mathbb{R} \rightarrow \mathbb{R}^{\sigma_1 + \cdots + \sigma_r + \rho}$$

und

$$H_i(u_i(t), \dot{u}_i(t), y_i(t), \dot{y}_i(t), x_i(t), \dot{x}_i(t), t) = 0, \quad t \geq t_0 \tag{2.4}$$

$$H_i : \mathbb{R}^{\sigma_i} \times \mathbb{R}^{\sigma_i} \times \mathbb{R}^{\sigma_i} \times \mathbb{R}^{\sigma_i} \times \mathbb{R}^{\pi_i} \times \mathbb{R}^{\pi_i} \times \mathbb{R} \rightarrow \mathbb{R}^{\sigma_i + \pi_i}$$

$$\text{für} \quad i = 1, 2, \ldots, r$$

und den Anfangswertbedingungen

$$w(t_0) = w^0,$$
$$u_i(t_0) = u_i^0, \quad y_i(t_0) = y_i^0, \quad x_i(t_0) = x_i^0, \tag{2.5}$$
$$\text{für} \quad i = 1, 2, \ldots, r.$$

Aus (2.3) und (2.4) folgt, daß die Anzahl der Gleichungen und der unbekannten Funktionen

$$n = \rho + \sum_{i=1}^{r} (2\sigma_i + \pi_i)$$

ist.

Es sei bemerkt, daß die Funktion $w(t)$ nur in (2.3) und die Funktionen $x_i(t)$ nur in (2.4) vorkommen, während die $u_i(t)$ und $y_i(t)$ in (2.3) und (2.4) auftreten.

Da in Anbetracht der Implizität, der Nichtlinearität und der Anzahl der Gleichungen eine analytische Lösung von (2.1), (2.2) bzw. des strukturierten Systems (2.3), (2.4), (2.5) im allgemeinen nicht durchführbar ist, werden numerische Verfahren benutzt.

Die prinzipiellen Schritte bei der numerischen Lösung des Systems (2.3) – (2.5) sind, man vergleiche [2], [3], [6], [11], [12]:

1. Diskretisierung von (2.3) und (2.4) mit Integrationsformeln,
2. Numerische Lösung der entstehenden nichtlinearen Gleichungssysteme mit Newton– bzw. anderen Iterationsverfahren,
3. Numerische Lösung der sich ergebenden linearen Gleichungssysteme mit i.a. schwachbesetzten Matrizen mit Sparse–Matrix–Techniken.

In den nachfolgenden Darlegungen wird auf die einzelnen Schritte detaillierter eingegangen.

3. Numerische Lösung

3.1. System von Algebro–Differentialgleichungen. Betrachtet wird das System (2.3) und (2.4) zum Zeitpunkt $t = t_{n+1}$.

Mit den rückwärtigen Differenzenformeln [4]

$$-h\dot{x}_{n+1} = \sum_{i=0}^{k} \alpha_i x_{n+1-i},$$

wobei $\dot{x}_{n+1}$ eine Approximation von $\dot{x}(t)$ zum Zeitpunkt $t = t_{n+1}$ und $x_n = x(t_n)$ bezeichnet und $h = t_{n+1} - t_n$ ist, ergibt sich das strukturierte System von nichtlinearen Gleichungen

$$F(u_1, y_1, \ldots, u_r, y_r, w) = 0 \tag{3.1}$$
$$F : \mathbb{R}^{\sigma_1} \times \mathbb{R}^{\sigma_1} \times \ldots \times \mathbb{R}^{\sigma_r} \times \mathbb{R}^{\sigma_r} \times \mathbb{R}^{\rho} \to \mathbb{R}^{\sigma_1 + \cdots + \sigma_r + \rho}$$

und

$$H_i(u_i, y_i, x_i) = 0 \tag{3.2}$$
$$H_i : \mathbb{R}^{\sigma_i} \times \mathbb{R}^{\sigma_i} \times \mathbb{R}^{\pi_i} \to \mathbb{R}^{\sigma_i + \pi_i}$$
$$\text{für} \quad i = 1, 2, \ldots, r$$

bzw. in der Form

$$\widehat{F}(x_1, u_1, \ldots, x_r, u_r, y_1, \ldots, y_r, w) = \begin{bmatrix} H_1(u_1, y_1, x_1) \\ \vdots \\ H_r(u_r, y_r, x_r) \\ F(u_1, y_1, \ldots, u_r, y_r, w) \end{bmatrix} = 0. \tag{3.3}$$

Es sei erwähnt, daß bei der Integration der Differentialgleichungen unter Nutzung verschiedener Kriterien eine Schrittweiten– und/oder Ordnungssteuerung benutzt wird.

Für die numerische Lösung von (3.1) und (3.2) werden unter besonderer Berücksichtigung der Struktur der Systeme Newton– und Gauß–Seidel–Verfahren eingesetzt.

3.2. Nichtlineare Gleichungssysteme.

Zweistufiges Newton–Verfahren, man vergleiche [6], [9], [12]. Es wird angenommen, daß aus (3.2)

$$y_i = G_i(u_i) \tag{3.4}$$

ableitbar ist, womit aus (3.1)

$$F(u_1, G_1(u_1), \ldots, u_r, G_r(u_r), w) = 0$$

folgt. Mit $u^T = (u_1, \ldots, u_r)$ ergibt sich die Newton–Korrektur des äußeren Verfahrens

$$\begin{bmatrix} u^{(1)} \\ w^{(1)} \end{bmatrix} = \begin{bmatrix} u^{(0)} \\ w^{(0)} \end{bmatrix} + \begin{bmatrix} \Delta u \\ \Delta w \end{bmatrix} \tag{3.5}$$

aus

$$\left[\frac{\partial F}{\partial u} + \left(\frac{\partial F}{\partial G_1} \cdot \frac{\partial G_1}{\partial u_1}, \ldots, \frac{\partial F}{\partial G_r} \cdot \frac{\partial G_r}{\partial u_r} \right) \frac{\partial F}{\partial w} \right] \begin{bmatrix} \Delta u \\ \Delta w \end{bmatrix} = -F \tag{3.6}$$

und die Korrekturen der inneren Verfahren

$$\begin{bmatrix} x_i^{(1)} \\ y_i^{(1)} \end{bmatrix} = \begin{bmatrix} x_i^{(0)} \\ y_i^{(0)} \end{bmatrix} + \begin{bmatrix} \Delta x_i \\ \Delta y_i \end{bmatrix}, \qquad i = 1, 2, \ldots, r \tag{3.7}$$

ergeben sich aus

$$\begin{bmatrix} \frac{\partial H_i}{\partial x_i} & \frac{\partial H_i}{\partial y_i} \end{bmatrix} \begin{bmatrix} \Delta x_i \\ \Delta y_i \end{bmatrix} = -H_i \qquad i = 1, \ldots, r. \tag{3.8}$$

Für die Berechnung der Jacobi–Matrizen des äußeren Verfahrens werden

$$\frac{\partial G_i}{\partial u_i}$$

benötigt, die wie folgt bestimmt werden. Mit der Annahme, daß aus (3.2)

$$x_i = G_i^*(u_i)$$

ableitbar ist und mit (3.4) kann unter Verwendung von (3.2) die Gleichung

$$\begin{bmatrix} \frac{\partial H_i}{\partial x_i} & \frac{\partial H_i}{\partial y_i} \end{bmatrix} \begin{bmatrix} \frac{\partial G_i^*}{\partial u_i} \\ \frac{\partial G_i}{\partial u_i} \end{bmatrix} = -\frac{\partial H_i}{\partial u_i}, \tag{3.9}$$

abgeleitet werden. Aus der letzten Gleichung ergibt sich die gesuchte Größe. Es sei bemerkt, daß die Koeffizienten–Matrizen von (3.8) und (3.9) übereinstimmen.

Eine andere Möglichkeit der Berechnung von $\frac{\partial G_i}{\partial u_i}$ wurde in [12] vorgeschlagen. Hierzu wird die Koeffizientenmatrix von (3.9) mit der Matrix M^T multipliziert und M so bestimmt, daß

$$M^T \cdot \begin{bmatrix} \frac{\partial H_i}{\partial x_i} & \frac{\partial H_i}{\partial y_i} \end{bmatrix} = (0, I)$$

ist, wobei 0 eine Nullmatrix und I eine Einheitsmatrix bezeichnen.

Die letzte Gleichung kann in der Form

$$\begin{bmatrix} \frac{\partial H_i}{\partial x_i} & \frac{\partial H_i}{\partial y_i} \end{bmatrix}^T \cdot M = \begin{bmatrix} 0 \\ I \end{bmatrix}$$

geschrieben werden. Es ist schließlich

$$\frac{\partial G_i}{\partial u_i} = -M^T \cdot \frac{\partial H_i}{\partial u_i}. \tag{3.10}$$

In [12] ist ein Konvergenzsatz für das zweistufige Newton–Verfahren bewiesen. Es wird gezeigt, daß es quadratisch konvergiert, falls das innere Verfahren gestoppt wird, wenn

$$\left\|(\Delta x_i, \Delta y_i)\right\| \leq \left\|(\Delta u, \Delta w)\right\|^2 \tag{3.11}$$

erfüllt ist.

Einstufiges Newton–Verfahren [6]

Mit

$$z = (x_1, u_1, \ldots, x_r, u_r, y_1, \ldots, y_r, w)$$

und (3.3) berechnet sich die Korrektur des einstufigen Newton–Verfahrens

$$z^{(1)} = z^{(0)} + \Delta z \tag{3.12}$$

aus

$$\begin{bmatrix} \frac{\partial H_1}{\partial x_1} & \frac{\partial H_1}{\partial u_1} & & & & \frac{\partial H_1}{\partial y_1} & & \\ & \ddots & & & & & \ddots & \\ & & \frac{\partial H_r}{\partial x_r} & \frac{\partial H_r}{\partial u_r} & & & & \frac{\partial H_r}{\partial y_r} \\ & \frac{\partial F}{\partial u_1} & \cdots & \frac{\partial F}{\partial u_r} & \frac{\partial F}{\partial y_1} & \cdots & \frac{\partial F}{\partial y_r} & \frac{\partial F}{\partial w} \end{bmatrix} \Delta z = -\widehat{F}. \tag{3.13}$$

Für die Lösung des linearen Systems (3.13) wird das Gaußsche Block–Eliminationsverfahren benutzt. Hierzu wird (3.13) in der folgenden Form geschrieben

$$\begin{bmatrix} A_1 & & & B_1 & & & \\ & \ddots & & & \ddots & & \\ & & A_r & & & B_r & \\ C_1 & \cdots & C_r & D_1 & \cdots & D_r & E \end{bmatrix} \begin{bmatrix} \Delta z_1 \\ \vdots \\ \Delta z_r \\ \Delta z_{r+1} \end{bmatrix} = \begin{bmatrix} b_1 \\ \vdots \\ b_r \\ b_{r+1} \end{bmatrix} \tag{3.14}$$

Das Gaußsche Block–Eliminationsverfahren für (3.14) erfordert die folgenden Schritte

1. Faktorisierung von $A_i = L_i \cdot U_i, \quad i = 1, 2, \ldots, r$
2. Berechnung von F_i^T unter Verwendung der Faktorisierung von 1.

$$U_i^T \cdot Z_i = C_i^T$$
$$L_i^T \cdot F_i^T = Z_i$$
$$i = 1, 2, \ldots, r$$

3. Berechnung der $\widehat{D}_i$ und $\hat{b}_i$ mit

$$\widehat{D}_i = D_i - F_i B_i$$
$$\hat{b}_i = F_i b_i$$
$$i = 1, 2, \ldots, r$$

4. Lösung des Gleichungssystems

$$(\widehat{D}_1 \cdots \widehat{D}_r E)\Delta z_{r+1} = b_{r+1} - \sum_{i=1}^{r} \hat{b}_i$$

5. Berechnung der Δz_i, $i = 1, \ldots, r$ mit

$$L_i \, u_i = b_i - B_i \, \Delta z_{r+1}$$
$$U_i \, \Delta z_i = u_i$$
$$i = 1, 2, \ldots, r$$

Gauß–Seidel–Verfahren

Es wird wiederum von (3.3) ausgegangen und auf dieses System ein Block–Gauß–Seidel–Verfahren angewendet. Die Iterationsvorschrift lautet

$$H_1(u_1^{(0)}, y_1^{(1)}, x_1^{(1)}) = 0$$
$$\ldots\ldots$$
$$H_r(u_r^{(0)}, y_r^{(1)}, x_r^{(1)}) = 0 \qquad\qquad (3.15)$$
$$F(u_1^{(1)}, y_1^{(1)}, \ldots, u_r^{(1)}, y_r^{(1)}, w^{(1)}) = 0.$$

Die Iterationsvorschrift (3.15) bedeutet eine Entkopplung des nichtlinearen Gleichungssystems (3.3) in Blöcken. Als nächstes müssen die in (3.15) vorhandenen nichtlinearen Gleichungssysteme gelöst werden.

Die nichtlinearen Systeme

$$H_i(u_i^{(0)}, x_i^{(1)}, y_i^{(1)}) = 0, \quad i = 1, 2, \ldots, r \qquad\qquad (3.16)$$

werden analog dem inneren Verfahren beim zweistufigen Newton-Verfahren behandelt. Für eine Vereinfachung der Schreibweise wird (3.16) notiert als

$$H_i(\tilde{u}_i, \tilde{x}_i, \tilde{y}_i) = 0,$$

womit die Iterationsvorschrift

$$\begin{bmatrix} \tilde{x}_i^{(1)} \\ \tilde{y}_i^{(1)} \end{bmatrix} = \begin{bmatrix} \tilde{x}_i^{(0)} \\ \tilde{y}_i^{(0)} \end{bmatrix} + \begin{bmatrix} \Delta \tilde{x}_i \\ \Delta \tilde{y}_i \end{bmatrix}, \quad i = 1, 2, \ldots, r \qquad\qquad (3.17)$$

lautet und die Newton-Korrekturen durch Lösung der Gleichungssysteme

$$\begin{bmatrix} \frac{\partial H_i}{\partial \tilde{x}_i} & \frac{\partial H_i}{\partial \tilde{y}_i} \end{bmatrix} \begin{bmatrix} \Delta \tilde{x}_i \\ \Delta \tilde{y}_i \end{bmatrix} = -H_i(\tilde{u}_i, \tilde{x}_i^{(0)}, \tilde{y}_i^{(0)}), \quad i = 1, 2, \ldots, r \qquad\qquad (3.18)$$

gefunden werden. Schließlich muß noch die Lösung der letzten Gleichung von (3.15) durchgeführt werden. Die Anwendung des Newton-Verfahrens, wobei die letzte Gleichung von (3.15) geschrieben wurde als

$$F(\tilde{u}_1, \tilde{y}_1, \ldots, \tilde{u}_r, \tilde{y}_r, \tilde{w}) = 0,$$

gibt die Iterationsvorschrift

$$
\begin{bmatrix} \tilde{u}_1^{(1)} \\ \vdots \\ \tilde{u}_r^{(1)} \\ \tilde{w}^{(1)} \end{bmatrix} = \begin{bmatrix} \tilde{u}_1^{(0)} \\ \vdots \\ \tilde{u}_r^{(0)} \\ \tilde{w}^{(0)} \end{bmatrix} + \begin{bmatrix} \Delta\tilde{u}_1 \\ \vdots \\ \Delta\tilde{u}_r \\ \Delta\tilde{w} \end{bmatrix}
$$

mit

$$
\begin{bmatrix} \frac{\partial F}{\partial u_1} \cdots \frac{\partial F}{\partial u_r} \frac{\partial F}{\partial w} \end{bmatrix} \begin{bmatrix} \Delta\tilde{u}_1 \\ \vdots \\ \Delta\tilde{u}_r \\ \Delta\tilde{w} \end{bmatrix} = -F(\tilde{u}_1^{(0)}, \ldots, \tilde{u}_r^{(0)}, \tilde{w}^{(0)}).
$$

Die Anwendung des Gauß–Seidel–Verfahrens auf (3.3) bereitet also keine Schwierigkeiten. Ein Nachteil des Gauß–Seidel–Verfahrens ist, daß es nur linear konvergiert. Der Wert des Gauß–Seidel–Verfahrens dürfte insbesondere in der Behandlung tiefer strukturierter nichtlinearer Gleichungssysteme liegen. Hiermit ist gemeint, daß (2.4) eine Struktur analog (2.3) und (2.4) hat. Für die Netzwerksimulation bedeutet dies, daß die Unternetzwerke wiederum Unternetzwerke sind. Ein so aufgebautes Netzwerk könnte in der äußeren Stufe durch Anwendung des Gauß–Seidel–Verfahrens und in den inneren Stufen durch die Benutzung des ein– bzw. zweistufigen Newtonverfahrens gelöst werden.

3.3. Lineare Gleichungssysteme. Bei der Anwendung des zweistufigen und einstufigen Newton–Verfahrens und beim Gauß–Seidel–Verfahren für die Lösung der nichtlinearen Gleichungssysteme sind lineare Gleichungssysteme zu lösen. Es ist

$$
\tilde{A}x = b \quad \text{mit } \tilde{A} \in \mathbb{R}^{n \times n} \quad \text{und } x, b \in \mathbb{R}^n. \tag{3.19}
$$

Bei der numerischen Simulation des elektrischen Verhaltens von Schaltkreisen unter Benutzung der eben erwähnten Verfahren haben die Matrizen der linearen Gleichungssysteme (3.19) die folgenden Eigenschaften:

$\tilde{A}$ unsymmetrisch, sparsam besetzt

n im allgemeinen groß.

Es sei erwähnt, daß sich die Besetztheit der Matrizen mit Nichtnull– bzw. Nullelementen während der Rechnung nicht ändert. Für die linearen Gleichungssysteme zeichnet sich demnach die folgende Aufgabenstellung ab. Es sind viele lineare Systeme mit unsymmetrischen Matrizen und gleicher Besetztheit mit Nichtnull– und Nullelementen zu lösen. Die Aufgabe wird mit dem Gaußchen Eliminationsverfahren behandelt, wobei dieses auf die genannte Problemstellung angepaßt wird.

Die Gleichung (3.19) wird also mit

$$
\begin{aligned}
P\tilde{A}Q &= L \cdot U \\
Ly &= Pb \\
UQ^{-1}x &= y
\end{aligned} \tag{3.20}
$$

gelöst, wobei P und Q Permutationsmatrizen und L eine untere und U eine obere Dreiecksmatrix sind. Die Hauptdiagonalelemente von L sind alle Eins.

Da die Matrizen von (3.19) sparsam besetzt sind, werden nur die Nichtnullelemente (NNE) gespeichert. Es wird das folgende Schema benutzt. Die Nichtnullelemente (NNE) von $\tilde{A}$ und deren Spaltenindizes werden in den Vektoren A und JA und die Indizes der Zeilenanfänge von $\tilde{A}$ in A im Vektor IA gespeichert. Es sind (m–Anzahl der NNE)

$$A \in \mathbb{R}^m, \quad JA \in \mathbb{N}^m \quad \text{und} \quad IA \in \mathbb{N}^{n+1},$$

wobei das letzte Element von IA gleich den um Eins erhöhten letzten Index von A bzw. JA ist. Die Matrix

$$\tilde{A} = \begin{bmatrix} 9 & & 2 & 1 \\ 1 & & 3 & & 5 \\ & 2 & & 4 & \\ 1 & & 7 & & 8 \\ & 5 & & 7 & 9 \end{bmatrix} \tag{3.21}$$

ist demnach wie folgt gespeichert:

$$A = [9, 2, 1, \ 1, 3, 5, \ 2, 4, \ 1, 7, 8, \ 5, 7, 9]^T$$
$$JA = [1, 4, 5, \ 1, 3, 5, \ 2, 4, \ 1, 3, 5, \ 2, 4, 5]^T$$
$$IA = [1, 4, 7, 9, 12, 15]^T.$$

Bei der Bestimmung der Permutationsmatrizen P und Q, d.h. der Festlegung einer Pivotstrategie, sind verschiedene Kriterien zu erfüllen. Die Strategie muß so sein, daß das Eliminationsverfahren numerisch stabil und die Anzahl der während der Elimination entstehenden Nichtnullelemente, d.h. das Fill-in, möglichst klein sind. Es sei erwähnt, daß sich die beiden Forderungen widersprechen können. Die Bestimmung des Pivotelementes läuft in jedem Eliminationsschritt wie folgt ab, wobei der Einfachheit wegen der 1. Schritt betrachtet wird.

Es sei

$$\tilde{A} = (a_{i,j}).$$

Zur Wahl als Pivotelement werden nur diejenigen Elemente zugelassen, die die sogenannte β–Bedingung erfüllen. Mit $\mathbb{I} = \{1, 2, \dots, n\}$ sei

$$\hat{a}_j = \max_{i \in \mathbb{I}} |a_{i,j}|, \quad \forall j \in \mathbb{I}.$$

Ein Matrixelement $a_{i,j} \neq 0$ erfüllt die β–Bedingung, falls zu einem vorgegebenen β mit $0 \leq \beta \leq 1$

$$\hat{a}_j \cdot \beta \leq |a_{i,j}|, \quad i, j \in \mathbb{I}$$

gilt. Entsprechend der Definition erfüllt in jeder Matrixspalte wenigstens ein Element die β–Bedingung.

Es bezeichne r_i die Anzahl der NNE in der i–ten Zeile und c_j die in der j–ten Spalte. Nach Markowitz [10] werden für jedes $a_{i,j} \neq 0$ Kosten nach der Beziehung

$$(r_i - 1)(c_j - 1)$$

definiert.

Für die Bestimmung eines Pivotelements werden die folgenden 4 Strategien betrachtet, wobei nur die die β–Bedingung erfüllenden Matrixelemente zur Pivotauswahl zugelassen werden. Falls mehrere Elemente die Bedingungen erfüllen, wird das betragsmäßig größte Element als Pivot genommen.

Strategie	
1	Pivot wird das Element mit den geringsten Kosten.
2	Es wird die Zeile mit minimaler Anzahl von Elementen, bei mehreren die erste, bestimmt. In den dadurch bestimmten Spalten wird Pivot das Element mit den geringsten Kosten.
3	Es wird die Spalte mit minimaler Anzahl von Elementen, bei mehreren die erste, bestimmt. In dieser dadurch bestimmten Spalte wird Pivot das Element mit den geringsten Kosten.
4	Es werden, falls mehrere vorhanden sind, alle Spalten mit minimaler Anzahl von Elementen bestimmt. In diesen Spalten wird Pivot das Element mit den geringsten Kosten.

Der numerische Aufwand für die einzelnen Strategien ist

Strategie	Operationen
1	$O(n \cdot m)$
2	$O(n \cdot n)$
3	$O(n \cdot n)$
4	$O(n \cdot n)$

Als Beispiel wurde ein lineares Gleichungssystem mit 2 904 Gleichungen und 58 142 Nichtnullelementen betrachtet, es wird im folgenden mit (LGS) bezeichnet.

Es ergab sich (VAX 4000–300, Zeitangaben in Sekunden)

Strategie	Fill–in	Zeit
1	10 100	533,56
2	16 246	24,08
3	18 870	9,12
4	16 738	26,71

Wie bereits oben dargelegt, sind bei der numerischen Lösung von strukturierten Systemen von Algebro–Differentialgleichungen viele lineare Gleichungssysteme mit der gleichen Besetztheit mit Nichtnullelementen zu lösen. Hierzu werden entweder ausführbare Maschinenprogramme generiert [7], deren Ausführung die Lösung der Gleichungssysteme mit

dem speziellen Gaußchen Eliminationsverfahren ist oder es wird ein sogenannter Pseudo–
Code erzeugt, dessen Interpretation die Lösung des Systems gibt. Die erste Methode ist
sehr schnell und vollständig vom Maschinencode des benutzten Computers abhängig. Das
zweite Verfahren kann unabhängig von einem Computer formuliert werden. Eine allge-
meine Darlegung wird an einer anderen Stelle gegeben. Das Vorgehen soll hier an einem
Beispiel erläutert werden. Gegeben sei wiederum die Matrix $\tilde{A}$ (vgl. (3.21)). Nach der
Bestimmung der Permutationsmatrizen P und Q ist

$$P\tilde{A}Q = \begin{bmatrix} 2 & 4 & & & \\ 5 & 7 & & & 9 \\ & 2 & 9 & & 1 \\ & & 1 & 7 & 8 \\ & & 1 & 3 & 5 \end{bmatrix} \tag{3.22}$$

Die Matrix $P\tilde{A}Q$ ist im Vektor A wie folgt gespeichert, wobei $\textcircled{i}$ das i–te Element in A
bezeichnet.

$$\begin{bmatrix} \textcircled{7} & \textcircled{8} & & & \\ \textcircled{12} & \textcircled{13} & & & \textcircled{14} \\ & \textcircled{2} & \textcircled{1} & & \textcircled{3} \\ & & \textcircled{9} & \textcircled{10} & \textcircled{11} \\ & & \textcircled{4} & \textcircled{5} & \textcircled{6} \end{bmatrix}$$

Die Fakorisierung von $\tilde{A}$ erfordert demnach die folgenden Operationen:

$$\begin{aligned}
A(12) &= A(12)/A(7) \\
A(13) &= A(13)-A(12)*A(8) \\
A(2) &= A(2)/A(13) \\
A(3) &= A(3)-A(2)*A(14) \\
A(9) &= A(9)/A(1) \\
A(11) &= A(11)-A(9)*A(3) \\
A(4) &= A(4)/A(1) \\
A(5) &= A(5)/A(10) \\
A(6) &= A(6)-A(4)*A(3)-A(5)*A(11)
\end{aligned}$$

Für die Arbeit mit dem Pseudo–Code werden von den obigen Ergibtanweisungen nur ein
Kennzeichen für die Art der Operation (es sind 6 ausreichend), die Indizes und die Länge
des Skalarproduktes gespeichert.

Bei der Vor– und Rückwärtsrechnung wird mit einem entsprechenden Pseudo–Code ge-
arbeitet.

Der Vorteil der Anwendung des Pseudo–Codes gegenüber der des genierten Maschinenprogramms besteht darin, daß er für seine Speicherung weniger Platz benötigt. Die Re-

chenzeit für die Interpretation des Pseudo–Codes ist allerdings größer als die für die Ausführung des generierten Maschinenprogramms.

Bei der numerischen Lösung des bereits angegebenen Gleichungssystems (LGS) mit einer etwas anderen Pivotstrategie ergaben sich die folgenden Werte (MicroVAX 3600, doppelte Genauigkeit, FORTRAN–Programme; $\beta = 10^{-6}$)

Fill–in	10 232	
Operationen Faktorisierung	271 133	
Pseudo–Code Faktorisierung	2 259 272	Byte
Pseudo–Code VR–Rechnung	581 840	Byte
Pivotstrategie	725,77	Sek.
Generierung Pseudo–Code	35,74	Sek.
Faktorisierung	4,17	Sek.
VR–Rechnung	0,48	Sek.
geschätzte Kondition	$1,14 \cdot 10^{12}$	

4. PARALLELISIERUNG DER NUMERISCHEN VERFAHREN

Betrachtet wird nun die Nutzung von Parallel– bzw. Vektorcomputern bei den angewendeten numerischen Verfahren. Sie können auf dem Niveau

- der Algebro–Differentialgleichungen,
- der nichtlinearen Gleichungssysteme,
- der linearen Gleichungssysteme

eingesetzt werden.

4.1. Niveau Algebro–Differentialgleichungen.
Es wird von der Formulierung der Aufgabenstellung nach (2.3) und (2.4) ausgegangen. Aus (2.4) ist unmittelbar ersichtlich, daß die Gleichungen unabhängig voneinander gelöst werden können. Angenommen, es können r Prozessoren benutzt werden. Dann bietet sich für die Parallelisierung der numerischen Lösung im Intervall $[t_0, t_1]$ der folgende Algorithmus an:

0. Man setze $k = 0$.
1. Man setze eine Näherung für

$$u_i^{(k)}(t), \quad i = 1, \dots, n, \quad t \in [t_0, t_1].$$

2. Man löse mit einem numerischen Verfahren auf jedem der r Prozessoren ein System der Form

$$H_i(u_i^{(k)}, \dot{u}_1^{(k)}, y_i^{(k)}, \dot{y}_i^{(k)}, x_i^{(k)}, \dot{x}_i^{(k)}, t) = 0 \qquad i = 1, \dots, r, \quad t \in [t_0, t_1)]$$

mit dem Ergebnis $x_i^{(k)}(t)$ und $y_i^{(k)}(t)$.

3. Man löse mit einem numerischen Verfahren auf einem Prozessor

$$F(u_1^{(k+1)}, \dot{u}_1^{(k+1)}, y_1^{(k)}, \dot{y}_1^{(k)}, \dots, w^{(k)}, \dot{w}^{(k)}, t) = 0, \qquad t \in [t_0, t_1]$$

mit dem Ergebnis $u_i^{(k+1)}(t), w^{(k)}(t)$.

4. Falls $\|u_i^{(k+1)}(t) - u_i^{(k)}(t)\| \leq \epsilon \quad \forall i$ (ϵ – vorgegebene Fehlerschranke), dann neues Zeitintervall und nach 0, sonst $k = k + 1$ und nach 2.

Der Vorteil dieses Zuganges liegt darin, daß im Schritt 2 jeder der Prozessoren viele arithmetische Operationen durchführen muß und der Algorithmus nur eine geringe Kommunikation erfordert.

4.2. Niveau nichtlinearer Gleichungssysteme. Eine Parallelisierung der Verfahren bietet sich entsprechend der Formulierung sowohl für die ein- und zweistufigen Newton-Verfahren als auch für das Gauß–Seidel–Verfahren unmittelbar an.

Für die einzelnen Verfahren ergibt sich:

Zweistufiges Newton–Verfahren

Der Algorithmus kann für einen festen Zeitschritt wie folgt formuliert werden:

0. Man setze $l = 0$.
1. Man setze eine Näherung für $u_i^{(l)}, w^{(l)}, \quad i = 1, \ldots, r$.
2. Man setze $k = 0$.
3. Man setze eine Näherung für $x_i^{(k)}, y_i^{(k)}, \quad i = 1, \ldots, r$.
4. Man löse auf jedem der r Prozessoren ein Gleichungssystem der Form (3.8) und führe einen Newtonschritt nach (3.7) mit dem Ergebnis $x_i^{(k+1)}, y_i^{(k+1)}$ durch.
5. Falls nach (3.11)

$$\|(\Delta x_i^{(k+1)}, \Delta y_i^{(k+1)}\| > \|(\Delta u^{(k)}, \Delta w^{(k)})\|^2$$

   ist, dann $k = k + 1$ und gehe nach 4.
6. Man berechne die Jacobi-Matrixen $\frac{\partial G_i}{\partial u_i}$ mit (3.9) bzw. (3.10).
7. Man löse auf einem Prozessor das lineare System (3.6) und führe einen Newtonschritt nach (3.5) mit dem Ergebnis $u_i^{(l+1)}$ und $w^{(l+1)}$ durch.
8. Falls $\|(\Delta u_i^{(l+1)}, \Delta w^{(l+1)}\| > \eta$ (η – vorgegebene Fehlerschranke) ist, setze $l = l + 1$ und gehe nach 2, sonst nächster Zeitschritt.

Es sei bemerkt, daß 4. – 6. parallel auf den Prozessoren durchgeführt werden können.

Einstufiges Newton–Verfahren

Bei einem Newtonschritt nach (3.12) kann die Berechnung von Δz unter Nutzung der parallelen Prozessoren vorgenommen werden.

Beim Gaußchen Block-Eliminationsverfahren für (3.14) können die Schritte 1., 2., 3. und 5. unter Nutzung von r Prozessoren parallel ausgeführt werden, während der Schritt 4. auf einem Prozessor ablaufen muß.

Das Gauß–Seidel–Verfahren kann ebenso parallelisiert werden. Die Parallelisierung ergibt sich unmittelbar aus (3.17) und (3.18).

4.3. Niveau linearer Gleichungssysteme. Unter Nutzung des in 3.3 eingeführten Pseudo–Codes können Algorithmen für Parallel– als auch für Vektorrechner formuliert werden. Hier werden nur die für Vektorrechner betrachtet.

Die Idee hierbei ist, bei den Anweisungen für den Pseudo–Code die zu erkennen, die voneinander unabhängig sind. Diese werden extrahiert und falls möglich zu entsprechenden Vektorbefehlen zusammengefaßt.

Eine Erkennung der entsprechenden Unabhängigkeiten ist mit dem Algorithmus von Yamamoto und Takahashi [13] möglich. Hierzu wird der Matrix

$$LU = P\tilde{A}Q$$

aus (3.20) die Matrix

$$M = (m_{i,j}) \qquad m_{i,j} \in \mathbf{N} \cup \{0\}$$

zugeordnet und $m_{i,j}$ als das Niveau der Unabhängigkeit bezeichnet. Falls einem Matrixelement das Niveau Null zugeordnet ist, sind bei der Faktorisierung keine Operationen erforderlich. Alle Matrixelemente mit gleichem Niveau können unabhängig voneinander berechnet werden. Bei der Faktorisierung sind zuerst alle Matrixelemente mit dem Niveau 1, dann mit Niveau 2 usw. zu bestimmen.

Nachfolgend ist der Algorithmus für die Berechnung der Elemente $m_{i,j}$ der Matrix M gegeben

```
M = 0
for I = 1,N-1 do
    for all {J: A(J,I )≠ 0 & J>I} do
        M(J,I) = 1+max(M(J,I),M(I,I))
        for all {K: A(I,K) ≠ 0 & K>I} do
            M(J,K) = 1+max(M(J,K),M(J,I),M(I,K))
        end
    end
end
```

Als Beispiel werde die Matrix $\tilde{A}$ (vgl. (3.21)) betrachtet. Es ergibt sich

$$M = \begin{bmatrix} 0 & 0 & & & \\ 1 & 2 & & & 0 \\ & 3 & 0 & & 4 \\ & & 1 & 0 & 5 \\ & & 1 & 1 & 6 \end{bmatrix}.$$

Die Faktorisierung wird nun in der folgenden Reihenfolge der Ergibtanweisungen vorgenommen:

Niveau	Anweisung
1	A(12) = A(12)/A(7) A(9) = A(9)/A(1) A(4) = A(4)/A(1) A(5) = A(5)/A(10)
2	A(13) = A(13) − A(12) *A(8)
3	A(2) = A(2)/A(13)
4	A(3) = A(3) − A(2) * A(14)
5	A(11) = A(11) − A(5) * A(3)
6	A(6) = A(6) − A(4) * A(3) − A(5) * A(11)

Für die Nutzung von Vektorcomputern müssen auf den einzelnen Niveaus Vektorbefehle erkannt werden. Es haben sich die folgenden bewährt:

Nr. Vektoroperation	Operation
1	Skalarprodukt
2	A(K) = 1/A(K)
3	A(K) = A(K) * A(L)
4	A(K) = (A(I) * A(J) + A(L) * A(M))A(M)

Die Schwierigkeit besteht hier darin, daß die Feldelemente indirekt adressiert werden. Im allgemeinen stehen für die einzelnen Vektorcomputer entsprechende Befehle zur Verfügung.

Für das bereits angegebene Gleichungssystem (LGS) wurden 171 Niveaus gefunden und folgendes festgestellt:

Niveau	Anzahl Anweisungen
0	20 565
1	1 952
2	18 399
3	1 362
4	4 238
5	5 124
⋮	⋮

Die größten Vektorlängen bei den Vektoroperationen waren:

Vektoroperation	Vektorlänge
2	1 952
3	18 399
3	357
3	639

Bei dem Gleichungssystem waren 91,2 % der Operationen vektorisierbar, wobei nur solche mit einer Vektorlänge größer als 3 gezählt worden sind.

LITERATUR

1. J. Borchardt, I. Bremer, F Grund, D. Horn, M Uhle, *MAGNUS - Mehrstufige Analyse großer Netzwerke auf der Basis von Untersystemen*, Proceedings, 3. Tagung Schaltkreisentwurf, Dresden, 1989, S. 148–152.

2. L.O Chua and P.-M. Lin, *Computer-Aided Analysis of Electronic Circuits*, Prentice–Hall, Inc , Englewood Cliffs, New Jersey, 1975

3. H. Elschner, A. Möschwitzer, A. Reibiger, *Rechnergestützte Analyse in der Elektronik*, Verlag Technik, Berlin, 1977.

4. C.W. Gear, *Numerical initial value problems in ordinary differential equations*, Prentice–Hall, Inc., Englewood Cliffs, New Jersey, 1971

5. F. Grund, *Modellierung nichtlinearer elektrischer Netzwerke*, Forschungsbericht, 26 S., Institut für Mathematik, Mohrenstraße 39, O – 1086 Berlin, 1981.

6 _____, *Nichtlineare Gleichungssysteme bei elektrischen Netzwerken mit Unternetzwerken*, Forschungsbericht, 24 S., Institut für Mathematik, Mohrenstraße 39, O – 1086 Berlin, 1981.

7. _____, *Numerische Lösung von Gleichungssystemen mit sparsam besetzten Matrizen*, Report R-Math-01/83, Tagung Numerische Lösung von Differentialgleichungen, Matzlow/Garwitz, 1982, S. 33–44.

8 C W Ho, A.E. Ruehli, and P C. Brennan, *The modified nodal approach to network analysis*, IEEE Trans. Circuits and Systems **CAS–22** (1975), 504–509.

9. W Hoyer and J.W. Schmidt, *Newton-type decomposition methods for equations arising in network analysis*, Z. Angew. Math Mech **64** (1984), 397–405.

10 H.M. Markowitz, *The elimination form of the inverse and its application to linear programming*, Management Sci. **3** (1957), 255–269.

11. N B. Rabbat and H.Y. Hsieh, *A latent macromodular approach to large-scale sparse networks*, IEEE Trans Circuits and Systems **CAS–23** (1976), 745–752.

12 N B G Rabbat, A L Sangiovanni-Vincentelli, and H.Y. Hsieh, *A multilevel Newton algorithm with macromodelling and latency for the analysis of large-scale nonlinear circuits in the time domain*, IEEE Trans. Circuits and Systems **CAS–26** (1979), 733–741

13. F Yamamoto and S Takahashi, *Vectorized LU decomposition algorithms for large-scale circuit simulation*, IEEE Trans. on Computer-Aided Design **CAD–4** (1985), 232–239

INSTITUT FÜR ANGEWANDTE ANALYSIS UND STOCHASTIK, HAUSVOGTEIPLATZ 5 – 7, O – 1086 BERLIN

E-mail address. grund@iaas–berlin dbp de

International Series of Numerical Mathematics, Vol. 117, © 1994 Birkhäuser Verlag Basel

Partitioning and Multirate Strategies in Latent Electric Circuits

Michael Günther* and Peter Rentrop
Mathematisches Institut
TU München
D-80290 München

Abstract. In highly integrated electric circuits only a few elements are stimulated by an input signal. The major part of basic elements remains latent. The latency can be exploited on the numerical simulation level by partitioning or by multirate strategies. Advantages and disadvantages of both approaches are discussed. It seems that multirate integration schemes, which use different step sizes for the active and the latent components, are more promising. Furthermore, a combination of partitioned and multirate schemes forms a base for a parallel evaluation. The inverter chain serves as a typical example for which partitioned Runge-Kutta methods and multirate Rosenbrock-Wanner schemes have been tested.

Key words: ordinary differential equations, partitioning and multirate strategies, MROW methods, PRK methods, parallel techniques, latency of electric circuits, inverter chain.

*The author is supported by FORTWIHR: The Bavarian Consortium for High Performance Scientific Computing

1   Introduction

In simulation packages the transient analysis of electric circuits is based on the modified nodal voltage analysis, see Bulirsch, Gilg [3], Denk, Rentrop [4], Feldmann et al. [5] and the cited literature there. Kirchhoff's laws lead to a system of implicit ordinary differential equations

$$F(t, U(t), \dot{Q}(t)) = 0$$

t: time; U(t): vector of voltages; Q(t): vector of charges

(1)

We are especially interested in the class of highly integrated circuits. They are marked by the *latency*, i. e. for some fixed time t only a small part of elements of the circuits is active, whereas the major part of elements remains passive. The possibly time varying latency can be used on the discretization and/or on the iteration level of numerical schemes in order to achieve a better efficiency. Since the modeling of large latent circuits often leads to an explicit formulation of (1), we will concentrate on methods for explicit autonomous initial value problems

$$\dot{U}(t) = f(U)$$

(2)

As usual, non autonomous problems can be made autonomous by introducing t as an additional variable.

In the following, one-step methods are studied, which exploit the latency on the discretization level. Partitioned Runge-Kutta methods, shortly PRKs, integrate – by using the same step size – the active components with an explicit method and the latent components with an implicit one. In part 2 their properties and the partitioning strategies are discussed. In part 3 a multirate Rosenbrock-Wanner method, shortly MROW, is introduced. Active and passive components are integrated with different step sizes. Two multirate strategies are presented. The performance of both approaches for an inverter chain are documented. Obviously the presented strategies can be used for an implementation on parallel machines. This is still being investigated.

2   Partitioned Runge-Kutta Methods

2.1   Partitioning Methods

For the numerical solution of an autonomous initial value problem of ordinary differential equations

$$\begin{aligned} \dot{y}(t) &= f(y) \quad , \quad y \in \mathbf{R}^n \\ y(t_0) &= y_0 \quad , \quad t > t_0 \end{aligned}$$

(3)

there exist several approved algorithms. Their reliability depends on the distribution of the time constants in the solution. A system with time constants of comparable size is called non stiff and can be solved by classical "non stiff" methods (explicit Runge-Kutta methods, multi step methods of Adams type or explicit extrapolation methods, see e. g. Stoer, Bulirsch [19], Hairer et al. [9]). Very different time constants in (3) lead to stiff systems. The resulting A-stability conditions require implicit or linearly implicit techniques ("stiff" methods).

The efficient application of algorithms depends on whether the given problem is classified as stiff or non stiff. The treatment of a stiff problem with a non stiff integrator leads to an extraordinary amount of computing time (up to a factor 1000), wrong results or failure of the method. The opposite situation, the solution of a non stiff problem with a stiff integrator, is less sensitive. However, depending on the problem, the computing time is increased by a factor which lies between 2 to 20.

In latent electric circuits active elements of (3) can be integrated by non stiff methods, whereas latent elements require stiff methods. Partitioning strategies try to exploit this inner structure of the system. There are two main types of partitioning:

- **Interval wise partitioning**
  An automatic local stiffness detection is performed for the whole system. The problem (3) is treated as non stiff or as stiff in subintervals. Since the latent part of an electric circuit introduces stiffness into (3), this approach seems not very promising in electric circuit simulation. This strategy is of less importance for parallel processors.

- **Dynamic subsystems partitioning**
  Often a splitting of (3) into stiff and non stiff components is known as a consequence of engineering arguments. In general, the classification of a component depends on time, i. e. a former stiff component may become non stiff and vice versa. Therefore an algorithm has to be derived, which automatically selects the stiff components in subintervals, i. e. splits (3) locally in stiff and non stiff subsystems.

In the following we investigate the latter type of partitioning: We assume that the y-vector can be partitioned into non stiff components y_N and stiff components y_S, see Rentrop [15], Weiner et al. [25]. After a suitable renumbering of the components of y, there holds

$$
\begin{aligned}
y &= (y_S, y_N)^T \\
y_S &= (y_1, \ldots, y_{ns}) \qquad \text{stiff components} \\
y_N &= (y_{ns+1}, \ldots, y_n) \qquad \text{non stiff components}
\end{aligned}
\tag{4}
$$

Then (3) leads to the partitioned form (5)

$$
\begin{array}{lllll}
\dot{y}_S(t) & = & f_S(y_S(t), y_N(t)), & y_S(t_0) & = & y_{S0} \\
\dot{y}_N(t) & = & f_N(y_S(t), y_N(t)), & y_N(t_0) & = & y_{N0}
\end{array}
\tag{5}
$$

where the following properties for the eigenvalues of the linearized system should be satisfied:

$$
\begin{array}{ll}
\min_{j=1,\ldots,n_S} \operatorname{Re}\left\{\lambda_j\left(\frac{\partial f_S}{\partial y_S}\right)\right\} \ll 0 & \text{(stiff part)} \\
\|\frac{\partial f_N}{\partial y_S}\|, \|\frac{\partial f_N}{\partial y_N}\| \le 1 & \text{(weak coupling of non stiff part)}
\end{array}
\tag{6}
$$

In numerical procedures this structure can be exploited on two levels:

- In system (5) the same implicit discretization method, e. g. BDF, is used for both parts - the stiff and the non stiff one. Only the iterative solution of the nonlinear equations is split. A modified Newton method is applied to the stiff components y_S, a functional iteration to the non stiff part y_N.

- A partitioned discretization is used for the solution of (5), i. e. the stiff subsystem is solved by a stiff method and the non stiff part by a non stiff method.

We will proceed on the latter level and propose a *Partitioned Runge-Kutta* (PRK)-method. For the stiff part a semi-implicit Rosenbrock-Wanner(ROW)-method and for the non stiff part a customary explicit Runge-Kutta method is used. Both methods are combined by so-called coupling conditions.

2.2   Definition of PRK-Methods

A partitioned Runge-Kutta method is defined by:

$$
\begin{pmatrix} y_S \\ y_N \end{pmatrix}^h (t_0 + h) = \begin{pmatrix} y_S \\ y_N \end{pmatrix}(t_0) + \sum_{i=1}^{s} c_i \cdot \begin{pmatrix} l_i \\ k_i \end{pmatrix}, \quad \text{with}
$$

$$
k_i = h f_N \left(y_{S0} + \sum_{j=1}^{i-1} \alpha_{ij} l_j, y_{N0} + \sum_{j=1}^{i-1} \alpha_{ij} k_j \right),
\tag{7}
$$

$$
(I - h\gamma Df_S) l_i = h f_S \left(y_{S0} + \sum_{j=1}^{i-1} \alpha_{ij} l_j, y_{N0} + \sum_{j=1}^{i-1} \alpha_{ij} k_j \right) + h Df_S \sum_{j=1}^{i-1} \gamma_{ij} l_j,
$$

for $i = 1, \ldots, s$, where $Df_S = \frac{\partial}{\partial y_S} f_S(y_S, y_N)$.

The method (7) can be briefly characterized:

- The non stiff part y_N is computed explicitly, the stiff part y_S semi-implicitly. For each integration step a system of linear equations of order $ns \le n$ must be solved, which requires s function evaluations and back substitutions. For $y_N \equiv y$ the method reduces to a Runge-Kutta method, for $y_S \equiv y$ one obtains a ROW method. For $\gamma = \gamma_{ij} = 0$ the ROW-method degenerates to a Runge-Kutta method.

- The standard programming of the right hand side $f(y)$ is assumed for the differential equation system. Therefore the coefficients α_{ij} for both parts y_S and y_N are chosen equally. For systems with a fixed partition and parallel processors a separate programming of the stiff and non stiff parts is convenient. In this case different values α_{ij} and $\tilde{\alpha}_{ij}$ for the computation of y_S and y_N are more efficient.

  Equally chosen weights c_i and stage number s reduce the number of equations of conditions and lead to a compact algorithm.

- A PRK method is well designed for stiffness detection, because a Runge-Kutta method is always embedded in a ROW method.

- A sub-splitting of y_S and y_N is necessary for a parallel implementation.

2.3   The Method PRK4

Embedded methods are important for the implementation of an efficient step size control. In order to obtain the necessary information about the local truncation error after one integration step, two solutions of different order are compared:

$$\begin{pmatrix} y_S \\ y_N \end{pmatrix}^h (t_0 + h) = \begin{pmatrix} y_S \\ y_N \end{pmatrix} (t_0) + \sum_{i=1}^{6} c_i \cdot \begin{pmatrix} l_i \\ k_i \end{pmatrix},$$

$$\begin{pmatrix} \hat{y}_S \\ \hat{y}_N \end{pmatrix}^h (t_0 + h) = \begin{pmatrix} y_S \\ y_N \end{pmatrix} (t_0) + \sum_{i=1}^{5} \hat{c}_i \cdot \begin{pmatrix} l_i \\ k_i \end{pmatrix}, \tag{8}$$

$$k_i, \quad l_i \quad i = 1, \dots, 6 \quad \text{see (7)}$$

with the local truncation errors

$$\begin{aligned} \tau(y_0, h) &= \frac{1}{h} \| y^h(t_0 + h) - y(t_0 + h) \| = \mathcal{O}(h^4) \\ \hat{\tau}(y_0, h) &= \frac{1}{h} \| \hat{y}^h(t_0 + h) - y(t_0 + h) \| = \mathcal{O}(h^3) \end{aligned} \tag{9}$$

In the algorithm PRK4 a third order method with $\hat{s} = 5$ is embedded in an A-stable fourth order method with $s = 6$.

Every integration step of PRK4 involves 6 function evaluations and the computation of the Jacobi-matrix Df_S of order $ns \leq n$. The computation of Df_S by numerical differentiation requires ns additional function evaluations. A linear equation system of order ns for 6 different right hand sides is solved in each step. The coefficient set of PRK4 can be found in [15].

2.4   Partitioning Strategies

The performance of a code for a stiff system cannot completely be described via the eigenvalues, see (6). Stiffness of a system can also be considered as a step size reduction, which is not caused by prescribed tolerances. In practice, the code performance is influenced by:

- **order and stability region of a (non stiff) method**
  Non stiff methods of high order and small stability region suffer even under mildly stiff problems, whereas low order methods are usually not so sensitive, because they do not allow large step sizes.

- **solution structure**
  Usually the solution of a stiff problem is characterized by boundary layers and asymptotic regions. Especially in the boundary layer the tolerance requirement can be more restrictive than the stability restrictions.

- **prescribed tolerance**
  High precision demands lead to step size restrictions which can be stronger than the stability restrictions.

The factors described above, which all influence the proposed step size, are essential for the decision, whether the stiffness of a problem causes step size restrictions or not. Therefore one main criterion for the stiffness detection and the switching strategy is the actual step size. Another main criterion is the analysis of the particular elements of the Jacobian.

The following five stiffness criteria are implemented in PRK4:

1. **Step size comparison for each component for two simultaneously applied methods**
   After $k = \max(n, 16)$ steps PRK4 simultaneously performs one synchronization step with the 6-stage ROW method of order 4 and with the underlying explicit Runge-Kutta method. Then the proposed step sizes are compared for each component. The stiff subsystem for the next k steps is defined by all components for which

$$1.4 \cdot h_{RK} < h_{ROW}$$

holds. For many problems this works well. In general, however, the reliability is not yet satisfactory. If one uses a stiff method, the error of the transient components is strongly damped. In this case, one step with a non stiff method does not increase the error at once. Therefore the test is not reliable enough. On the other hand, a more frequent comparison of h_{RK} and h_{ROW} is too expensive.

2. **Asymptotic behavior**
   Integrating the asymptotic regions of a stiff problem by an explicit method yields step sizes which are too small, whereas the solution is harmless. The step size strategy restricts the quotient of two successive successful step sizes by the interval $[0.5, 1.5]$. If the quotient of the function values of 16 successful step sizes lies between 0.3 and 3, the component fixing the step size is declared as stiff. A similar criterion was proposed by Shampine [20].

3. **Successful step size prediction**
   PRK4 controls the number of successful and failed steps. Normally, the step size prediction is very reliable. Therefore a component is declared as stiff, if more than four step sizes during eight integration steps are rejected.

By use of a stiff integration code the "iteration matrix"

$$E = I - h\gamma Df$$

is available, and its spectral radius $\rho(E)$ can be estimated via the matrix norms, which are used for the pivoting strategy:

$$\rho(E) \leq 1 + \gamma h \|Df\|$$

This gives another possibility to check the stiffness. If

$$h\|Df\| \leq c,$$

where c stands for the negative real limit of the stability region of the non stiff method, a non stiff, otherwise a stiff method is used.

So far, all arguments are valid for a switching within the whole system. If one intends to switch only some components, heuristic arguments lead to

4. **Stability check**
   With $Df = (d_{ij})$, one examines the products $h \cdot d_{ij}$, $i, j = 1, \ldots, n$:

   - *diagonal elements:* If $h \cdot d_{ii} < 0$ and $|h \cdot d_{ii}| > c$, the i-th component is declared as stiff.

- *non diagonal elements:* If $|h \cdot d_{ij}| > c$ with $i \neq j$ and $d_{ji} \neq 0$, both the i-th and j-th component are declared as stiff.

The following criterion is applied in multirate strategies, too (cf. ch. 3).

5. **Extrapolated values**

Let $y_i^h(t_0 + h), i \in \{1, \ldots, ns\}$, be computed by a stiff method and $\tilde{y}_i^h(t_0 + h)$ be the extrapolated value, based on $y_i^h(t_0)$. If

$$|y_i^h(t_0 + h) - \tilde{y}_i^h(t_0 + h)| \leq TOL \cdot \|SK_i\|$$

$$
\begin{array}{rcl}
\text{with} \quad TOL & : & \text{required precision} \\
SK_i & := & \max_{t_0 \leq t_j \leq t_{end}} \{C, |y_i^h(t_j)|\}, \\
C & : & \text{scaling factor}
\end{array}
$$

holds, then the i-th component remains stiff, otherwise it is declared as non stiff.

3 Multirate Rosenbrock-Wanner Methods

3.1 Local Time Step Control

The circuit simulator TITAN of the SIEMENS AG, a further development of SPICE2 of the University of California, Berkeley, includes a static circuit partitioning which allows the application of a local time step control, see Schultz [17]. We will derive the connection between the local time step control in TITAN and the multirate methods.

Fig. 1a gives a schematic example for a hierarchical subcircuit structure. The circuit part A is the nominal circuit part and carrier of two subcircuits B and C. Furthermore, the subcircuit C consists of the circuit parts D and E. Dormant blocks, i. e. C, D and E in Fig. 1b, are computed in the so-called extrapolation mode: for each external branch of the block a simple substitute circuit is chosen, consisting of a resistor and a voltage source.

For a mathematical interpretation of the extrapolation mode, the implicit Euler scheme is applied to a constant, linear capacitor:

$$Q = C \cdot U \quad \Rightarrow \quad C \cdot \dot{U} = I$$

$$C \cdot \frac{U_{n+1} - U_n}{h} = I_{n+1} \quad \Rightarrow \quad U_{n+1} = U_n + \frac{h}{C} \cdot I_{n+1}$$

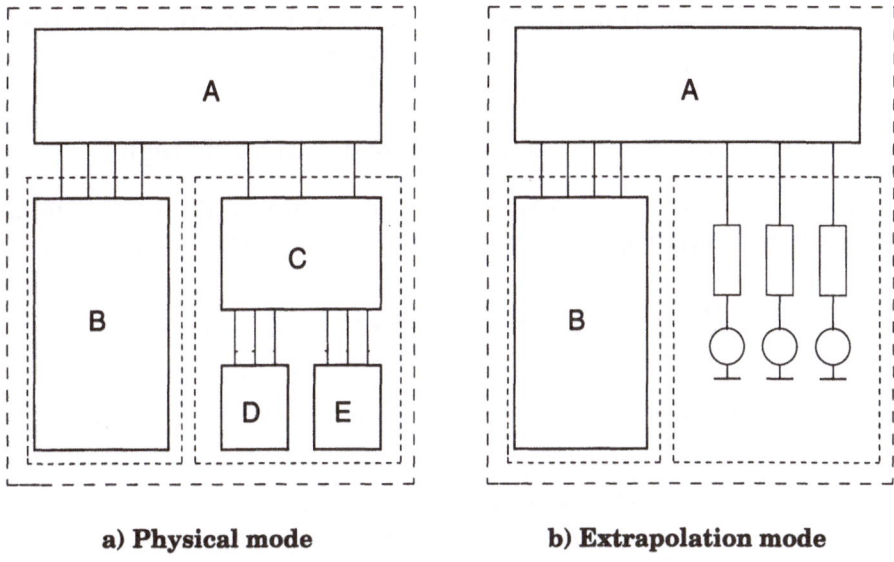

a) Physical mode   **b) Extrapolation mode**

Fig. 1: Hierarchically partitioned circuit

This yields the following substitute circuit, the so-called companion model for the implicit Euler method:

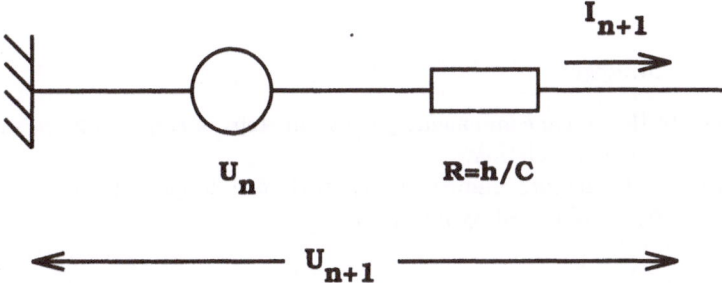

Fig. 2: Companion-model of the implicit Euler scheme

In electrical engineering, the local truncation error of the implicit Euler method can be interpreted as the difference between the capacitor C and the substitute circuit.

The model of Schultz uses fixed resistances R and sources U_n in the extrapolation mode. The step size h is the only degree of freedom left. Therefore, the so-called physical mode for the circuit blocks corresponds to the integration of differential equation components with different step sizes, which results in *multirate* methods.

Latency is used in the local time step control on the iteration level of the Newton iterations in [17], too. The following multirate investigations concentrate on the discretization level.

3.2   Slowest First and Fastest First Strategies

According to dynamic subsystem partitioning, the solution of (3) is split into active components $y_S(t)$, hopefully a very small subset of y(t), and latent components $y_L(t)$:

$$y = \begin{pmatrix} y_S \\ y_L \end{pmatrix} \quad , \quad y_S \in \mathbb{R}^{ns} \quad , \quad y_L \in \mathbb{R}^{nl} \quad , \quad ns + nl = n \quad , \tag{10}$$

This leads to a split system

$$\begin{aligned} \dot{y}_S(t) &= f_S(y_S, y_L) \quad , \quad y_S(t_0) = y_{S0} \\ \dot{y}_L(t) &= f_L(y_S, y_L) \quad , \quad y_L(t_0) = y_{L0} \end{aligned} \tag{11}$$

One method with two step sizes is used for the numerical integration of (11).The active components y_S are integrated with the small step size h, the latent components y_L are integrated with the large step size H. A synchronization of the *micro steps h* and the *macro steps H* is performed after each macro step. There are two possible strategies for a time integration step from t_0 to $t_0 + H$, see Gear,Wells [7].

- *Fastest First Strategy*

  - integrate the active components y_S with m steps of step size h; y_L-values are obtained by extrapolation.
  - integrate the latent components y_L with one large step H; y_S-values in $[t_0, t_0 + H]$ are obtained by interpolation.

- *Slowest First Strategy*

  - integrate the latent components y_L with one large step H; y_S-values in $[t_0, t_0 + H]$ are extrapolated.
  - integrate the active components y_S with m steps of step size h; y_L-values are interpolated.

Whereas the fastest first strategy seems quite evident, the justification of the slowest first strategy is the decoupling of active and latent parts. In view of step size control and back up the slowest first strategy looks more promising. It is possible to replace numerical interpolation by extrapolation, too.

3.3   Definition of MROW methods

Due to the desired steep voltage courses, which correspond to different time scales, electric circuits lead to stiff models. Therefore our multirate investigations concentrate on ROW-methods. An MROW-method for the numerical solution of system (11) for $t \in [t_0, t_0 + H]$ is defined:

- $h = \frac{1}{m} \cdot H$,   H: macro step, h: micro step   (12a)

- latent components y_L (12b)

$$y_L^H(t_0 + H) \;=\; y_{L0} + \sum_{i=1}^{s} c_i \cdot k_i$$

$$\left(I - H \cdot \gamma \cdot \frac{\partial f_L}{\partial y_L}(y_{S0}, y_{L0})\right) \cdot k_i =$$

$$H \cdot f_L \left(\hat{y}_S(t_0 + \alpha_i H), y_{L0} + \sum_{j=1}^{i-1} \alpha_{ij} k_j\right) +$$

$$+ H \frac{\partial f_L}{\partial y_L}(y_{S0}, y_{L0}) \sum_{j=1}^{i-1} \gamma_{ij} k_j, \qquad i = 1, \ldots, s$$

where $\hat{y}_S(t)$ is an *extrapolated* value for $y_S(t)$.

- active components y_S (12c)

$$y_S^h(t_0 + (\lambda + 1) \cdot h) \;=\; y_S^h(t_0 + \lambda h) + \sum_{i=1}^{s} c_i \cdot l_i$$

$$\left(I - h \cdot \gamma \cdot \frac{\partial f_S}{\partial y_S}\left(y_S^h(t_0 + \lambda h), \tilde{y}_L(t_0 + \lambda h)\right)\right) \cdot l_i =$$

$$h \cdot f_S \left(y_S^h(t_0 + \lambda h) + \sum_{j=1}^{i-1} \alpha_{ij} l_j, \tilde{y}_L(t_0 + \lambda h + \alpha_i h)\right) +$$

$$+ h \cdot \frac{\partial f_S}{\partial y_S}\left(y_S^h(t_0 + \lambda h), \tilde{y}_L(t_0 + \lambda h)\right) \sum_{j=1}^{i-1} \gamma_{ij} l_j, \qquad i = 1, \ldots, s$$

using $\alpha_i = \sum \alpha_{ij}$ with $\alpha_{ij} = 0$ for $i \le j$, $i, j = 1, \ldots, s$

for $\lambda = 0, 1, \ldots, m - 1$, where $\tilde{y}_L(t)$ is an *extrapolated* value for $y_L(t)$.

- rational (1,1)-extrapolation scheme (implemented in MROW(2)3)   (12d)

  With $\bar{y}(t_0 + \tilde{h}) := \hat{y}(t_0 + \tilde{h})$ or $\tilde{y}(t_0 + \tilde{h})$, resp., and $x \circ y := \sum_{l=1}^{n} x_l y_l$, $\bar{y}_i$ is defined for $i = 1, \ldots, ns$ or $i = ns + 1, \ldots, n$, resp., as follows:

$$\bar{y}_i(t_0 + \tilde{h}) := y_i(t_0) + \frac{2 \cdot \tilde{h} \cdot f_i(t_0)^2}{2 \cdot f_i(t_0) - \tilde{h} \cdot (Df_i(t_0) \circ f(t_0))}$$

If the value of the denominator falls below a given tolerance, the rational extrapolation is replaced by the Taylor expansion of y_i up to order 2.

Remarks:

- We consider the coupling of active and latent components only by their function values which is justified by the behavior of electric circuits. Another method of decoupling active and latent parts before integrating the system is given by the waveform relaxation, see Lory [13].

- The (1,1)-rational extrapolation was implemented, since a polynomial extrapolation could destroy the A-stability property, see Gear, Wells [7]. Even an interpolation scheme for the computation of the active components was not necessary, since the rate of variation of the latent components is very small.

- Since we do not rely on interpolation schemes, the latent and active components can be computed independently at the same time. Therefore a parallel implementation of MROW(2)3 on two processors is possible. In order to achieve a sufficient speed-up by parallelization, a further sub-splitting of y_S and y_L is necessary.

A-stability considerations are similar as for PRK-methods, see [10]. If the active components are integrated with the A-stable semi-implicit Euler scheme, the passive components do not destroy A-stability, provided that a rational (1,1)-extrapolation is used. For detailed investigations see Andrus [1], Skelboe [22], Skelboe, Anderson [23] and Gear, Wells [7].

3.4   The method MROW(2)3

Special methods for a given order p are constructed according to Strehmel, Weiner [24] or Stoer, Bulirsch [19]. Since MOSFET-models in electric circuits introduce discontinuities into the right hand side of (3), it makes no sense to develop schemes of arbitrarily high order. Therefore we constructed a third order method with an embedded second order method for step size control.

Lemma: MROW(2)3
There exist MROW-methods of order p=3 with stage number s=4 and three function evaluations per step. For step size control a method of order p=2 with s=3 is embedded. The parameters $\gamma, \alpha_2, \alpha_3, \alpha_{32}, \beta_{32}, \beta_{43}$ and c_4 are free. The parameter γ leads to A-stable schemes, if $0.39434 \leq \gamma \leq 1.28057$.

Remarks:

- Without embedding technique an MROW3 method can be achieved with s=3 and two function evaluations per step.

- An A-stable MROW2(3)-method with stage number s=3 is not possible, since the equations of conditions lead to: $6 \cdot p_1 \cdot p_2 - 3 \cdot p_3 = 0$ with $p_1 = \frac{1}{2} - \gamma$, $p_2 = \frac{1}{6} - \frac{1}{2} \cdot \gamma$ and $p_3 = \frac{1}{6} - \gamma + \gamma^2$, which is only satisfied for $\gamma = 0$.

In Table 1, a coefficient set for MROW(2)3 , with $\tilde{\gamma}_{ij} = \gamma_{ij}/\gamma$, is listed, which has performed very well in practical examples.

$\gamma = 0.395$	$\tilde{\gamma}_{21} = -3.98945316978$
$\tilde{\gamma}_{31} = 0.975871843730$	$\tilde{\gamma}_{32} = 0$
$\tilde{\gamma}_{41} = 0.443240634322$	$\tilde{\gamma}_{42} = -1.841436891$
$\tilde{\gamma}_{43} = -0.582949001513$	
$\alpha_{21} = 1.0$	
$\alpha_{31} = 0.125$	$\alpha_{32} = 0.375$
$\hat{c}_1 = 0.332571347358$	$\hat{c}_2 = 0.332571347358$
$\hat{c}_3 = 0.334857305285$	
$c_1 = 0.166666666667$	$c_2 = 0.166666666667$
$c_3 = 0.333333333333$	$c_4 = 0.333333333333$

Tab. 1: Coefficients of MROW(2)3

The remaining free parameters were chosen as: $\alpha_3 = 1/2$, $\beta_{32} = \alpha_{32} = 3/8$, $c_4 = 1/3$ and $\beta_{43} = 1/(c_4 \cdot \beta_{32}) \cdot (p_8/2 - p_3 p_2)/(p_1 - p_3^2/p_8)$.

3.5 Technical Details of Implementation

- As a step size control for the active and passive components a modification from Stoer, Bulirsch [19] was chosen. A proposal for the new macro step size $\mathbf{H}_{L,new}$ is defined by

$$\mathbf{H}_{L,new} := 0.5 * max\{\mathbf{h}_{new}\} \qquad (13)$$

The individual step size proposals $\mathbf{h}_{new}$ for the latent components are based on the old macro step size, the ones for the active components on the last micro step size in the old macro step. To prevent a step size control which is highly zigzag in character, $\mathbf{H}_{L,new}$ is limited by

$$0.5 * \mathbf{H}_L \leq \mathbf{H}_{L,new} \leq 1.5 * \mathbf{H}_L \qquad (14)$$

A proposal for the first micro step size in the new macro step $\mathbf{H}_{S,new}$ is given by

$$\mathbf{H}_{S,new} := min(min\{\mathbf{h}_{new}\}, 0.5 * \mathbf{H}_{L,new}) \qquad (15)$$

As a step size control for the remaining micro steps the method of [19] is chosen. If a macro step size control fails, the macro step size is applied only to the subset of the successful latent components. The failed components are declared as active.

- As a multirate strategy the slowest first method was applied. In the case of a failed step size prediction this technique shows great advantages for the macro step, see Gear, Wells [7]:

  - In failed steps the fastest first technique demands the integration of all active components before integrating the subset of latent components. By applying the slowest first method, however, one only has to repeat the macro step with a new set of passive components.
  - In the case of the slowest first strategy the backup can be done provided that one additional value is stored for all variables.

  Since the coupling of active and passive components must be very weak, the extrapolation of the active components over a huge interval leads to acceptable errors.

- The partitioning strategy is based both on numerical and technical information:

  - The individual step size proposal h_{new} for each component in a new macro step serves as a partitioning criterion: If $h_{new} \geq H_{L,new}$, then the component is treated as passive, otherwise as active. Since the step size prediction of an active component depends on the former micro step size (which is in general much smaller than the macro step size), a deactivation according to the proposed partitioning strategy can fail. Therefore a switching of active to passive may also be released by comparing the approximated solution and a linear extrapolated value: If the difference between these two values is limited by a given tolerance in all micro steps of the preceding macro step, then the component is regarded as latent.
  - Additionally, including information about the neighborhood of active components in the partitioning strategy minimizes their rate of rejection. The inverter chain (cf. chapter 4) can serve as an example: The runtime of a signal is delayed by passing through the circuit, i. e. a window of active elements passes through the circuit in time. A partitioning strategy can exploit this information by generating a window of active components $[n_{min}, n_{max}]$ with n_{min} or n_{max}, resp., the minimum or maximum number, resp., of an active element. To increase the reliability of the partitioning strategy, the neighbors of active components could be activated, i. e. the window could be enlarged. In other respects technical information can be exploited: To avoid a leap over of an active phase, the maximum macro step size has to be limited by the length of active phases.

4 Test Example: Inverter Chain

4.1 The Modeling of a MOSFET

In electric circuit simulation, models for semiconductor devices are used which imitate physical reality to a great extent, see Sieber et al. [18]. The main modeling principles are the following:

- A model consists of the five basic elements resistor, capacitor, inductor, voltage and current sources.

- The nonlinear controlled voltage or current source and resistors describe the static behavior.

- The nonlinear controlled capacitors or inductors describe the dynamic behavior.

The inverter chain consists of voltage sources, capacitors, resistors and MOSFETs (*metal oxide semiconductor field effect transistors*). For a modeling of the basic elements see Kampowski et al. [11]. In general, a MOSFET model is based on the voltage-current characteristics. Its electric performance is described by the poles gate G, drain D, source S and bulk B, as it is shown in Fig. 3. Drain and source are positive doped zones in the negative bulk.

Gate G controls the current I_{DS} from D to S by the potential difference between gate and source. The gate is isolated from the channel DS by a thin SiO_2-layer: this means that no or only a very small stationary gate current is present, which is independent of the bias applied to the gate.

One distinguishes between two classes of MOSFETs: the n-channel and the p-channel type. The n-channel FET is characterized by the following property: the smaller the potential difference U_{GS}, the smaller the drain current I_{DS} (cf. the characteristics below). For the p-channel FET it obviously holds: the smaller U_{GS}, the larger I_{DS}. The characteristics of an n-channel transistor can be transformed into those of a p-channel transistor by reversing the circuit's voltage and all currents.

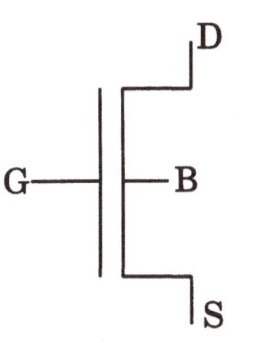

Fig. 3: MOSFET-Symbol

Independent of the channel type one furthermore distinguishes between enhancement and depletion MOSFETs:

- *n-channel enhancement MOSFET*

  It has the following characteristics:

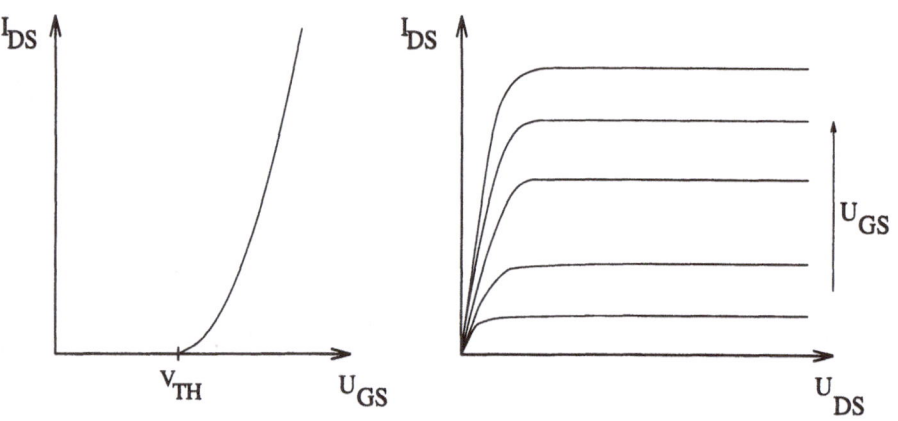

U_{TH} represents a threshold for the drain current I_{DS}, see left graph: $I_{DS} > 0$ holds only for $U_{GS} > U_{TH} > 0$. Since no drain current is present for $U_{GS} = 0$, the enhancement MOSFET is called self-locking.

- *n-channel depletion MOSFET*

  The transistor is called self-conducting: because of $U_{TH} < 0$, there is a drain current even for $U_{GS} = 0$. The characteristics are as follows:

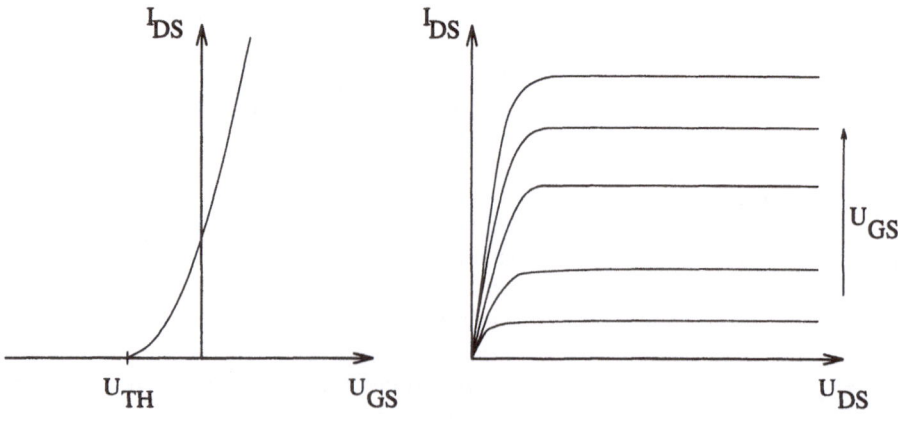

Remark: The two mentioned types of n-channel FETs are physically characterized by different p-doping degrees of the channel region. Enhancement means concentration, i. e. a strong p-doping, depletion means reduction, i. e. a weak p-doping.

In a simple model, a MOSFET transistor can be described by a voltage controlled current source to fit the drain current with respect to the gate voltage, see Fig. 4a. It holds

$$I_{GS} = I_{GD} = I_{GB} = 0$$
$$I_{BD} = I_{BS} = 0$$
$$I_{DS} = f(U_G, U_D, U_S),$$

where the function $f = f(U_G, U_D, U_S)$ is piecewise defined by

$$f = K \cdot \max\{(U_G - U_S - U_T), 0\}^2 - K \cdot \max\{(U_G - U_D - U_T), 0\}^2. \qquad (16)$$

K and U_T are given technical parameters, e. g. $U_T = 1V$ and $K = 2 \cdot 10^{-4} AV^{-2}$. More complex models are shown in Fig. 4b-e.

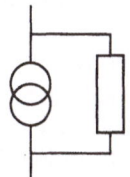
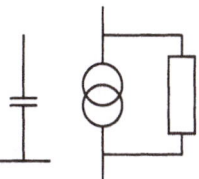

Fig. 4a: Level 1   **Fig. 4b:** Level 2   **Fig. 4c:** Level 3

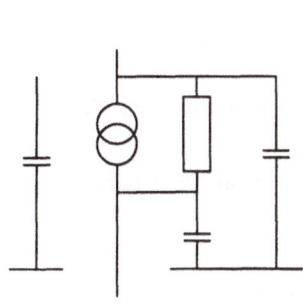
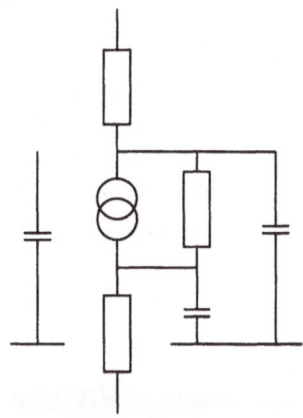

Fig. 4d: Level 4   **Fig. 4e:** Level 5

Remarks to Figure 4:

4a: Simple MOSFET model: voltage controlled current source

4b: A parallel switched resistor is added to the voltage controlled current source . This resistor models the so-called leakage current, since the $I_{DS} - U_{DS}$ − characteristic is never exactly 0.

4c: The gate is modeled by an additional capacitor: the capacitor voltage controls the voltage source,too.

4d: Two more capacitors model the capacitance between the n^+-doping at drain or source, resp. , and the bulk pole.

4e: The two additional resistors describe the path resistance, i. e. the flux in the current, through the n^+-doped area at drain or source, respectively.

A much more complicated and extensive MOSFET model has been derived by Shichman and Hodges [21]. For the descriptions see Günther, Rentrop [6], Schmidt [16].

The modeling of the inverter chain

A MOSFET inverter transforms a given signal from the logical value 0 to 1 and vice versa. Status 1 is represented by a constant operating voltage U_{op}, status 0 is the ground voltage $U_0 = 0V$. U_T represents a threshold such that any signal U_{in} greater than $U_{op} - U_T$ stands for the logical value 1, otherwise 0. In Fig. 5 the circuit of a MOSFET inverter is presented.

Using the simple MOSFET model of Fig. 4a, the first Kirchhoff's law applied at node 1 yields the differential equation

$$\frac{1}{R} \cdot (U_1 - U_{op}) + C(\dot{U}_1 - \dot{U}_0) + f(U_{in}, U_1, U_0) = 0$$

or rewritten

$$\dot{U}_1 = \frac{1}{RC} \cdot (U_{op} - U_1) - \frac{1}{C} \cdot f(U_{in}, U_1, U_0) \tag{17}$$

This is an explicit ODE with discontinuities in the derivatives of the right hand side, see the definition of f in (16).

Choosing $R = 5 \cdot 10^3 \Omega$ and $C = 0.2 \cdot 10^{-12} F$ and measuring the time in nsec$=10^{-9}$sec, (16) and (17) lead to

$$
\begin{aligned}
\dot{U}_1 &= U_{op} - U_1 - \tilde{f}(U_{in}, U_1, U_0) \\
\text{with} \quad \tilde{f}(U_G, U_D, U_S) &= \frac{1}{K} \cdot f(U_G, U_D, U_S)
\end{aligned}
\tag{18}
$$

With $U_{op} = 5V$, the response U_1 to a given input signal U_{in} is shown in Fig. 6.

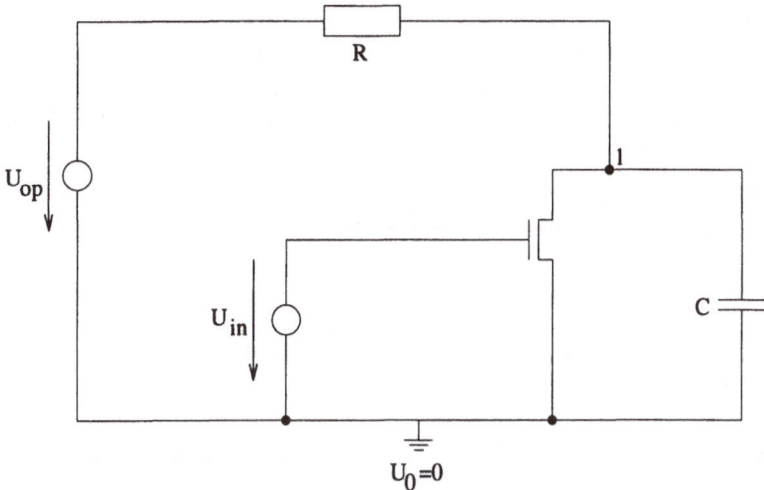

Fig. 5: Circuit of single inverter

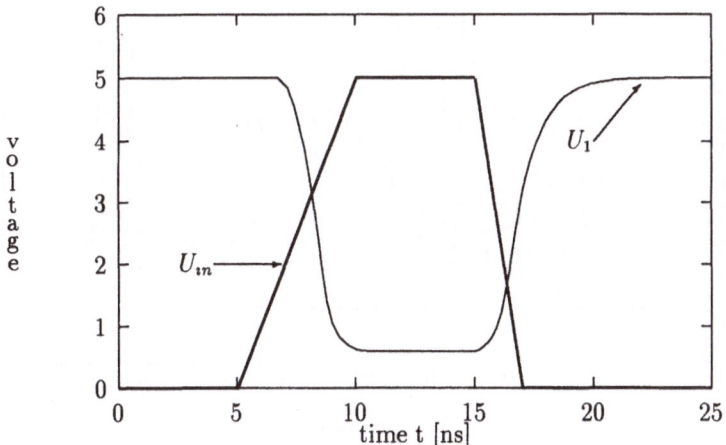

Fig. 6: Response U_1 to input signal U_{in}

A series of n coupled inverters, the inverter chain, delays the runtime of a signal, if n is even. Its circuit diagram is given in Fig. 7.

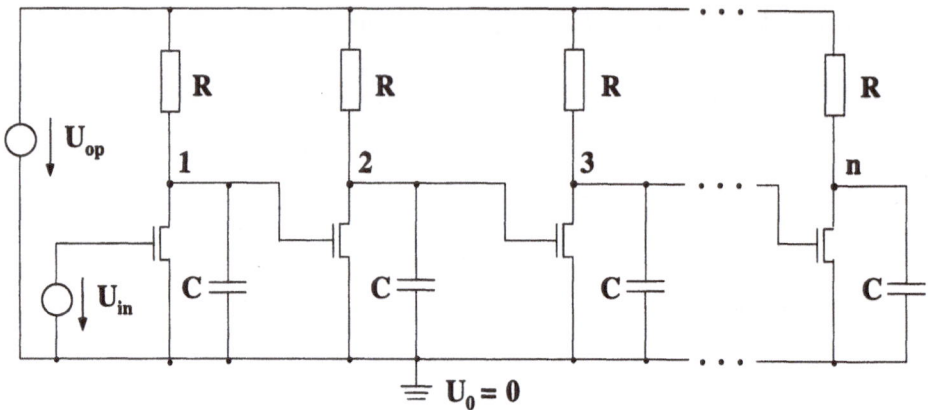

Fig. 7: Circuit of inverter chain

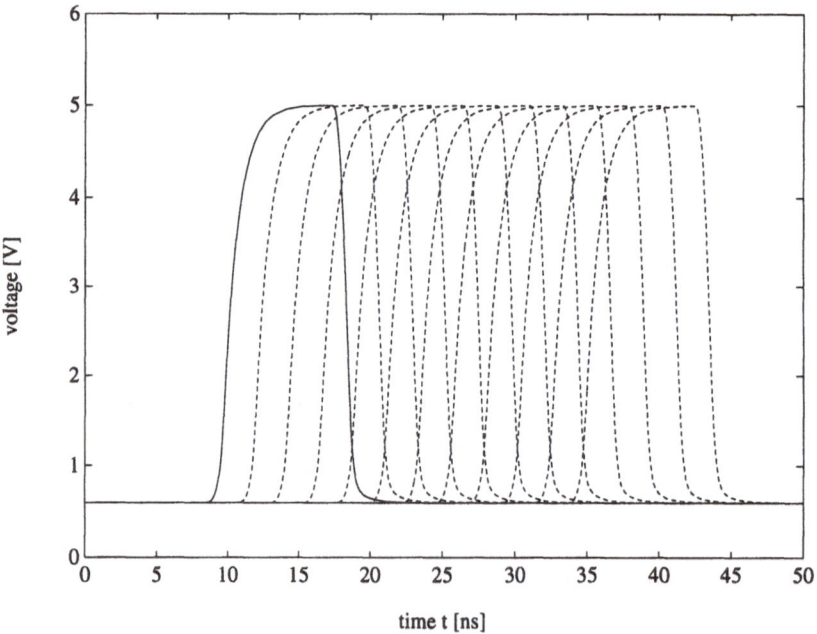

Fig. 8: Phase skip of signals of inverter chain

With $I_{GD} = I_{DS} = 0$ for our simplified transistor model, (18) immediately leads to the following system of ODEs for the unknown nodal voltages $U_1, \ldots, U_n$:

$$\begin{aligned}
\dot{U}_1 &= U_{op} - U_1 - \tilde{f}(U_{in}, U_1, U_0) \\
\dot{U}_k &= U_{op} - U_k - \tilde{f}(U_{k-1}, U_k, U_0), \quad k = 2, \ldots, n
\end{aligned} \tag{19}$$

Fig. 8 shows the nodal voltages $U_2, U_6, U_{10}, \ldots$ for the given input signal U_{in}, cf. Fig. 6: The input signal passes through the inverter chain: therefore it is delayed. For a given time only a small part of the circuit is active. The major part of the inverter chain remains latent.

If the output of an inverter chain is used as the input voltage, a ring oscillator is created. Its models and properties are discussed in Kampowsky et al. [11] and in Schultz [17].

The inverter chain and the ring oscillator can be used to build large circuits with regular structure and a predictable solution. Both circuits possess a large amount of latency which can be exploited by partitioned methods, e. g. Weiner et al. [25], or by multirate methods, e. g. Schultz [17], Günther, Rentrop [6].

4.2 Numerical Results

Since the right hand side of (19) is discontinuous in the first derivative, a consequence of the MOSFET model (16), each inverter introduces several discontinuity points caused by its modeling. Due to our experience, it does not make sense to detect and localize these points in order to restart the integration, i. e. to apply the so-called switching point technique from optimal control, see e. g. Bulirsch [2]. Therefore we prefer to integrate with low order methods, e. g. MROW(2)3, and to restart integration only at the edges of the input signal for t = 5ns, 10ns, 15ns and 17 ns, cf. Fig. 6.

PRK4 and MROW(2)3 were tested for the inverter chain, described by equation (19), with a different number of inverters. The A-stable one-step method GRK4A of Kaps, Rentrop [12] served as method of comparison. For all methods the initial step size HI=1.E-2 for the prescribed precision TOL=1.E-3 was used. The resulting linear equation systems were first solved by full Gaussian elimination techniques.

The following speed-ups for PRK4 and MROW(2)3 in comparison with GRK4A for 10 and 40 inverters are obtained:

#Inverter	10	40
Speed-up of PRK4	0.6	0.8
Speed-up of MROW(2)3	0.8	2.0

Tab. 2: Speed-up factors for TOL = 1.E-3

This rather daunting result for PRK4 can easily be explained: A large speed-up factor for PRK4 can only be expected, if the problem is strongly stiff and the number of stiff components remains small compared to the dimension of the problem. The inverter chain, however, is only weakly stiff and possesses a large amount of latency, i. e. only a few components are active. So the stiffness criterion 5 (cf. ch. 2. 5) handles these latent components as stiff also in the non stiff, asymptotic phase. The overhead from Linear Algebra remains dominant. For that reason we skip PRK4 in the following tests.

On the other hand, MROW(2)3 is allowed to use very large step sizes for most of the components: these savings compensate the overhead costs.

In Fig. 9 the successful switching strategy, both for the stiff – non stiff decision in PRK4 and for the latent – active decision in MROW(2)3, is evident.

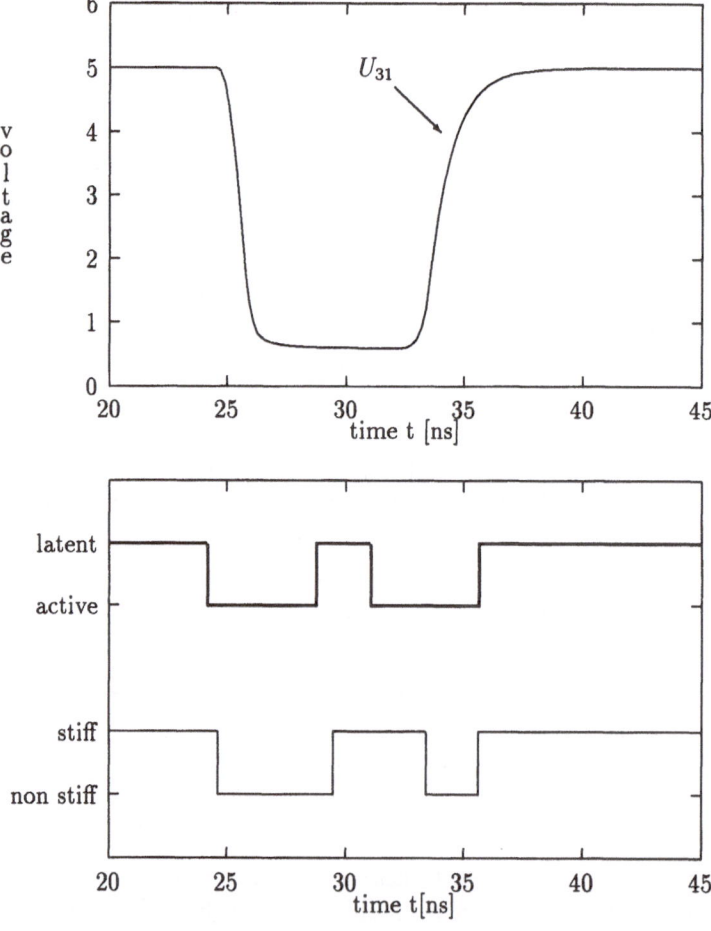

Fig. 9: Partitioning of the U_6-component in PRK4 and MROW(2)3

The maximum macro step size HMAXMAKRO of MROW(2)3 was limited by 1.145ns: this corresponds to the interval in which two successive inverters of even index are activated. The partitioning strategy of MROW(2)3 exploits this engineer's knowledge as follows: if the i-th inverter is recognized as active, the two immediate neighbors in signal flow direction, component i+1 and i+2, are activated, too. This reflects the behavior of the inverter chain which delays the runtime of an input signal, see Fig. 8. An application of the window technique described in ch. 3.5 saves computational costs by an additional factor 2.

Since the Jacobian is only a band matrix with lower bandwidth 1 and upper bandwidth 0, we also applied the band solver DECB/SOLB of Hairer and Wanner [10]. This reduces the costs for the Linear Algebra from $\mathcal{O}(n^3)$ to $\mathcal{O}(n)$. Since a monotone partitioning does not destroy this band structure, a slightly modified band solver can be used in MROW(2)3, too. For matrices with irregular sparse structure, dynamic sparsing methods can be applied, see Nowak [14]. Therefore the number of operations both in GRK4A and MROW(2)3 are of order $\mathcal{O}(n)$. Consequently, the possible speed-up, compared to an application of full Gaussian elimination, is lowered. The results are listed in Table 3 and 4.

The following abbreviations are used:

NSTEP:	Number of steps (only for GRK4A)
NSTEPREJ:	Number of rejected steps (only for GRK4A)
NSTEPMAK:	Number of macro steps (only for MROW(2)3)
NSTEPMAKREJ:	Number of rejected macro steps (only for MROW(2)3)
NSTEPMIK:	Number of micro steps (only for MROW(2)3)
NSTEPMIKREJ:	Number of rejected micro steps (only for MROW(2)3)
ERROR:	Maximum error of solution components at the end of the interval
TIME:	Total computing time in seconds for the solution of the problem.

Computations were performed in FORTRAN single precision on a SUN-workstation Sparc ELC, using the optimization option **-fast** in the FORTRAN compiler f77.

#Inverter	50	100	200
NSTEP	308	508	909
NSTEPREJ	94	168	321
ERROR	$1 \cdot 10^{-6}$	$1 \cdot 10^{-6}$	$1 \cdot 10^{-6}$
TIME	0.9	3.3	15.9

#Inverter	400	800	1600
NSTEP	1709	3309	6508
NSTEPREJ	621	1220	2419
ERROR	$1 \cdot 10^{-6}$	$1 \cdot 10^{-6}$	$3 \cdot 10^{-6}$
TIME	88.4	576.8	4132.2

Tab. 3: Test results of GRK4A for the inverter chain, TOL = 1.E-3

#Inverter	50	100	200
NSTEPMAK	99	124	175
NSTEPMAKREJ	5	5	5
NSTEPMIK	416	571	1074
NSTEPMIKREJ	81	149	265
ERROR	$4 \cdot 10^{-7}$	$4 \cdot 10^{-7}$	$3 \cdot 10^{-7}$
TIME	0.9	2.5	8.5

#Inverter	400	800	1600
NSTEPMAK	275	474	875
NSTEPMAKREJ	5	6	9
NSTEPMIK	2074	4068	7720
NSTEPMIKREJ	543	1085	2221
ERROR	$3 \cdot 10^{-7}$	$2 \cdot 10^{-7}$	$7 \cdot 10^{-7}$
TIME	34.9	181.1	1103.4

Tab. 4: Test results of MROW(2)3 for the inverter chain, TOL = 1.E-3

MROW(2)3 solves the test examples as reliably and precisely as GRK4A. According to the lower order of MROW(2)3, the number of NSTEPMIK is slightly larger than NSTEP. The reliability of the proposed dynamical partitioning strategy is shown by the small number of rejected macro steps. In general, the step size rejections for micro steps are up to 30%. This is the price to be paid for integrating the discontinuities. The latency is well reproduced by the multirate method: Fig. 10 shows the individual step sizes for each component in MROW(2)3 for n=100. One can see two bands of activity passing through the components, corresponding to the increasing and decreasing steep voltage drops of the signal. Only twenty components were integrated as active per time step: all other components are integrated with maximum macro step size, see Fig. 11.

As we have seen, the use of a band solver yields an operation count of only $\mathcal{O}(n)$ both in GRK4A and MROW(2)3. Therefore, an upper bound for the possible speed-up by multirating can be estimated by

$$MSPUP_n := \frac{\#\text{step sizes}(\text{GRK4A})}{\#\text{step sizes}(\text{MROW}(2)3)} \tag{20}$$

where $\#\overline{\text{step sizes}}(\text{MROW}(2)3)$ denotes the average number of macro and micro step sizes necessary for integrating one component by MROW(2)3. Since one has to deal with a computational overhead, caused by multirate strategy and the extrapolation costs, this value is only of theoretical interest.

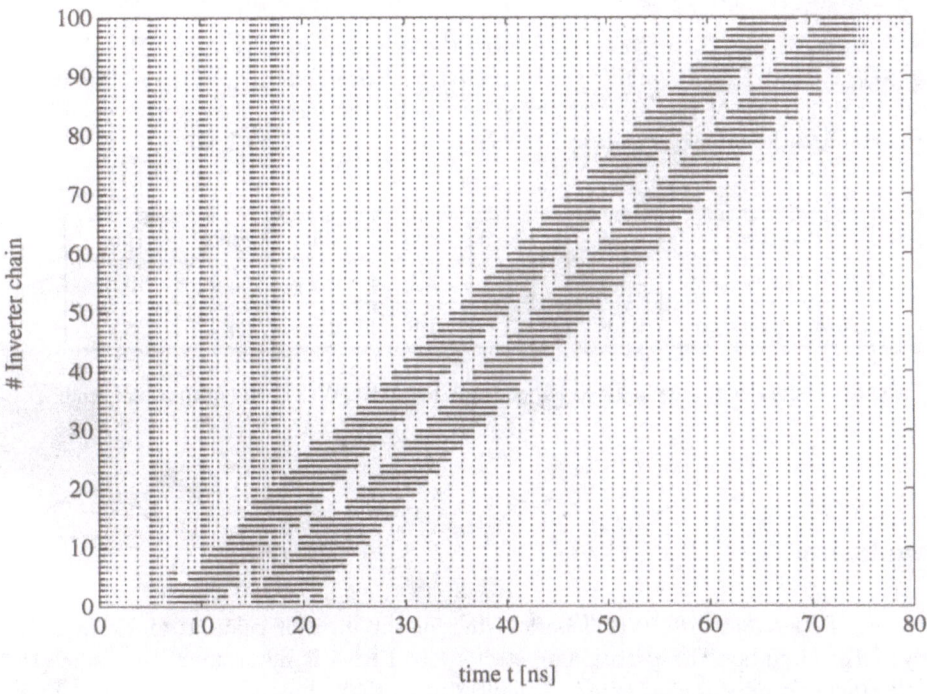

Fig. 10: Individual step sizes for the inverter chain, n=100

The restarting of integration at the edges of the input signal is reflected in the interval $t \in [0, 17ns]$.

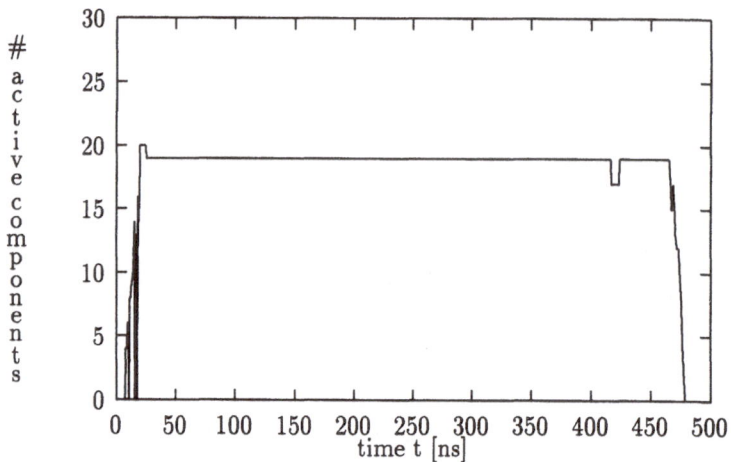

Fig. 11: Number of active components in MROW(2)3 for 800 inverters

Tab. 5 shows $MSPUP_n$ in comparison with the achieved speed-up $SPUP_n$: the ratio of $MSPUP_n$ and $SPUP_n$ marks the computational overhead. Note that MROW(2)3 exploits the latency of the inverter chain very well: the larger the inverter chain, the more extended the latency, the larger the speed-up.

#Inverter	50	100	200	400	800	1600
$MSPUP_n$: theory	1.5	2.4	3.1	4.7	5.9	6.6
$SPUP_n$: practice	1.0	1.3	1.9	2.5	3.2	3.8

Tab. 5: Theoretical and achieved speed-ups for MROW(2)3 in comparison with GRK4A

Conclusion

The codes PRK4 and MROW(2)3 were tested for the inverter chain as an example for (very) large circuits. The partitioning strategy of PRK4 is not competitive, since the tested circuit is only weakly stiff. A multirate method, however, saves computation time, since the latent circuits allow large step sizes for most components. The multirate strategies must be based both on numerical information and on circuit information, e. g. neighborhood of active elements, which is available via the data structures in electric circuit simulation packages.

Acknowledgment

We are indebted to the Numerical Analysis group of R. Bulirsch of the TU München for helpful discussions. We thank very much U. Feldmannn and A. Gilg from the "Zentrale Forschung und Entwicklung" of the SIEMENS AG, München.

References

[1] J. F. Andrus: Stability of a multi-rate method for numerical integration of ode's, *Computers Math. Applic. 25 (1993), pp. 3-14*

[2] R. Bulirsch: Die Mehrzielmethode zur numerischen Lösung von nicht linearen Randwertproblemen und Aufgaben der optimalen Steuerung. Lecture at the course in "Flugbahnoptimierung" of the Carl Cranz Gesellschaft, October 1971

[3] R. Bulirsch and A. Gilg: Effiziente numerische Verfahren für die Simulation elektrischer Schaltungen. In H. Schwärtzel (ed.), Informatik in der Praxis, *Springer Verlag, 1986, pp. 3-12*

[4] G. Denk and P. Rentrop: Mathematical models in electric circuit simulation and their numerical treatment. NUMDIFF5, ed. K. Strehmel, *Teubner, Leipzig, Bd. 121 (1991), pp. 305-316*

[5] U. Feldmann, U. A. Wever, Q. Zheng, R. Schultz and H. Wriedt: Algorithms for modern circuit simulation, *Archiv für Elektronik und Übertragungstechnik 46 (1992), pp. 274-285*

[6] M. Günther, P. Rentrop: Multirate Rosenbrock-Wanner methods and latency of electric circuits, *Appl. Numer. Math. 13 (1993), pp. 83-102*

[7] C. W. Gear and R. R. Wells: Multirate linear multistep methods, *BIT 24 (1984), pp. 484-502*

[8] E. Hairer: Order conditions for numerical methods for partitioned ordinary differential equations, *Numer. Math. 36 (1981), pp. 431-445*

[9] E. Hairer, S. P. Nørsett and G. Wanner: Solving ordinary differential equations I. Nonstiff problems, *Springer Verlag, Berlin, 1987*

[10] E. Hairer, G. Wanner: Solving ordinary differential equations II. Stiff problems, *Springer Verlag, Berlin, 1991*

[11] W. Kampowsky, P. Rentrop and W. Schmidt: Classification and numerical simulation of electric circuits, *Surveys Mathematics Industry 2 (1992), pp. 23-65*

[12] P. Kaps and P. Rentrop: Generalized Runge-Kutta methods of order four with stepsize control for stiff ordinary differential equations, *Numer. Math. 33 (1979), pp. 55-68*

[13] P. Lory: Simulation of VLSI-circuits: relaxation techniques. In: J. Manley, S. McKee, D. Owens (eds.): Proceedings ECMI 3, Glasgow 1988, *Teubner-Kluwer, Stuttgart, pp. 409-414 (1990)*

[14] U. Nowak: Dynamic sparsing in stiff extrapolation methods, *Preprint SC 92-21, Konrad-Zuse-Zentrum für Informationstechnik Berlin, 1992*

[15] P. Rentrop: Partitioned Runge-Kutta methods with stiffness detection and stepsize control, *Numer. Math. 47 (1985), pp. 545-564*

[16] W. Schmidt: Limit cycle computation of oscillating electric circuits, *In this Proceedings*

[17] R. Schultz: Local timestep control for simulating electrical circuits. In R. Bank et al., Mathematical Modelling and Simulation of Electrical Circuits and Semiconductor Devices, *ISNM 93 (1990), pp. 73-84*

[18] E. R. Sieber, U. Feldmann, R. Schultz and H. Wriedt: Timestep control for charge conserving integration in circuit simulation, *In this Proceedings*

[19] J. Stoer, R. Bulirsch: Introduction to numerical analysis, 2nd ed. , *Springer Verlag, New York, 1993*

[20] L. F. Shampine: Stiffness and nonstiff differential equation solvers, II: detecting stiffness with Runge-Kutta methods, *ACM-TOMS 3 (1977), p. 44-53*

[21] H. Shichman and D. A. Hodges: Insulated-gate field-effect transistor switching circuits, *IEEE J. Solid State Circuits, SC-3 (1968), pp. 285-289*

[22] S. Skelboe: Stability properties of backward differentiation multirate formulas, *Appl. Numer. Math. 5 (1989), pp. 151-160*

[23] S. Skelboe and P. U. Anderson: Stability properties of backward Euler multirate formulas, *SIAM J. Sci. Stat. Comp. 10 (1989), pp. 1000-1009*

[24] K. Strehmel and R. Weiner: Linear-implizite Runge-Kutta-Methoden und ihre Anwendungen, *Teubner, Leipzig, Bd. 127, 1992*

[25] R. Weiner, M. Arnold, P. Rentrop, K. Strehmel: Partitioning strategies in Runge-Kutta Type methods, *IMA J. Numer. Anal. (1993) 13, pp. 303-319*

Dipl.-Math. M. Günther and Prof. Dr. P. Rentrop, Mathematisches Institut, TU München, D-80290 München

International Series of Numerical Mathematics, Vol. 117, © 1994 Birkhäuser Verlag Basel

Circuit Simulation –
an Application for Parallel ODE Solvers? [1]

Martin Kiehl[2]

Abstract. Parallel ODE solvers were mentioned long before parallel computers were available. But now the interest partly seems to decrease. One reason is that we cannot expect much progress from the theory. This is shown for arbitrary explicit one-step methods and implicit Runge-Kutta methods, i.e.: full implicit Runge-Kutta (IRK) methods, diagonal implicit Runge-Kutta (DIRK) methods, singly diagonal implicit Runge-Kutta (SDIRK) methods, semi-implicit Runge-Kutta (SIRK) methods and Rosenbrock-Wanner (ROW) methods.

But due to communication overhead parallelism is not always economical. The construction of special parallel methods turns out to be useful only for very special applications. Circuit Simulation is one of the most interesting candidates.

AMS Subject Classification: 65L05, 65L06, 65Y05

Key words: one-step methods, Runge-Kutta methods, minimal stages, parallel ODE, extrapolation

1. Introduction

In (1895) *Runge* [15] constructed his 2 order method for the solution of initial value problems

$$y'(x) = f(x, y) \quad y(x_0) = y_0 \quad y \in \mathbf{R}^n.$$

Since that time many integration schemes with higher order p were constructed, with as few evaluations of f as possible.

---

[1]This work was partly supported by:
  the Bavarian Consortium on High Performance Scientific Computing (FORTWIHR)
[2]Mathematisches Institut, TU München, D-80290 München, Germany
  e-mail: kiehl@mathematik.tu-muenchen.de

Butcher found that the number of evaluations s called stages for sequential methods was restricted by

$$s \geq \begin{cases} p & \text{for } p \geq 5 \\ p+1 & \text{for } p \geq 7 \\ p+2 & \text{for } p \geq 8 \end{cases}.$$

Further restrictions were given by many authors for different type of methods regarding imbedded methods or methods of special stability property.

On parallel computers, the evaluation of the right-hand side f with different argument can in theory be computed in parallel on different processors (we call this 'parallelism in the method' or 'parallel methods') in the same time as one single evaluation on one processor.

Therefore we call a group of evaluations of f within a given integration scheme a parallel stage if all arguments are independent of each other, and if all evaluations within the parallel stage number r only use information from the first $r - 1$ parallel stages. A method which allows parallel evaluations of f is called a parallel methods (parallelism in the method).

In the following we assume that nearly all the computing time is spent on the evaluation of f,
that no parallelism is used to evaluate f,
that the communication overhead is negligable and
that each evaluation of f takes approximately the same time regardless of the arguments. If s_p is the number of parallel stages then we find that each integration step takes a s_p multiple of one execution time of f, and it seems very natural to look for methods of a given order p but with minimal s_p.

In the case of $f(x,y) = f(x)$ of a simple integration problem, we can choose any integration formula, where $f(x_i)$ is to be evaluated on some points x_i ; $i = 1, \ldots, k$. This can be done simultaneously on k processors and we get a theoretical speed-up close to factor k, which in practice actually can be obtained.

It is an interesting question if similar results can be found in the general case, where f also depends on y.

Communication and overhead also plays an important role.

Each component of y' can be evaluated in parallel independent from the others. We call this 'parallelism in the right-hand side' or 'parallelism in the problem'. But this reduces the size of the parallel tasks and is efficient only for problems with high dimension n. If we neglect the overhead the speed-up factor comes close to the number of processors r. r thereby can be very large.

This cannot be expected if parallel methods are' used which are not competitive as sequential mathod. But the parallel tasks are at least as big as f and not only a part of it. Thus parallel methods can become interesting if the overhead cannot be neglected or the user does not want to parallelize the right-hand side.

2.   Embedded Formulas

For the sake of a sophisticated step-size control, it is of great advantage not only to construct one approximation formula of high order p, but also to find an additional approximation formula with order $p - 1$ which does not require too many additional evaluations of f. This leads to additional order conditions and to methods with only few additional stages.

If at least 2 processors are available, it is clear, that the approximation of lower order can be computed on the additional processor. Thus we are not restricted to embedded formulas and can use any approximation formulas of order p with minimal stages and can still use usual step-size strategies.

If we suppose that enough processors are available, the theory can be concentrated on single order approximations without regard to embedding. This significantly reduces the number of order conditions.

3.   Minimal Stages for Explicit One-Step Methods

It was investigated in [12] that a simple test differential equation allows to destinate the minimal number of parallel stages of explicit one-step methods.

We just have to consider the differential equation

$$y'(x) = Ay = \begin{pmatrix} 0 & 0 & \cdots & \cdots & 0 \\ 1 & 0 & \ddots & & \vdots \\ 0 & 1 & \ddots & \ddots & \vdots \\ \vdots & \ddots & \ddots & \ddots & 0 \\ 0 & \cdots & 0 & 1 & 0 \end{pmatrix} y(x) \quad ; \quad y(0) = \begin{pmatrix} 1 \\ 0 \\ \vdots \\ \vdots \\ 0 \end{pmatrix} \in \mathbf{R}^{p+1} \tag{3.1}$$

respectively

$$y_1(x) = 1 \quad ; \quad y_{i+1}(x) = \int_0^x y_i(t)\mathrm{d}t \text{ for } i \geq 1.$$

Thus we have $y_{i+1}(x) = x^i/i!$.

It is easy to see that a Runge-Kutta method with s sequential and s_p parallel stages is of the form

$$\eta(x_0 + h) = y(x_0) + h\sum_{j=1}^{s} b_j k_j \tag{3.2}$$

$$k_0 = y_0 \tag{3.3}$$

$$k_i = f(c_i h, k_0 + h\sum_{j=1}^{s(0)} a_{i,j} k_j) \quad 0 = s(0) < i \leq s(1) \tag{3.4}$$

$$k_i \;=\; f(c_i h, k_0 + h \sum_{j=1}^{s(1)} a_{i,j} k_j) \quad s(1) < i \leq s(2) \tag{3.5}$$

$$\vdots \tag{3.6}$$

$$k_i \;=\; f(c_i h, k_0 + h \sum_{j=1}^{s(s_p-1)} a_{i,j} k_j) \quad s(s_p - 1) < i \leq s(s_p) = s \,. \tag{3.7}$$

$$\tag{3.8}$$

If applied to (3.1), it produces $\eta_{s_p+1}(x_0 + h) = 0$. Therefore it cannot approximate $y_{s_p+1}(x_0 + h) = h^{s_p}/s_p!$ with an order $p > s_p$.

This has already been obtained by *Nørsett* and *Simonsen*[14], and in [12] the result was generalized for arbitrary explicit one-step methods.

4. Implicit Runge-Kutta Methods

Big problems are of course the main application of parallel methods, which in most cases are stiff. Therefore implicit methods are of special interest. There an approximation of the Jacobian f_y is used, which can be obtained by difference approximation.

$$f_{y_j}(x, y) \doteq \frac{f(x, y + \varepsilon e_j) - f(x, y)}{\varepsilon} \quad j = 1, \dots, n. \tag{4.1}$$

This approximation is of course very well suited for parallel computation, but in practice the dimension n of the problems which require parallel computation will mostly be much bigger than the number of processors. Then f_y should be computed as seldom as possible, and the characteristic numbers for the efficiency of an implicit method is again the number of parallel stages, also in the case where parallelism is regarded..

Let us assume that f_y is evaluated only once in each step, for example at the starting point $(f_{y0} := f_y(x_0, y_0))$, and consider the diagonal implicit Runge-Kutta (DIRK) method.

$$\eta(x_0 + h) \;=\; y(x_0) + h \sum_{j=1}^{s} b_j k_j \tag{4.2}$$

$$k_i \;=\; f(x_0 + c_i h, y(x_0) + h \sum_{j=1}^{i-1} a_{i,j} k_j + h a_{i,i} k_i) \tag{4.3}$$

where each k_i is computed by an iteration process

$$k_i^{(0)} \;=\; f(x_0 + c_i h, y(x_0) + h \sum_{j=1}^{i-1} a_{i,j} k_j)$$

$$k_i^{(l)} \;=\; f(x_0 + c_i h, y(x_0) + h \sum_{j=1}^{i-1} a_{i,j} k_j + h a_{i,i} k_i^{(l-1)}) + f_{y0}(h a_{i,i} k_i^{(l)} - h a_{i,i} k_i^{(l-1)})$$

$$\Longleftarrow (\quad I \quad -ha_{i,i}f_{y_0})k_i^{(l)} = f(x_0 + c_ih, y(x_0) + h\sum_{j=1}^{i-1} a_{i,j}k_j + ha_{i,i}k_i^{(l-1)}) - f_{y_0}(ha_{i,i}k_i^{(l-1)})$$

This iteration terminates in the case of a linear differential equation $y' = Ay$ with a constant matrix A. Then DIRK methods are equivalent to semi-implicit Runge-Kutta (SIRK) methods

$$(I - ha_{i,i}f_{y_0})k_i = f(x_0 + c_ih, y(x_0) + h\sum_{j=1}^{i-1} a_{i,j}k_j), \qquad (4.4)$$

which are therefore also called linear implicit Runge-Kutta Methods.

If, on the other hand, the iteration does not terminate after the first step, we can still regard DIRK methods as linear implicit Runge-Kutta methods, if we regard all iterations $k_i^{(l)}$ as one stage.

Both methods fit in the more general scheme of Rosenbrock-Wanner (ROW) methods.

$$(I - h\gamma_{i,i}f_{y_0})k_i = f(x_0 + c_ih, y(x_0) + h\sum_{j=1}^{i-1} a_{i,j}k_j) + f_{y_0}h\sum_{j=1}^{i-1} \gamma_{i,j}k_j. \qquad (4.5)$$

Lemma : A ROW method with order $2p$ has at least $s_p \geq p$ parallel stages.

This was derived in [12] similar to explicit methods by just considering the differential equation

$$y'(x) = \begin{pmatrix} 0 & 0 & \cdots & \cdots & 0 \\ x & 0 & \ddots & & \vdots \\ 0 & x & \ddots & \ddots & \vdots \\ \vdots & \ddots & \ddots & \ddots & 0 \\ 0 & \cdots & 0 & x & 0 \end{pmatrix} y(x) \quad ; \quad y(0) = \begin{pmatrix} 1 \\ 0 \\ \vdots \\ \vdots \\ 0 \end{pmatrix} \qquad (4.6)$$

respectively

$$y_1(x) = 1 \quad ; \quad y_{i+1}(x) = \int_0^x xy_i(t)\mathrm{d}t \text{ for } i \geq 1.$$

General implicit Runge-Kutta methods

$$\eta(x_0 + h) = y(x_0) + h\sum_{j=1}^{s} b_j k_j \qquad (4.7)$$

$$g_i = y(x_0) + h\sum_{j=1}^{s} a_{i,j}f(x_0 + c_ih, g_j) \quad i = 1, \ldots, s \qquad (4.8)$$

require Newton iterations of the implicit equations. If we use $f_y(x_0, y_0)$ as an approximation of all $f_y(x_0 + c_ih, g_j)$, these methods also become Rosenbrock-Wanner methods.

In the case of linear differential equations $y'(x) = Ay(x)$, we again find existing sequential methods which achieve this limit ([3], [6], [4]). The s stage Gauß method is of order $2s$ and A-stable. For odd $p = 2s - 1$, we can use *Lobatto* or *Radau* methods. In the case of a constant matrix A each stage only requires one parallel evaluation of f

For non-linear problems there are ROW methods with $s_p = p - 1$ stages at least for small orders p ([11], [2]) and semi-implicit extrapolation methods with order $p = s$ for any arbitrary order p ([1], [5]).

A small restriction of this method class yields to implicit Runge-Kutta methods where the coefficients in the diagonal of the tableau all have the same single value $a_{i,i} = a$. These singly diagonal implicit Runge-Kutta (SDIRK) methods require only one decomposition of a matrix $(I - ha_{i,i}f_y(x_0, y_0))$ and are therefore very well suited for high dimensional problems, which are of special interest for parallel computation because of their long computing time.

They fit into the restricted class of ROW methods where $\gamma_{i,i} = \gamma$.

In the case $y'(x) = Ay(x)$ these methods become

$$\eta(x_0 + h) \;=\; \sum_{j=0}^{s} b_j k_j$$

$$(I - \gamma h A)k_i \;=\; hA \sum_{j=0}^{i-1} \beta_{i,j} k_j,$$

with $\beta_{i,j} := a_{i,j} + \gamma_{i,j}$ and $k_0 := y_0$.

For this class the following can be shown ([12]): **A ROW method with $\gamma_{i,i} = \gamma$ and order $p \leq 32$ needs at least $s = p - 1$ parallel stages.**

The same restrictions hold for linear singly implicit Runge-Kutta methods and singly implicit Runge-Kutta methods ([12]).

On the other hand there are already ROW methods with order $p = s$ respectively $p = s+1$ at least for small values of s ([7] [11], [2]). Thus the restriction is optimal.

5.   Communication, Parallelism in the Problem

The theory of minimal stages has turned out to be much easier for parallel than for sequential one-step methods.

We do not need to look for embedded codes. Two independent not-embedded approximations can be calculated on two proccessors.

If not-embedded Runge-Kutta methods are used instead of embedded Runge-Kutta methods, we can expect a speed-up factor of about 1.5 for high order.

The calculation of the Jacobian by (4.1) naturally speeds up nearly by a factor r if r processors are used.

These speed-ups can be obtained very simply with low analytical effort.

Of course we might additionally improve the stability [10] and we can construct more general parallel methods, but it seems that there is not much hope to find speed-ups in the class of explicit one-step methods, nor for ROW, SDIRK or SIRK methods, which are really exceedingly better than the speed-up we obtain from extrapolation.

Extrapolation codes seem to be optimally suited for parallel computation. These codes additionally allow to calculate approximations with different stepsizes independently before the extrapolation starts. The communication overhead, which has not been regarded in this paper up to now thereby is reduced to two messages between one processor and all the others. The speed-up which can be expected is then very close to factor r if r processors or more are available. r thereby has to be $r \geq p/2$. Similar results can be found for fully implicit methods in [8] and for $r < p/2$ in [9].

The following table shows the Runge-Kutta tableau of an explicit extrapolated Euler method of order 6 arranged to run on 3 processors in the usual form $\dfrac{c \mid A}{B}$ where c and $A = (a_{i,j})$ define the coefficients of (3.5), and where the i'th row of B corresponds to the vector b of (3.2) used to obtain an order i approximation.

c	1	2	3	4	5	6	7	8	9	10	11	12	13	14	15	16	17	18
0	0																	
$\frac16$	$\frac16$	0																
$\frac26$	$\frac16$	$\frac16$	0															
$\frac36$	$\frac16$	$\frac16$	$\frac16$	0														
$\frac46$	$\frac16$	$\frac16$	$\frac16$	$\frac16$	0													
$\frac56$	$\frac16$	$\frac16$	$\frac16$	$\frac16$	$\frac16$	0												
0	0	0	0	0	0	0	0											
$\frac15$	0	0	0	0	0	0	$\frac15$	0										
$\frac25$	0	0	0	0	0	0	$\frac15$	$\frac15$	0									
$\frac35$	0	0	0	0	0	0	$\frac15$	$\frac15$	$\frac15$	0								
$\frac45$	0	0	0	0	0	0	$\frac15$	$\frac15$	$\frac15$	$\frac15$	0							
$\frac12$	0	0	0	0	0	0	$\frac12$	0	0	0	0	0						
0	0	0	0	0	0	0	0	0	0	0	0	0	0					
$\frac14$	0	0	0	0	0	0	0	0	0	0	0	0	$\frac14$	0				
$\frac24$	0	0	0	0	0	0	0	0	0	0	0	0	$\frac14$	$\frac14$	0			
$\frac34$	0	0	0	0	0	0	0	0	0	0	0	0	$\frac14$	$\frac14$	$\frac14$	0		
$\frac13$	0	0	0	0	0	0	0	0	0	0	0	0	$\frac13$	0	0	0	0	
$\frac23$	0	0	0	0	0	0	0	0	0	0	0	0	$\frac13$	0	0	0	$\frac13$	0
	1	0	0	0	0	0	0	0	0	0	0	0	0	0	0	0	0	0
	-1	0	0	0	0	0	0	0	0	0	0	2	0	0	0	0	0	0
	$\frac12$	0	0	0	0	0	0	0	0	0	0	-4	0	0	0	0	0	$\frac92$
	$-\frac16$	0	0	0	0	0	0	0	0	0	0	4	0	0	0	$\frac{32}{3}$	0	$-\frac{27}{2}$
	$\frac{1}{24}$	0	0	0	0	0	0	0	0	0	$\frac{625}{24}$	$-\frac83$	0	0	0	$-\frac{128}{3}$	0	$\frac{81}{2}$
	$-\frac{1}{120}$	0	0	0	0	$\frac{648}{10}$	0	0	0	0	$-\frac{3125}{24}$	$\frac43$	0	0	0	$\frac{256}{3}$	0	$-\frac{81}{4}$

All diagonal blocks can be computed independently. We call this 'high-level parallelism' and can characterize the level of a parallel method by the number of parallel block stages.

For extrapolation methods the block-stage number s_b is one. Therefore extrapolation codes are not only optimal with respect to s_p but also with respect to communication.

The first rows of each diagonal block refer to the fact that $f(x_0, y_0)$ is evaluated by each of the three processors. This allows to send only x_0 and y_0. The numerical approximation is only dependent on very few k_i, as $b_i = 0$ for most of the b_i. This also reduces the information flow back to the central processor.

Also for other orders p and a given number of processors r a Runge-Kutta tableau based on an extrapolated Euler method with r diagonal blocks of nearly the same size can be obtained if only $p \geq 2r$. The speed-up then comes very close to r.

The speed-up obtained from other methods which are specially designed for parallel computation, and are not competitive on sequential computers must of course be much smaller than the number of the processors used.

This is important as the evaluation of

$$y'(x) = \begin{pmatrix} y_1'(x) \\ y_2'(x) \\ \vdots \\ y_n'(x) \end{pmatrix} = f(x, y) = \begin{pmatrix} f_1(x, y) \\ f_2(x, y) \\ \vdots \\ f_n(x, y) \end{pmatrix} \tag{5.1}$$

can clearly be done in parallel on n processors. In general we have only r processors with $r << n$ and the work can be well balanced between all the processors. This is true especially for big problems were computing time is large and parallelization is necessary.

The disadvantage of this type of parallelism is that the subtasks distributed to the different processors are just a fraction of a single evaluation of f. If the communication time is not negligible, this can severely reduce the speed-up.

As a rough approximation we have

$$c(n) = (c_0 + c_w n)/t_{float}$$

with $c_0 :=$ start-up time to initiate a message, $c_w :=$ constant factor, $n :=$ number of words of the message, $c(n) :=$ message time and $t_{float} :=$ floating-point-operation time.

On the other hand we can then use that sequential method which is best adapted to our problem, and if the communication time is negligible we obtain a speed-up that approaches the number of available processors.

Therefore we can divide all problems in three classes.

1.) Small problems, where communication time for sending the arguments x, y to another processor takes nearly the same time as computing $f(x, y)$ (Nowadays small means less then 1000 floating point operations in f), or where sequential computers can compute the solution fast and cheap (less than 1 Gflop for the whole problem).

2.) Very big problems, where communication time is negligible ($c_0 << c_w n \Rightarrow n$ big and $c_f >> c_w n$, $c_f :=$ number of operations in f) and parallelism in (5.1) leads to optimal speed-up and no parallel ODE-solver is required. Because each subtask is only a fraction of a single evaluation of f, communication is negligible only if f is of high complexity. (Nowadays high complexity means more than about 10 000 floating point operations in f.)

3.) Mid-size problems, where communication time is not negligible but not dominant. (between 1000 and 10 000 floating point operations in f.) Then extrapolation codes seem to be best, as they allow a high level parallelism, where each subtask is a multiple of a single evaluation of f.

Of course the boundaries of the classification depend on the hardware but not the classification itself. We generally find that parallel ODE-solvers are only economical for the mid-size problems, whatever mid-size means. For this class of problems high-level parallelism is recommended.

A further disadvantage of parallelizing (5.1) is that it has to be done by the user. This is very important if only one problem of a special type is solved. But then the time for the formulation of the problem by far exceeds the computation time. Speeding up the latter does not seem very important. On the other hand we often have to solve problems of the same type again and again. In this case the parallelization of (5.1) is reasonable.

6.   Application to Circuit Simulation

However, f is not always given in a form like (5.1). For example f can be implicitly generated by program packages, i.e., f can be evaluated, but not analytically examined. Then splitting of f is difficult and parallel IVP solvers may be applicable also for very big problems.

Program packages which do not allow an easy insight into the equations are used in many important application field of industry like circuit simulation (SPICE), TITAN (SIEMENS)), multi-body systems (MEDYNA (DLR), ADAMS (automobile industry), SIMPACK (MAN)) or reaction kinetics. As the right-hand side f is not even given, it is of course difficult to split it.

For example SPICE allows to simulate electrical circuits by defining the circuit in a modular way. This is done by using models which represent basic components of the circuit (diodes, transistors etc.). Each component of the circuit is then defined by the basic type, some characterizing parameters and the numbers of the nodes to which they are connected. These definitions are then interpreted by SPICE at runtime.

Instead of generating analytic expressions of $f(x, y_1, \ldots, y_n)$, SPICE computes $f(x, y) = g_1(x, y) + g_2(x, y) + \cdots + g_m(x, y)$. The g_j represent the contribution of circuit component number j, and y represents the vector of all node voltages and branch currents of the circuit. Each g_j depends only on the y_i of those components which are listed in the definition of circuit component number j and it also contributes only to these components. So each g_j can be evaluated in parallel but the processor requires the necessary node values.

As there is no rule which node values are needed to compute g_j, this means a big communication overhead. On the other hand the modules g_j are very small, so that the overhead can easily become dominant.

If r processors are available, it is therefore preferable to split the whole circuit in r parts of similar size $f(x, y) = G_1(x, y) + G_2(x, y) + \cdots + G_r(x, y)$. Each G_j is then treated on one processor. The problem is, however, to split the right hand side which requires to intervene into the package, and to choose the parts in a way that all G_j only depend on a small subset of the node values. If this is not possible a parallel IVP solver is advantageous.

However, both problems have been solved in the meantime, so that splitting f is now possible. This was basicly done to run SPICE on vector computers. It is a typical phenomenon that applications, which are important enough to develop program packages, are also important enough to reformulate the implementation of the right-hand side.

SPICE now runs on sequential and vector computers. The latter sometimes also have a few parallel processors. Under this condition it can become preferable to use parallel extrapolation codes where up to 4 processors allow a speedup of nearly factor 4. The comunication thereby is less than for splitting f and the parallel tasks are well balanced. Therefore extrapolation methods can be of interest to speed up the simulation of circuits on 2 or 4 processor vector computers. However, for big circuits or if IVP solvers other then extrapolation codes are used splitting f seems to be the best choice as the overhead is negligable in this case.

Conclusion:

We have assumed that the computing time is basicly spent on the evaluation of f. Then it is generally not necessary to construct new parallel methods if the problems are of large dimension. These problems can be solved and should be solved by using vectorization and parallelization of the right-hand side if many processors are available, or by using parallel extrapolation if not more then 4 processors are available.

Parallel IVP solvers can therefore only be used for mid-size problems and are therefore of limited interest. Then the speedup cannot be much better than that already obtained from parallel extrapolation. (For very special problem classes, however, it is of course possible to construct special parallel methods and obtain exceeding speed-ups [13].)

Only for applications where f cannot be splitted parallel IVP solvers others then extrapolation methods may be useful also for very big problems.

However, in many applications the differential equations are stiff. A main part of the computations is then spent in solving a large scale linear equation system. If this is the dominating part of the computation then of course parallel linear algebra routines can be applied. But this does not require a specially designed integration routine.

References

[1] Bader, G., Deuflhard, P.: A semi-implicit mid-point rule for stiff systems of ordinary differential equations. *Numer. Math., 41*, 373-398 (1983).

[2] Burrage, K.: A special family of Runge-Kutta methods for the solving stiff differential equations. *BIT, 18*, 22-41 (1978).

[3] Butcher, J.C.: Integration processes based on Radau quadrature formulas. *Math. Comp., 18*, 233-244 (1964).

[4] Chipman, F.H.: A-stable Runge-Kutta processes. *BIT, 11*, 384-388 (1971).

[5] Deuflhard, P.: Recent progress in extrapolation methods for ordinary differential equations. *SIAM Review, 27*, 505-535 (1985).

[6] Ehle, B.L.: On Padè approximations to the exponential function and A-stable methods for the numerical solution of initial value problems. *Research report CSRR 2010, Dept. AACS, Univ. of Waterloo, Ontario, Canada*, (1969).

[7] Hairer, E.,, Wanner, G.: Solving ODE II. *Springer*, (1991).

[8] van der Houwen, P.J., Sommeijer, B.P.: Parallel iteration of high-order Runge-Kutta methods with stepsize control. *J. Comp. Appl. Math., 29*, 111-127 (1990).

[9] van der Houwen, P.J., Sommeijer, B.P.: Iterated Runge-Kutta methods on parallel computers. *SIAM J. Sci. Stat. Comp., 12 No.5*, 1000-1028 (1991).

[10] Iserles, A., Nørsett, S.P.: On the theory of parallel Runge-Kutta methods. *IMA J. Num. Anal., 10*, 463-488 (1990).

[11] Kaps, P., Rentrop, P.: Generalized Runge-Kutta methods of order four with stepsize control for stiff ordinary differential equations. *Numer. Math., 33*, 55-68 (1979).

[12] Kiehl, M.: Parallel Integration Methods for Initial Value Problems with Minimal Stages. *Technische Universität München, Mathematisches Institut, Report M-9210*, (1992).

[13] Kiehl, M.: Parallel Multiple Shooting for the solution of Initial Value Problems. *Technische Universität München, Mathematisches Institut, Report M-9211*, (1992).

[14] Nørsett, S.P., Simonsen, H.H.: Aspects of parallel Runge-Kutta methods. *Springer Lecture Notes in Mathematics 1386*, 103-117 (1989).

[15] Runge, C.: Über die numerische Auflösung von Differentialgleichungen. *Math. Ann., 44*, 167-178 (1895).

International Series of Numerical Mathematics, Vol. 117, © 1994 Birkhäuser Verlag Basel

Numerical stability criteria for differential-algebraic systems

R. März

Humboldt-Universität zu Berlin

Abstract: In this paper we transfer classical results concerning Lyapunov stability of stationary solutions x_* to the classes of DAEs being most interesting for circuit simulation, thereby keeping smoothness as low as possible. We formulate all criteria in terms of the original equation. Those simple matrix criteria for checking regularity, Lyapunov stability etc. are easily realized numerically.

Key words: Differential algebraic systems, Lyapunov stability

1 Introduction

This paper deals with autonomous quasilinear differential-algebraic equations (DAEs)

$$(1.1) \qquad A(x)x' + g(x) = 0,$$

where the leading coefficient matrix $A(x)$ is everywhere singular but has constant rank. ¿From a geometric point of view, (1.1) should induce a smooth vectorfield on a certain state manifold. However, if it does so, the vectorfield as well as the manifold are given implicitly only, and they are not available in practice for higher index DAEs except for interesting case studies. This is why we insist on terms of (1.1) for numerical stability criteria.

In this paper we transfer classical results concerning Lyapunov stability of stationary solutions of regular ordinary differential equations (ODEs) to the case of DAEs (1.1).

Due to analytic techniques we keep the smoothness as low as possible while the concept of understanding DAEs as differential equations on manifolds supposes more smoothness than it seems to be natural.

For instance, the semi-explicit DAE

$$(1.2) \qquad u' - \varphi(u, v) = 0,$$

$$(1.3) \qquad \Psi(u, v) = 0,$$

with C^1 functions φ, Ψ has index 1 if $\Psi'_v(u, v)$ remains nonsingular. Clearly, due to the Implicit Function Theorem, exactly one solution of (1.2), (1.3) passes through each consistent initial point u_0, v_0, that is, $\Psi(u_0, v_0) = O$. Locally, (1.2), (1.3) is equivalent with

$$(1.4) \qquad\qquad u' = \varphi(u, f(u)),$$

$$(1.5) \qquad\qquad v = f(u),$$

whereby $f(u)$ and $\varphi(u, f(u))$ depend continuously differentiably on u.
On the other hand, the geometric concept understands (1.2), (1.3) to be the linearly implicitly given vectorfield

$$(1.6) \qquad\qquad u' - \varphi(u, v) = 0,$$

$$(1.7) \qquad\qquad \Psi'_u(u, v)u' + \Psi'_v(u, v)v' = 0,$$

on the manifold

$$\mathcal{M} := \{(u^\mathsf{T}, v^\mathsf{T})^\mathsf{T} \in I\!\!R^m : \Psi(u, v) = 0\}.$$

To arrive with (1.6)), (1.7) at a C^1 vectorfield again, we should assume $\Psi \in C^2$. Moreover, the explicit regular ODE resulting from (1.6), (1.7), namely

$$(1.8) \qquad\qquad u' = \varphi(u, v),$$

$$(1.9) \qquad\qquad v' = -\Psi'_v(u, v)^{-1}\Psi'_u(u, v)\varphi(u, v),$$

is called the underlying ODE of the DAE (1.2), (1.3). Considering this ODE on the whole space $I\!\!R^m$ instead on the manifold $\mathcal{M}$ would not be helpful for answering stability questions since the asymptotics of (1.8), (1.9) on the whole of $I\!\!R^m$ might show a different behaviour than its restriction to $\mathcal{M}$. So a stationary solution being stable on $\mathcal{M}$ may become unstable on $I\!\!R^m$.

Sometimes it is easier to deal with DAEs having even a constant leading coefficient matrix, say

$$(1.10) \qquad\qquad \tilde{A}\tilde{x}' + \tilde{g}(\tilde{x}) = 0.$$

Hence, instead of considering the original DAE (1.1) we may turn to the enlarged systems

$$(1.11) \qquad\qquad Px' - y = 0,$$

$$(1.12) \qquad\qquad A(x)y + g(x) = 0,$$

or

$$(1.13) \qquad\qquad x' - y = 0,$$

(1.14) $$A(x)y + g(x) = 0,$$

which have obviously the form (1.10). Using system (1.11),(1.12) we always assume $A(x)$ to have a constant nullspace $N = \ker A(x)$, and P stands for any projection matrix with $\ker P = N$. If $\ker A(x)$ depends on x (that is it rotates with varying x) we use (1.13), (1.14).

It is well known that enlarging (1.1) to (1.11), (1.12) leaves the index invariant. However, in contrary, the index of system (1.13), (1.14) becomes higher than that of the original DAE (1.1)

At this place it should be mentioned that we are basing on the tractability index (e.g. [1], [2]) defined in terms of the Jacobians of the functions A, g and $\tilde{A}, \tilde{g}$, respectively. Recall that the tractability index represents a generalization of the Kronecker index. Moreover, for the DAEs being discussed in the following, the tractability index is shown to coincide with the differentiation index as well as with the geometric one, supposed the latter exists. This is why we use simply the notion of an *index*.

In circuit simulation the charge oriented modelling leads to DAEs of the form

(1.15) $$A\frac{d}{dt}C(x) + g(x) = 0,$$

or, equivalently, to the system

(1.16) $$Ay' + g(x) = 0,$$

(1.17) $$y - C(x) = 0.$$

Thereby, A is a constant matrix. The system (1.16), (1.17) is somewhat easier to integrate than

(1.18) $$AC'(x)x' + g(x) = 0$$

since one can perform Newton iterations without second derivatives of C.

Obviously, system (1.16), (1.17) has a constant leading matrix, i.e. it is of the form (1.10).

Moreover, the enlarged DAE (1.16), (1.17) has index 1 iff (1.18) has so, supposed the condition

(1.19) $$im \, A \equiv im \, AC'(x)$$

is satisfied. Often $C'(x)$ is nonsingular, hence (1.19) is given trivially.

In the following we provide stability criteria for DAEs of the form (1.10) in terms of $\tilde{A}$ and $\tilde{g}'(\tilde{x}_*)$, where $\tilde{x}_*$ denotes the stationary solution under discussion. Clearly, all those results can be traced back for (1.1) resp. (1.15) immediately.

2   Basic linear algebra

Given two matrices $A, B \in L(\mathbb{R}^m)$ we form the matrix chain

$$A_0 := A, \quad B_0 := B,$$

(2.1) $$A_{j+1} := A_j + B_j Q_j, \quad B_{j+1} := B_j P_j, \ j \geq 1.$$

Thereby, $Q_j \in L(\mathbb{R}^m)$ stands for any projector onto $\ker A_j$, and $P_j := I - Q_j$.

Lemma 2.1 *The matrix pencil $\lambda A + B$ is regular with index μ if and only if A_μ is nonsingular but $A_j, j = 0, \ldots, \mu - 1$ are not*

The proof is referred to in [3].

Lemma 2.2 *For given regular index μ pencil $\lambda A + B$ the projections $Q_0, \ldots, Q_{\mu-1}$ can be chosen such that*

(2.2) $$P_0 \cdots P_{\mu-1} = \hat{A}^D \hat{A}$$

becomes true, where $\hat{A}^D$ denotes the Drazin inverse of $\hat{A} := (cA + B)^{-1} A$.

The proof is given in [4].
Recall (e.g. [5]) that $\hat{A}^D \hat{A}$ respresents the spectral projection onto the finite eigenspace of the pencil along the infinite one.
As a consequence (cf. [5]) of Lemma 2.2, the linear constant coefficient DAE

(2.3) $$Ax' + Bx = 0$$

is equivalent with

(2.4) $$(P_0 \cdots P_{\mu-1} x)' + M P_0 \cdots P_{\mu-1} = 0,$$

(2.5) $$x = P_0 \cdots P_{\mu-1} x,$$

where

(2.6) $$M := P_0 \cdots P_{\mu-1} A_\mu^{-1} B.$$

Any finite eigenvalue of the pencil is an eigenvalue of M simultaneously. The corresponding eigenspaces belong to $\operatorname{im} P_0 \cdots P_{\mu-1}$. Moreover, M has the zero eigenvalue corresponding to $\ker P_0 \cdots P_{\mu-1} \subset \ker M$.
So, asking for the stability of (2.3) we check the eigenvalues of M or, equivalently, those of the pencil.
The matrix chain (2.1) can be computed numerically without special difficulties. It may be considered as a practical tool for index checking and regularity tests e.g. during the numerical integration. In particular, transforming

$$(2.7) \qquad\qquad HA\Pi = \begin{pmatrix} R_{11} & R_{12} \\ 0 & 0 \end{pmatrix}, \quad R_{11} \text{ nonsingular,}$$

say, by a Housholder matrix H and a permutation matrix Π, we know

$$Q = \Pi \begin{pmatrix} 0 & -R_{11}^{-1} R_{12} \\ 0 & I \end{pmatrix} \Pi^{-1}$$

to be a projection onto $\ker A$.

Thus, the main work we have to do is performing (2.7).

There is some good experience with realizing those tests ([8]). Note that in the framework of index reduction techniques (e.g. [5], [7], [3]) one has to make similar efforts at each step.

3 Lyapunov stability

First of all let us recall the famous Theorem of Lyapunov that we will generalize for DAEs.

Consider the regular ODE

$$(3.1) \qquad\qquad x' + g(x) = 0,$$

which is assumed to have the stationary solution x_*, i.e. $g(x_*) = 0$.

Theorem 3.1 *(e.g. [9]):* Let $g \in C^2(\mathcal{D}, I\!\!R^m), \mathcal{D} \subset I\!\!R^m$ *open,* $x_* \in \mathcal{D}, g(x_*) = 0$. *If all eigenvalues of the matrix* $-B, B := g'(x_*)$, *have negative real parts, the equilibrium point* x_* *is asymptotically stable.*

In particular, stability in the sense of Lyapunov inlcudes the solvability of all initial value problems with initial values $x_0 \in B(x_*, \tau), \tau > 0$ sufficiently small, and all those solutions have continuations up to infinity.

How to apply this result to the nonlinear index 1 and index 2 DAEs

$$(3.2) \qquad\qquad Ax' + g(x) = 0$$

we are interested in?

Again, x_* is a stationary solution iff

$$(3.3) \qquad\qquad g(x_*) = 0.$$

However, now we have to compare them with consistent intial values only, that is, with $x_0 \in B(x_*, \tau) \cap \mathcal{M}$, where $\mathcal{M}$ denotes the corresponding state manifold.

If (3.2) has index 1 on $\mathcal{D}, \mathcal{D} \subset I\!\!R^m$ open, it simply holds

$$\mathcal{M} = \{x \in \mathcal{D} : g(x) \in im\, A\}.$$

In that case, the nullspace $\ker A$ and the tangent space

$$S_0(x_*) := T_{x_*}\mathcal{M} = \{z \in I\!\!R^m : g'(x_*)z \in im\, A\}$$

intersect trivially, thus

$$\ker A \oplus S_0(x_*) = I\!\!R^m.$$

This makes clear that the right initial conditions may be stated e.g. by means of the projector P_0 onto $S_0(x_*)$ along $\ker A$.

Theorem 3.2 ([10], [11]): *Let $g \in C^2(\mathcal{D}, I\!\!R^m), \mathcal{D} \subset I\!\!R^m$ open, $x_* \in \mathcal{D}$, $g(x_*) = 0$, $g'(x_*) =: B$. Let the pencil $\lambda A + B$ be regular with index 1, and all its eigenvalues have negative real parts.*
Then there are a $\tau > 0$, and $\delta(\varepsilon) > 0$ to each $\varepsilon > 0$ such that

(i) *all IVPs for (3.2) with*

$$P_0(x(0) - x^0) = 0, \quad |P_0x^0 - P_0x_*| \leq \tau$$

have unique solutions on $[0, \infty)$,

(ii) $|P_0x^0 - P_0x_*| \leq \delta(\varepsilon)$ *implies*

$$|x(t; x^0) - x_*| \leq \varepsilon, \quad t \geq 0, \quad \text{and}$$

(iii) $|x(t; x^0) - x_*| \longrightarrow 0 \quad (t \to \infty).$

It should be mentioned that this assertion remains true if P_0 is any projector with $\ker P_0 = \ker A$.

Comparing with Lyapunov's Theorem above we should take into account that the eigenvalues of the regular zero index pencil $\lambda I + B$ are exactly the eigenvalues of the matrix $-B$.

Clearly, as for regular ODEs, all terms characterizing stability, that is, A, B, P_0, are available numerically.

The proof follows the classical lines (e.g. [9]) via linearization at the stationary solution. For the linear part, the decoupling by means of the projector technique sketched in Section 2 has been applied.

Further, let us stress once more that $g \in C^2$ is assumed even in the classical case.

It seems to be straightforward to propose an adequate Caratheodory theory for index 1 DAEs (cf. [13] for first aspects).

Unfortunately, to obtain a similar result also for the index 2 case, we need some structural condition to be sure that the DAE has index 2 around x_* indeed provided the pencil $\lambda A + g'(x_*)$ has.

Let us illustrate this problem by the following example (cf. [12], [6]), which describes a simple nonlinear resistor circuit. The system

(3.4)
$$\left.\begin{array}{l} x_1' - \alpha(x_3) = 0 \\ x_2' - \beta(x_3) = 0 \\ x_1 + x_2 x_3 + x_3^3 = 0 \end{array}\right\}$$

with given smooth functions $\alpha, \beta; \mathbb{R} \to \mathbb{R}$, is easily checked to have index 1 all over $\{x \in \mathbb{R}^3 : x_2 + 3x_3^2 \neq 0\}$. The surface given by the constraint in (3.4) has a fold. On the fold curve

$$\mathcal{F} := \{x \in \mathbb{R}^3 : x_1 + x_2 x_3 + x_3^3 = 0, x_2 + 3x_3^2 = 0\}$$

the pencils $\lambda A + g'(x)$ are regular but they have index 2. Different choices of the functions α, β may lead to bifurcations and impasse points, respectively. Hence, that problem (3.4) represents rather an index 1 DAE having singularities on the fold curve. It should not be considered to be an index 2 DAE.

By Example (3.4), the linearization at points on the fold curve is shown to make no sense at all in general. So we need certain additional structure to be sure that the linearization makes sense and the information at the only point x_* will suffice.

Theorem 3.3 ([6]) : *Let $g \in C^2(\mathcal{D}, \mathbb{R}^m), \mathcal{D} \subset \mathbb{R}^m$ open, $x_* \in \mathcal{D}, g(x_*) = 0, g'(x_*) =: B$. Let the pencil $\lambda A + B$ be regular with index 2, and all its eigenvalues have negative real parts.*
Additionally, let the condition

(3.5) $(I - AA^+)\{g(x) - g(P_0 x)\} \in im\,(I - AA^+)BQ_0, x \in B(x_*, \sigma),$

be given for certain $\sigma > 0$.
Then there are a $\tau > 0$, and $\delta(\varepsilon) > 0$ to each $\varepsilon > 0$ such that

(ı) all IVPs for (3.2) with

$$P_0 P_1(x(0) - x^0) = 0, \quad |P_0 P_1 x^0 - P_0 P_1 x_*| \leq \tau$$

have unique solutions on $[0, \infty)$,

(ıı) $|P_0 P_1 x^0 - P_0 P_1 x_*| \leq \delta(\varepsilon)$ implies

$$|x(t; x^0) - x_*| \leq \varepsilon, \quad t \geq 0, \quad and$$

(ııı) $|x(t; x^0) - x_*| \longrightarrow 0 \quad (t \to \infty)$.

Thereby, P_0, P_1 are projectors, $\ker A = \ker P_0, \ker(A + BQ_0) = \ker P_1$.

Condition (3.5) means, roughly speaking, that the derivative free part $(I - AA^+)g(x)$ within (3.2) should depend on the nullspace component $Q_0 x$ only linearly. In the case of Hessenberg form equations this is given trivially. Therefore, (3.5) covers both linear equations and Hessenberg form ones. Some further generalization is possible but much more technical.

Again, the characteristic terms A, B, P_0, P_1 are available, the index of the pencil as well as its regularity may be checked numerically.

Information on the pencil eigenvalues may be obtained directly, but also via the matrix (2.6) by usual methods.

Finally, turning back to our original equation (1.1) and its enlarged form (1.13),(1.14), we find the following statement that applies to index 1 DAEs whose leading coefficient has a nullspace varying with x.

Corollary 3.4 *(cf.* [10]*): Let* $A \in C^2(\mathcal{D}, L(I\!R^m)), g \in C^2(\mathcal{D}, I\!R^m), \mathcal{D} \subset I\!R^m$ *open,* $x_* \in \mathcal{D}, g(x_*) = 0,$

$$A(x_*) =: A_0, \quad g'(x_*) =: B,$$

P_0 *be a projector,* $\ker P_0 = \ker A_0$.
Additionally, let

(3.6) $im\, A(x) = im\, A(x_*), \quad x \in B(x_*, \sigma).$

Let the pencil $\lambda A_0 + B$ *be regular with index 1, and all its eigenvalues have negative real parts.*

Then the assertions (i) - (iii) of Theorem 3.2 become true if we replace (3.2) by (1.1).

Note that Corollary 3.4 is obtained by applying Theorem 3.3 to the enlarged system (1.13), (1.14) and tracing back the result to (1.1). Thereby, condition (3.6) appears as the reflection of (3.5). It should be stressed further that the projector P_0 onto $\ker A_0 = \ker A(x_*)$ obviously depends on x_*.

In case of a constant nullspace $\ker A(x) \equiv N$ we may use a projector independent of the linearization point and drop the structural condition (3.6) again.

Related results on index 3 equations are given in [10], however this applies to constrained multibody systems rather than to circuit simulation.

Considering limit circles makes some more difficulties since now one cannot work locally around a single point, that is, restrict the problem to a single chart. The linearization result given in [14] is hoped to be the right tool to overcome these difficulties.

References

[1] März, R.(1992) Numerical methods for differential-algebraic equations. Acta Numerica 1992, 141-198

[2] Griepentrog, E., Hanke, M. and März, R. (1992) Towards a better understanding of differential-algebraic equations. Seminarberichte Humboldt-Universität Berlin, Fachbereich Mathematik Nr. 92-1, 2-13

[3] Griepentrog E. and März, R. (1989) Basic properties of some differential-algebraic equations. Z. für Anal. u. ihre Anwendungen 8, 25-40

[4] März, R. (1993) Canonical projectors for linear differential-algebraic equations. Preprint 93-17, Humboldt-Universität Berlin, Fachbereich Mathematik

[5] Lewis, F.L. (1986) A survey of linear singular systems. Circuits Systems Signal Process 5(1), 3-36

[6] März, R. (1992) On quasilinear index 2 differential-algebraic equations. Seminar-bericht (cf. [2]) 92-1, 39-60

[7] Gear, C.W. and Petzold, L.R. (1984) ODE methods for the solution of differential/algebraic systems. SIAM J. Numer. Anal. 21, 716-728

[8] Lamour, R. (1992) personal communication

[9] Pontryagin, L.S. (1961) Ordinary differential equations. FIZMATGIZ, Moskow (in Russian)

[10] März, R. (1991) Practical Lyapunov stability criteria for differential algebraic equations. Preprint 91-28 (cf. [4]), to appear in Banach Center Publications

[11] Tischendorf, C.(1991) On stability of solutions of autonomous index-1 tractable and quasilinear index-2-tractable dae's. Preprint 91-25 (cf. [4]), to appear in Circuits Systems Signal Process

[12] Chua, L.O. and Deng An-Chang (1989) Impasse points. J. of Circuit Theorey and Applications 17, 213-235

[13] Dolezal, V. (1986) Generalized solutions of semistate equations and stability. Circuits Systems Signal Process 5(4), 391-403

[14] März, R. and Tischendorf, C. (1993) Solving more general index 2 differential algebraic equations. Preprint 93-6 (cf.[4]), Computers and Mathematics with Applications, to appear

Prof. R. März, FB Mathematik, Humboldt-Universität Berlin, Unter-den-Linden 6
D-10099 Berlin, Germany

International Series of Numerical Mathematics, Vol. 117, © 1994 Birkhäuser Verlag Basel

Analysis of Linear Time-invariant Networks in the Frequency Domain

Wolfgang Mathis
University of Wuppertal
Department of Electrical Engineering, D-5600 Wuppertal 1, Fuhlrottstr. 10
FAX: +49-202-3008; E-Mail: mathis at cyber.URZ.UNI-WUPPERTAL.DBP.DE

1. Introduction and Formulation of the Problem

It is well-known that linear time-invariant electrical networks represent successful models in nearly all branches of electrical engineering. Therefore networks consisting of linear resistors, capacitors and inductors as well as dependent and independent sources – so-called active RLC networks – are of much interest till now in order to describe linear analog circuits. Studying electrical networks we have to remember the fundamental concept of this class of systems. The main idea is to build systems consisting of relative simple subsystems which are cooperate by means of a connection subsystem. Although this principle is used in many areas of engineering and natural sciences these 'network concepts' differ in their mathematical structures and the meaning of its variables. In electrical network theory systems are described by currents and voltages. Therefore we need equations which govern the simple subsystems and the connection subsystem; in this area these simple subsystems are denoted as network elements. First considerations of electrical networks were published by Ohm in 1828 [18]. His studies were continued by Kirchhoff, Helmholtz, Maxwell and others; for a historical overview see Wunsch [29]. The description of electrical networks is based on the following three classes of equations:

1) Network elements characterized by currents and voltages
   (Collected in an affine-type equation)

$$\mathbf{Y}\mathbf{u} + \mathbf{Z}\mathbf{i} = \mathbf{c}, \tag{1}$$

2) network elements characterized by differential of integral relationships
   (in active RLC networks the capacitors and inductors)

$$L_k \frac{di_L^k}{dt} = u_L^k, \qquad C_l \frac{du_C^l}{dt} = i_C^l, \qquad k, l = 1, 2, \ldots \tag{2}$$

3) the connection subsystem

$$\mathbf{A}\mathbf{i} = \mathbf{0}, \qquad \mathbf{B}^T \mathbf{u} = \mathbf{0}, . \tag{3}$$

where $(\mathbf{A}, \mathbf{B})$ is exact. The exactness is defined by $R(\mathbf{B}) = N(\mathbf{A})$ where $R(\mathbf{B})$ is the range of $\mathbf{B}$ and $N(\mathbf{A})$ is the null-space of $\mathbf{A}$. An equivalent definition of an exact pair of matrices as well as some conclusions of this property are contained in Mathis [14]. It follows from 1) – 3) that these equations for describing networks form a system of algebraic and differential equations; as an abbreviation DAE is used. There are several methods available in order to formulate differential equations of the form (see e.g. Chua, Lin [3])

$$\mathbf{S}\dot{\mathbf{x}} = \mathbf{A}\mathbf{x} + \mathbf{B}_1\mathbf{u}(t) + \mathbf{B}_2\dot{\mathbf{u}}(t) + \cdots, \tag{4}$$

where $u(t)$ consists of the independent sources or input quantities of the network. If there is $\mathbf{S} = \mathbf{1}$ these equations are called 'state space equations' in applications (control theory, system and network theory). For the mathematical analysis of the general case it is useful to associate a 'generalized state space' to equations (4); therefore Dziurla and Newcomb [6] as well as Mathis [12] denoted these type of equations as 'semistate space equations' and 'generalized state space equations', respectively. In control theory equations (4) are called 'singular systems' (Campbell [2]) or 'descriptor systems' (see e.g. Cobb [4]). For the description of linear systems and networks the general case ($\mathbf{S} \neq \mathbf{1}$) can be reduced to state space equations. This is useful because theory as well as numerics of these explicit differential equations is well established (e.g. Zwillinger [32]). Unfortunately the formulation of these equations for a network using the fundamental network equations (1) – (3) is often very tedious; an approach was presented by Pottle [19]. Because it is very simple to formulate generalized state space equations in this paper we restrict us to this type of equations.

Beside currents and voltages in $\mathbf{x}$ – the state variables or generalized state variables – we have to choose other network variables which can be interpreted as output variables of the network. In linear time-invariant networks output variables $\mathbf{y}$ yield as linear combinations of state variables $\mathbf{x}$ and input variables $\mathbf{u}$; one obtains

$$\mathbf{y} = \mathbf{Cx} + \mathbf{Du}(t). \tag{5}$$

Restricting us to those networks with a single input and a single output the generalized state space equations (4) and the output equations (5) result in a reduced form

$$\mathbf{S\dot{x}} = \mathbf{Ax} + \mathbf{b}_1 u(t) + \mathbf{b}_2 \dot{u}(t) + \cdots, \tag{4'}$$

$$y = \mathbf{c}^T \mathbf{x} + d \cdot u(t) \tag{5'}$$

for the description of the class of linear time-invariant networks. In the following sections we are interested in the transfer behavior of these systems.

2. Transfer Functions for Generalized State Space Equations

For the analysis of (4') with a large number of equations several numerical methods are available. If $\mathbf{S}$ is a singular matrix equations (4') are implicit differential equations or so-called differential algebraic equations (DAE). Equations of this type and methods for its numerical solution were studied intensively since the early eighties by Campbell, Gear, Petzold and many others; see e.g. the monograph of Brenan, Campbell, Petzold [1] for further references. On the other hand main results of the linear case with constant coefficients were published much earlier by Gantmacher [9]. First studies were published in control and network theory by Dziurla et al. [6], Verghese et al. [26], Cobb [4] and some others.

For the analysis of linear differential equations with constant coefficients it is suitable to apply an operational solution method; beside the classical Laplace transform (see e.g. Desoer, Kuh [5]) another suitable method for this type of equations was presented by Mathis and Marten [11] (see also Mathis [14]) based on results of Yosida [30]; this method uses ideas from commutative algebra is called HMY calculus (Heaviside-Mikunsinki-Yosida).

Beside of the conceptual clarity this calculus avoids in contrast to the Laplace transform the introduction of complex currents and voltages and the Dirac's function and its derivatives are included in a well-defined way. Using the HMY calculus (see Mathis [14]) the equations (4') and (5') can be formulated as follows ($\star$ is the convolution operation of on $I\!R_+$ defined functions)

$$\mathbf{S}s \star \mathbf{x} + \mathbf{p}_1(s) = \mathbf{A}\mathbf{x} + \mathbf{b}_1\{u(t)\} + \mathbf{b}_2 s \star \{u(t)\} + \cdots + \mathbf{p}_2(s); \tag{6}$$
$$y = \mathbf{c}^T\mathbf{x} + d \cdot \{u(t)\} \tag{7}$$

where the polynomials $\mathbf{p}_1$ and $\mathbf{p}_2$ include the initial conditions. The unknown functions are so-called hyperfunctions which extends the continuous functions on $I\!R_+$ (see Yosida [30]). Wunsch [28] and Mathis [14] (using the HMY calculus) showed that these polynomials are of no interest for the transfer behavior of systems and networks. Therefore we can omit these terms in the following. If we solve (6) with respect to $\mathbf{x}$ and eliminate this network variable in (7); we obtain

$$y = \left(\mathbf{c}^T(s\mathbf{S} - \mathbf{A})^{-1}\mathbf{b}(s) + d\right) \star \{u(t)\}$$

where $\mathbf{b}(s) := \mathbf{b}_1 + s\mathbf{b}_2 + \cdots$. The transfer function of a network formulated above is defined

$$H(s) := \mathbf{c}^T(s\mathbf{S} - \mathbf{A})^{-1}\mathbf{b}(s) + d. \tag{8}$$

So and Sandberg [23] proved that (8) can be reduced (index r) to a quotient of two determinants

$$H(s) = \frac{\det(s\tilde{\mathbf{1}} - \tilde{\mathbf{A}}_r)}{\det(s\mathbf{1} - \mathbf{A}_r)}. \tag{9}$$

Related results were given by Fettweis [8] and Sorensen [22]. Unfortunately these approaches can be instable from a numerical point of view. For this reason it seems to be better to use another form of $H(s)$

$$H(s) = \frac{\det(s\tilde{\mathbf{S}}(s) - \tilde{\mathbf{A}}(s))}{\det(s\mathbf{S} - \mathbf{A})} \tag{10}$$

which can be determined from the linear inhomogeneous equations (7) and (8) by means of Cramer's rule. Note, that in general the coefficients of the numerator matrix are polynomials in s. One of the main reasons for calculating $H(s)$ is that we can establish the stability of networks by considering the poles of these functions. Obviously Laplace's evaluation formula is not suitable for determining the determinants because the number of terms generated terms is very large (in essential $n!$. If we interpret the denominator determinant as the characteristic polynomial of a generalized eigenvalue problem

$$\mathbf{M}\mathbf{x} = \lambda\mathbf{N}\mathbf{x}$$

an efficient algorithm for this problem is necessary in order to determine transfer functions (10). With respect to special eigenvalue problem ($\mathbf{N} = \mathbf{1}$) we distinguish (just like Zurmühl, Falk [31]) these problems by its pairs of matrices: generalized eigenvalue problem

$(\mathbf{M}, \mathbf{N})$ and the special eigenvalue problem $(\mathbf{M}, \mathbf{1})$. In contrast to the latter case a $(\mathbf{M}, \mathbf{N})$ eigenvalue problem includes not only finite but also infinite eigenvalues (zero eigenvalues of the reversed matrix pair $(\mathbf{N}, \mathbf{M})$) if $\mathbf{M}$ is singular. Note, that the numerator determinant cannot be interpreted as the characteristic polynomial of a generalized eigenvalue problem in general cases because the matrices are depend on s. In the next section we solve this problem by means of a network theoretical consideration.

3. Two-Sets-of-Eigenvalues Approach and Network Analysis

As So and Sandberg [23] formulated their approach no efficient algorithms for the $(\mathbf{M}, \mathbf{N})$ eigenvalue problem were available. In 1973 Moler and Stewart [17] generalized the powerful QR algorithm for the $(\mathbf{M}, \mathbf{1})$ problem; it is called QZ algorithm. Their algorithm together with some additional improvements is contained in the EISPACK package (see Garbow et al. [10]). First remarks for applying it to the evaluation of network determinants were already given by Chua, Lin ([3], p.570). Unfortunately first tests with the QZ algorithm showed that in realistic filter networks (e.g. high pass filters) infinite eigenvalues appear. These eigenvalues lead to erroneous numerical values and, therefore, a stability of the network cannot be decided (for an example see Mathis [12]). Based on his developed method for a computational calculation of Kronecker's canonical form [24] of a singular pair of matrices (denoted as pencil matrices) van Dooren considered applications to linear system theory [25]. The latter results were used and extended by Mathis [12] to the analysis of linear time-invariant networks. A first version based on network equations of Schwartz [21] which are formulated with nodal potentials. The main features of the algorithm can be given as follows

a) Construction of the network matrices
b) Application of scaling strategies from network theory (e.g. Desoer, Kuh [5])
c) Permutation of rows and columns in order to find eigenvalues without numerical calculations because of internal structures of sparse network matrices (Ward [27])
d) Reduction of zero and infinite eigenvalues (van Dooren [24])
e) Balancing of the pair of matrices (Ward [27])
f) Ordering of rows and columns with respect to its norms (Ward [27])
g) Application of the QZ algorithm (Moler and Stewart [17])

This combination algorithm shows that several preprocessing steps have to be performed in order to get reliable results. A second implementation of this algorithm based on the network equations of ANP3 (Sorensen [22]). This approach transforms a general linear time-invariant RLC network with transformers, an independent input source and controlled sources into a RC network with an input sources and only a voltage controlled current source. For this reason nodal potential equations can be reformulated in the form

$$\mathbf{S}s \star \mathbf{x} = \mathbf{A}\mathbf{x} + \mathbf{b}_1\{u(t)\} + \mathbf{b}_2 s \star \{u(t)\}; \tag{12}$$
$$y = \mathbf{c}^T \mathbf{x} + d \cdot \{u(t)\} \tag{13}$$

with the associated transfer function

$$H(s) = \frac{\det(s\tilde{\mathbf{S}} - \tilde{\mathbf{A}})}{\det(s\mathbf{S} - \mathbf{A})} \tag{14}$$

where the matrices $\tilde{S}$ and $\tilde{A}$ are constant matrices. This demonstration program system CIRCDES includes further below mentioned features. The last implementation based on the modified nodal approach (see e.g. Mathis [14]). Menzel [16] implemented the main parts of the combination algorithm in the circuit simulator TITAN of the Siemens AG (Feldmann et al. [7]); TITAN is designed only for the internal usage. These versions were tested by more than hundred examples. All these tests were performed as white-box as well as black-box tests (Schmitz et al. [20]) and results showed the robustness of this algorithm. Because the poles and zeros with additional constants for numerator and denominator determine the transfer function in a unique manner derived quantities can be calculated. For example, above mentioned program system CIRCDES calculates Nyquist plots and root loci of the zeros and poles in dependence of an arbitrary network parameter; these curves can be useful in analog circuit design. Furthermore the determination of Bode plots is possible but these plots can be calculated in another way, too. In a forthcoming paper (Mathis [15]) we discuss some more details about applications of these quantities of linear time-invariant networks.

4. An Example

In the last section we mentioned that the constructed combination algorithm was tested with a large number of examples. Especially the implementation in TITAN opened the possibility to test our algorithm with practical circuits because linear networks can be generated by means of a linearization process from nonlinear transistor models. In this section some results of an operational circuit μA 709 (Fig.1) with 15 transistors, 19 resistors and 2 capacitors are discussed where the used bipolar transistor model consists of further elements; these results were worked out by Menzel [16]. In this case one obtains 33 zeros and 34 poles. In order to verify the accuracy of the algorithm Bode plots were calculated and compared with Bode plots which were generated with the AC analysis mode of TITAN; Menzel obtained very small differences between the results (correspondence more than 4 digits). Furthermore Menzel observed that AC analysis of TITAN which calculates Bode plots pointwise for each frequency needs nearly twice as much time as the presented approach.

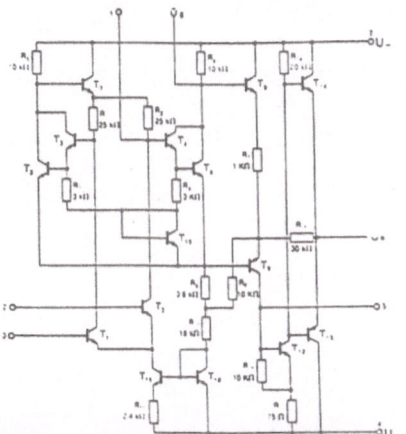

Fig.1: Operational Amplifier μA 709

POLSTELLEN

FEHLERMARKE	= 6
NENNERGRAD	= 34
NENNERANTIGRAD	= 0
NENNERKONSTANTE	= 0.000000000000000000E+00

	REALTEIL	IMAGINAERTEIL
P(1)=	-8.2653502907089050D+10 +J*	0.0000000000000000D+00
P(2)=	-4.9460407108297040D+10 +J*	0.0000000000000000D+00
P(3)=	-2.9604352332495801D+10 +J*	0.0000000000000000D+00
P(4)=	-2.0926003006252550D+10 +J*	0.0000000000000000D+00
P(5)=	-1.9752354892648849D+10 +J*	0.0000000000000000D+00
P(6)=	-1.1410893854880008D+10 +J*	0.0000000000000000D+00
P(7)=	-9.4123008428024225D+09 +J*	0.0000000000000000D+00
P(8)=	-6.3011048815019131D+09 +J*	0.0000000000000000D+00
P(9)=	-3.8961493632018124D+09 +J*	0.0000000000000000D+00
P(10)=	-3.6106602146682551D+09 +J*	0.0000000000000000D+00
P(11)=	-3.1394173449840803D+09 +J*	0.0000000000000000D+00
P(12)=	-2.9857854605973778D+09 +J*	0.0000000000000000D+00
P(13)=	-2.9223114808796690D+09 +J*	0.0000000000000000D+00
P(14)=	-1.7082990934684106D+09 +J*	1.3274024978982334D+09
P(15)=	-1.7082990934684107D+09 +J*	1.3274024978982337D+09
P(16)=	-1.4060822211347676D+09 +J*	-6.5736788815199003D+07
P(17)=	-1.4060822211347676D+09 +J*	6.5736788815199003D+07
P(18)=	-7.5008065989649159D+08 +J*	-6.0387653303385235D+07
P(19)=	-7.5008065989649141D+08 +J*	6.0387653303385276D+07
P(20)=	-4.9033427121137494D+08 +J*	0.0000000000000000D+00
P(21)=	-3.3881590047449267D+08 +J*	0.0000000000000000D+00
P(22)=	-2.7811235892298478D+08 +J*	0.0000000000000000D+00
P(23)=	-2.7574888274277687D+08 +J*	0.0000000000000000D+00
P(24)=	-2.1408942638423871D+08 +J*	0.0000000000000000D+00
P(25)=	-1.1907378845467049D+08 +J*	-6.2426447379334930D+07
P(26)=	-1.1907378845457049D+08 +J*	8.2426447379334904D+07
P(27)=	-1.5148946644357456D+08 +J*	0.0000000000000000D+00
P(28)=	-1.7375099182395247D+08 +J*	0.0000000000000000D+00
P(29)=	-7.4824252415880571D+07 +J*	0.0000000000000000D+00
P(30)=	-3.4612661481900368D+07 +J*	0.0000000000000000D+00
P(31)=	-1.3217928682509143D+07 +J*	-5.4900000143603173D+06
P(32)=	-1.3217928682509143D+07 +J*	5.4900000143603152D+06
P(33)=	-4.3496225345188577D+06 +J*	0.0000000000000000D+00
P(34)=	-6.0259755922616050D+03 +J*	0.0000000000000000D+00

NULLSTELLEN

FEHLERMARKE	= 6
ZAEHLERGRAD	= 33
ZAEHLERANTIGRAD	= 0
ZAEHLERKONSTANTE	= 0.000000000000000000E+00

	REALTEIL	IMAGINAERTEIL
P(1)=	-2.5615275289087870D+10 +J*	-8.9982280733919891D+10
P(2)=	-2.5615275289087806D+10 +J*	8.9982280733919937D+10
P(3)=	-5.2762957659245061D+10 +J*	0.0000000000000000D+00
P(4)=	-4.2861140598170612D+10 +J*	0.0000000000000000D+00
P(5)=	-2.1412192461044805D+10 +J*	0.0000000000000000D+00
P(6)=	-1.7871976257178439D+10 +J*	0.0000000000000000D+00
P(7)=	-1.1223038488915916D+10 +J*	0.0000000000000000D+00
P(8)=	-8.9816958556663446D+09 +J*	0.0000000000000000D+00
P(9)=	-3.7139692717272674D+09 +J*	0.0000000000000000D+00
P(10)=	-3.5016967508759404D+09 +J*	0.0000000000000000D+00
P(11)=	-3.0101286667980450D+09 +J*	0.0000000000000000D+00
P(12)=	-2.9382027958443033D+09 +J*	-4.4734784572063684D+07
P(13)=	-2.9382027958443034D+09 +J*	4.4734784572063684D+07
P(14)=	-1.6808393535028014D+09 +J*	-1.3149815928976561D+09
P(15)=	-1.6808393535028017D+09 +J*	1.3149815928976563D+09
P(16)=	-1.3192567990664427D+09 +J*	0.0000000000000000D+00
P(17)=	3.2978885745664787D+08 +J*	0.0000000000000000D+00
P(18)=	-8.0212475063929802D+08 +J*	0.0000000000000000D+00
P(19)=	-1.2765864546561094D+08 +J*	-2.2849487485293099D+08
P(20)=	-1.2765864546561098D+08 +J*	2.2849487485293096D+08
P(21)=	-3.7562920575001609D+08 +J*	0.0000000000000000D+00
P(22)=	-3.0294431958761019D+08 +J*	-3.9950757675900735D+07
P(23)=	-3.0294431958761019D+08 +J*	3.9950757675900768D+07
P(24)=	-3.3929969741034716D+08 +J*	0.0000000000000000D+00
P(25)=	-1.6337717557570257D+08 +J*	-8.7807935410095785D+07
P(26)=	-1.6337717557570257D+08 +J*	8.7807935410095785D+07
P(27)=	-1.6981084565595175D+08 +J*	0.0000000000000000D+00
P(28)=	-5.8136321031562410D+07 +J*	-3.0935899409081601D+06
P(29)=	-5.8136321031562388D+07 +J*	3.0935899409081598D+06
P(30)=	-1.3293408318012710D+07 +J*	0.0000000000000000D+00
P(31)=	-1.2518127748655946D+07 +J*	0.0000000000000000D+00
P(32)=	2.7334697314885173D+05 +J*	0.0000000000000000D+00
P(33)=	-6.6037116378291948D+03 +J*	0.0000000000000000D+00

Fig.2: Poles and Zeros of the OP μA 709

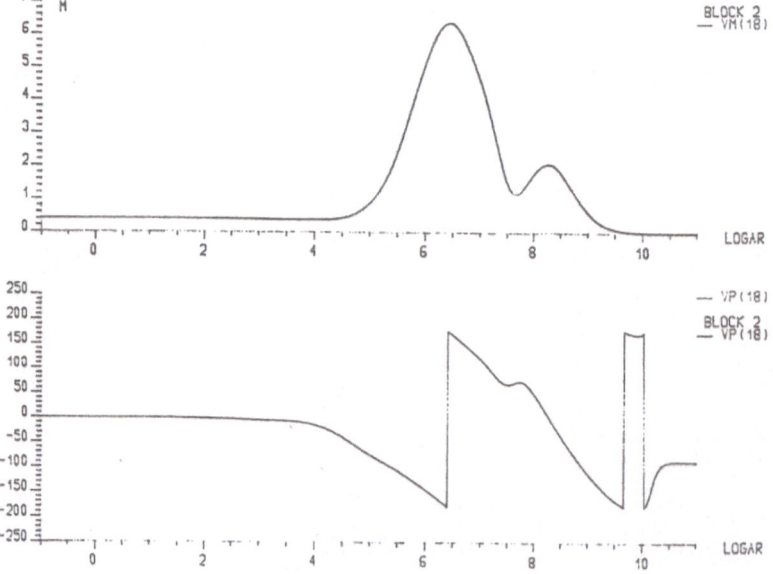

Fig.3: Bode Plots of the OP μA 709

References

[1] Brenan, K.E.; S.L. Campbell; L.R. Petzold: Numerical Solution of Initial-Value Problems in Differential-Algebraic Equations. North-Holland, New York-Amsterdam 1989

[2] Campbell, S.L.: Singular Systems of Differential Equations. Pitman, San Francisco 1980

[3] Chua, L.O.; P.-M.Lin: Computer-Aided Analysis of Electronic Circuits. Prentice-Hall, Inc., Englewoods Cliffs (NJ) 1975

[4] Cobb, D.: Feedback and Pole Placement in Descriptor Variable Systems. Intern.Journ. Contr. 33(1981)1135-1146

[5] Desoer, C.A.; E.S. Kuh: Basic Circuit Theory. McGraw-Hill Book Comp., New York 1967

[6] Dziurla, B.; R. Newcomb: The Drazin Inverse and Semistate Equations. Proc. 4th Intern. Symp. MTNS, Delft (The Netherlands), pp.283-289, 1979

[7] Feldmann, U.; U.A. Wever; Q. Zheng; R. Schultz; H. Wriedt: Algorithms for Modern Circuit Simulation. AEÜ 46(1992)274-285

[8] Fettweis, A.: On the Algebraic Derivation of the State Equations. IEEE CT-16(1969)171-175

[9] Gantmacher, F.R.: Matrizenrechnung I + II. VEB Deutscher Verlag der Wissenschaften, Berlin 1971

[10] Garbow, B.S.; J.M. Boyle; J.J. Dongarra; C.B. Moler: Matrix Eigenvalue Routines – EISPACK Guide Extension. Springer-Verlag, Berlin-New York 1977

[11] Marten, W.; W. Mathis: New Algebraic Methods in Linear Time-invariant System Theory. Proc. ECCTD'87, Paris 1987

[12] Mathis, W.: Zur Theorie und Numerik verallgemeinerter Zustandsgleichungen im Frequenzbereich und deren Anwendung bei der Netzwerkanalyse. Dissertation, Braunschweig 1984

[13] Mathis, W.: Bestimmung von Übertragungsfunktionen linearer Netzwerke als 2-faches verallgemeinertes Eigenwertproblem. 4. Symp. Simulationstechnik, Proc. of ASIM 87: Halin, J.(Hrsg.): Simulationstechnik. Springer-Verlag, Berlin 1987

[14] Mathis, W.: Theorie nichtlinearer Netzwerke. Springer-Verlag, Berlin 1987

[15] Mathis, W.: Issues Towards CAD Environments for Analog Circuits. Will be pushlish in the journal 'VLSI Design', 1993

[16] Menzel, A.: Anpassung und Implementation eines Verfahrens zur Pol/Nullstellen-Analyse in einen industriellen Schaltkreis-Simulator. Diploma Thesis, Techn. University Braunschweig, 1990

[17] Moler, C.B.; G.W. Stewart: An Algorithm for Generalized Matrix Eigenvalue Problems. SIAM J. Num. Anal. 10(1973)241-256

[18] Ohm, G.S.: Die Galvanische Kette, mathematisch bearbeitet (Nachdruck von 1827). VEB Deutscher Verlag der Wissenschaften, Berlin 1989

[19] Pottle, C.: State-Space Techniques for General Active Network Analysis. In: Kuo, F.F.; J.F. Kaiser: System Analysis by Digital Computer. John Wiley&Sons, Inc., New York 1966

[20] Schmitz, P.; H. Bons; R. von Mege: Software – Qualitätssicherung – Testen im Software-Lebenszyklus (2. Edition). Friedr. Vieweg&Sohn, Braunschweig-Wiesbaden 1983

[21] Schwartz, E.: Symbolic Analysis of Active RC-Networks with a Minicomputer. AEÜ 32(1978)456-462

[22] Sorensen, E.V.: A Linear Semisymbolic Circuit Analysis Program Based in Algebraic Eigenvalue-Technique (ANP3) Report: Inst. Circuit Theory and Telecom., Techn. Univ. Denmark, 288 Lyngby, Oct. 1972

[23] So, H.C., I.W. Sandberg: A Two-Sets-of-Eigenvalue Approach to the Computer Analysis of Linear Systems. IEEE CT-16(1969)509-517

[24] van Dooren, P.M.: Computation of Kronecker's Canonical Form of a Singular Pencil. Linear Algebra Appl. 27(1979)103-140

[25] van Dooren, P.M.: The Generalized Eigenstructure Problem in Linear System Theory. IEEE AC-26(1981)111-129

[26] Verghese, G.C.; B.C. Lévy; T. Kailath: A General State-Space for Singular Systems. IEEE AC-26(1981)811-831

[27] Ward, C.R.: Balancing the Generalized Eigenvalue Problem. SIAM J.Sci.Stat.Comp. 2(1981)141-152

[28] Wunsch, G.: Moderne Systemtheorie. Akademische Verlagsgesellschaft Geest u. Portig, Leipzig 1962

[29] Wunsch, G.: Geschichte der Systemtheorie. Akademie-Verlag, Berlin 1985

[30] Yosida, K.: Operational Calculus. Springer-Verlag, New York-Berlin 1984

[31] Zurmühl, R.; S. Falk: Matrizen 1. Grundlagen. Springer-Verlag, Berlin 1992 (6.Edition)

[32] Zwillinger, D.: Handbook of Differential Equations. Academic Press, Boston-San Diego 1989

Acknowledgements: The author would like to thank J. Krehnke, Heiko Schiller, Günther Wegener and Andreas Menzel for their engagement during the preparation of their diploma thesis and studying projects at the Techn. University of Braunschweig. Furthermore it is my pleasure to thank Siemens AG (Munich) for support and, in particular, Dr.rer.nat. Uwe Feldmann and his research and development group (ZFE) at Siemens for many discussions and their hospitality.

International Series of Numerical Mathematics, Vol. 117, © 1994 Birkhäuser Verlag Basel   91

Limit Cycle Computation of Oscillating Electric Circuits *

W. Schmidt

Mathematisches Institut, Technische Universität München

Postfach D-80290 München, Germany

Abstract

Limit cycle computation of oscillating electric circuits without external input can be formulated as a two point boundary-value problem. An efficient way to solve it is developed, based on inner coupling of a stiff integration method and multiple shooting. Furthermore, if the structure of this specific boundary-value problem is taken into account, this algorithm can be accelerated and stabilized by some analytical calculations. To test the method, a complete set of equations, including parameters, is given, which describes a ring oscillator, consisting of a series of N MOS-FET inverters. The MOS-FET transistors are substituted by a circuit invented by Shichman and Hodges.

1. Formulation of a Boundary-Value Problem

Oscillating electric circuits without external input are often represented by an autonomous system of ordinary differential equations

$$(1) \qquad \dot{x}(t) = f(x(t)), \qquad x(0) = x_0, \qquad f : \mathbb{R}^n \to \mathbb{R}^n.$$

Here, periodic solutions of (1) are of special interest. They are determined by the period T and a set of initial values x_0. Seydel (1981) formulates a nonlinear $n+1$-dimensional boundary-value problem to compute these unknowns T and $x_0 \in \mathbb{R}^n$:

$$(2) \qquad \begin{pmatrix} \dot{x}(t) \\ \dot{T}(t) \end{pmatrix} = \begin{pmatrix} T \cdot f(x(t)) \\ 0 \end{pmatrix}, \qquad x(0) = x(1).$$

In (2), the integration interval $[0, T]$ is normalized to the unit interval $[0, 1]$. Of course, system (2) has no unique solution, since there are $n+1$ differential equations and only n

---

*Dedicated to my father to his 65th birthday

corresponding boundary conditions. Several ways to obtain uniqueness are possible: for example, the phase shift can be fixed by an additional boundary condition:

$$(3) \qquad r(x(0), x(1)) := x_i(0) - \alpha = 0, \quad r : \mathbb{R}^n \times \mathbb{R}^n \to \mathbb{R}, \quad \alpha \in \mathbb{R}.$$

In other words, an arbitrary component $x_i(0)$, $1 \le i \le n$, is prescribed by condition (3). Other choices of r are practicable (refer to Seydel (1981)). With the following abbreviations

$$Y(t) := \begin{pmatrix} x(t) \\ T(t) \end{pmatrix}, \qquad R(Y(0), Y(1)) := \begin{pmatrix} x(1) - x(0) \\ r(x(0), x(1)) \end{pmatrix},$$

and

$$F(Y(t)) := \begin{pmatrix} T \cdot f(x(t)) \\ 0 \end{pmatrix}, \qquad F : \mathbb{R}^{n+1} \to \mathbb{R}^{n+1},$$

the expanded boundary-value problem (2), (3) has the form:

$$(4) \qquad \dot{Y}(t) = F(Y(t)), \qquad R(Y(0), Y(1)) = 0.$$

The solution of boundary-value problems of this kind is a standard task in numerical analysis. In the next section the multiple shooting method is adapted to the special structure of equation (4).

2.   Multiple shooting and periodic solutions

The multiple shooting method is a well-proved technique for the numerical solution of two-point boundary-value problems. It is completely described in Stoer, Bulirsch (1979). Here, only the main principles are outlined.

In the first step the multiple shooting method divides the integration interval $[0, 1]$ into $m - 1$ subintervals given by

$$0 = t_1 < t_2 < \cdots < t_{m-1} < t_m = 1.$$

At each node t_j, $1 \le j \le m$, a $(n + 1)$-dimensional starting vector S_j is prescribed and in each subinterval $[t_j, t_{j+1}], 1 \le j \le m - 1$, an initial value problem

$$(5) \qquad \dot{Y}(t) = F(Y(t)), \qquad Y(t_j) = S_j$$

is integrated. Since the solution of (5) depends on both t and S_j, we write $Y_j(t; S_j)$ instead of $Y_j(t)$ alone. Because the solution of the differential equation (4) should at least be continuous, we obtain a system of nonlinear equations for the starting vectors $S_1, S_2, \ldots, S_m$:

(6)
$$\Phi(S_1, S_2, \ldots, S_m) = \begin{pmatrix} Y_1(t_2; S_1) - S_2 \\ Y_2(t_3; S_2) - S_3 \\ \vdots \\ Y_{m-1}(t_m; S_{m-1}) - S_m \\ R(S_1, S_m) \end{pmatrix} = \begin{pmatrix} 0 \\ 0 \\ \vdots \\ 0 \\ 0 \end{pmatrix}.$$

A well-known way to solve system (6) is the damped Newton method. It requires the computation of the Jacobian

(7)
$$\Phi_{S_1, S_2, \ldots, S_m} = \begin{pmatrix} G_1 & -I & & \\ & \ddots & \ddots & \\ & & \ddots & I \\ & & & G_{m-1} \\ A & & & B \end{pmatrix}.$$

The $(n+1) \times (n+1)$ submatrices A, B, and G_j, $1 \le j \le m-1$ are defined by

$$A := \frac{\partial R(S_1, S_m)}{\partial S_1}, \qquad B := \frac{\partial R(S_1, S_m)}{\partial S_m}, \qquad G_j := \frac{\partial Y_j(t_{j+1}, S_j)}{\partial S_j}.$$

The calculation of A and B does not cause any trouble. Because of the simple structure of the boundary condition R, it is sufficient to compute them only in the very first step of Newton's iteration. The crucial point in the implementation of shooting methods is the computation of the $m-1$ matrices G_j, i.e. the differentiation of the trajectories with respect to the initial vector S_j. In standard codes they are calculated by numerical differentiation, but this is a very expensive method: in each subinterval $n+1$ additional initial value problems with slightly modified initial values must be solved per Newton step. Even if standard update techniques like Broyden's approximation are used, this overhead is too large. Taking into account that the differential equation (5) requires an integration routine for stiff problems, here a more efficient path can be taken: analysis tells us that the matrices G_j satisfy the linear matrix-differential equation

(8)
$$\dot{G}_j(t) = F_Y(Y(t)) \cdot G_j(t), \qquad G_j(t_j) = I, \qquad t \in [t_j, t_{j+1}],$$

where $I \in \mathbb{R}^{n+1, n+1}$ denotes the identity matrix. The differential equation (8) can be integrated with moderate additional amount by an inner coupling of the multiple shooting method and the stiff integrator. This means that instead of numerical differentiation, only matrix multiplications must be done. In the case of BDF-integrators, this technique is completely described in Kampowsky, Rentrop, and Schmidt (1992).

So far, no use has been made of the special structure of equation (4). But the coupling of the trivial differential equation $\dot{T} = 0$ to the original system $\dot{x}(t) = f(x(t))$ must not be done explicitly; it is only of a formal nature. Moreover, the corresponding parts of the

Jacobi matrix $\Phi_{S_1,S_2,...,S_m}$ can be computed analytically with a few additional operations. To simplify notation, we will limit ourselves to an arbitrary subinterval $[t_j, t_{j+1}]$ and suppress the index j. In other words, we solve the $(n+1) \times (n+1)$ matrix differential equation

(9) $$\dot{G}(t) = F_Y(Y(t)) \cdot G(t), \qquad G(t_j) = I,$$

in the interval $[t_j, t_{j+1}]$, instead of (8).

When this is written without abbreviations, we obtain a block differential equation

(10) $$\begin{pmatrix} \dot{g}_{11}(t) & \dot{g}_{12}(t) \\ \dot{g}_{21}(t) & \dot{g}_{22}(t) \end{pmatrix} = \begin{pmatrix} T \cdot f_x(x(t)) & f(x(t)) \\ 0 & 0 \end{pmatrix} \cdot \begin{pmatrix} g_{11}(t) & g_{12}(t) \\ g_{21}(t) & g_{22}(t) \end{pmatrix},$$

with the initial value $G(t_j) = I$. The submatrices are dimensioned by:

(11) $$g_{11} \in \mathbb{R}^{n,n}, \quad g_{12} \in \mathbb{R}^{n,1}, \quad g_{21} \in \mathbb{R}^{1,n}, \quad g_{22} \in \mathbb{R}.$$

For the particular elements g_{11}, g_{21}, and g_{12} of the initial value problem (10) the solution can be explicitly calculated in each interval $[t_j, t_{j+1}], 1 \leq j \leq m-1$.

(12) $$\dot{g}_{22}(t) = 0, \quad g_{22}(t_j) = 1 \quad \Longrightarrow \quad g_{22}(t) = 1 \in \mathbb{R},$$
(13) $$\dot{g}_{21}(t) = 0, \quad g_{21}(t_j) = 0 \quad \Longrightarrow \quad g_{21}(t) = 0 \in \mathbb{R}^{1,n},$$

(14) $$\dot{g}_{12}(t) = T \cdot f_x(x(t))g_{12}(t) + f(x(t)), \quad g_{12}(t_j) = 0$$
$$\Longrightarrow \quad g_{12}(t) = (t - t_j) \cdot f(x(t)) = \tfrac{1}{T}(t - t_j) \cdot \dot{x}((t)) \in \mathbb{R}^{n,1}.$$

Of course, each integration scheme evaluates the right side of a differential equation, thus $\dot{x}(t) \in \mathbb{R}^{n,1}$ is already known. Therefore the computation of g_{12}, g_{21}, and g_{22} requires only three additional operations. In practical terms, the trivial equation $\dot{T} = 0$ is only formally coupled to the original equation (1), which does not really increase the dimension of the problem. Finally, the most difficult question, the integration of the remaining matrix differential equation

(15) $$\dot{g}_{11}(t) = T \cdot f_x(x(t))g_{11}(t), \qquad g_{11}(t_j) = I, \in \mathbb{R}^{n,n}$$

can be solved by the connection of multiple shooting and a stiff integrator, mentioned above (Kampowsky et al. (1992)).

3.   Circuit of the ring oscillator

The ring oscillator, shown in figure 1, is a typical example of an oscillating, autonomous circuit. It consists of a series of $N-1$ coupled MOS-FET transistors $T_1, T_3, T_4, T_5 \ldots, T_N$. The output signal of the last transistor T_N is reinjected through the additional inverter T_2 to the first one. To simplify our considerations, we assume that all N inverters are identical. Thus all resistors are equal to R and all capacitors are equal to C. To apply nodal voltage analysis to this circuit, the occurring MOS-FET devices are replaced by appropriate equivalent circuits . For example, a first approach with a low level MOS-FET model is given in Kampowsky et al. (1992). Zheng and Dellnitz (1990) discuss these equations from both the analytical and the numerical point of view. This model simulates the qualitative behaviour of a MOS-FET transistor quite well, but for more precise analysis models of higher levels should be preferred.

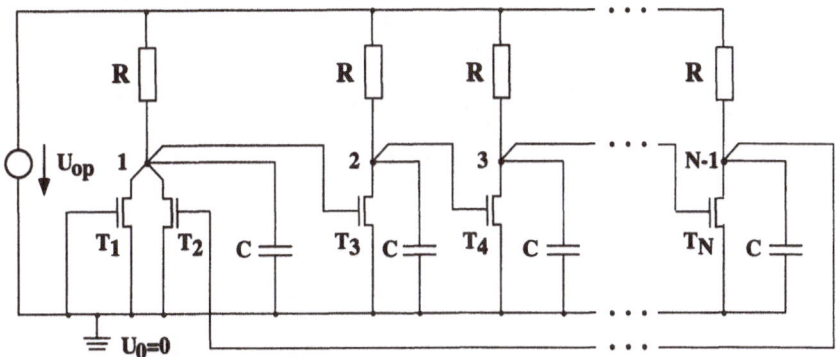

Figure 1: MOS-FET ring oscillator

Shichman and Hodges (1968) developed one of these more sophisticated models. They replaced the MOS-FETs, given schematically in figure 2, by a combination of resistors, voltage dependent capacitors, diodes, and a current source. The voltage - current dependency between the four nodes gate (G), drain (D), source (S), and bulk (B) is modelled by the following set (16), ..., (19) of highly nonlinear equations.

The two substrate junctions are represented by the ideal diode equations

$$(16) \qquad I_{BS} = I_{BS}(V_{BS}) = \begin{cases} -I_S \cdot (\exp{(\frac{V_{BS}}{V_{TH}}} - 1)), & V_{BS} \leq 0, \\ 0, & V_{BS} > 0. \end{cases}$$

$$(17) \qquad I_{BD} = I_{BD}(V_{BD}) = \begin{cases} -I_S \cdot (\exp{(\frac{V_{BD}}{V_{TH}}} - 1)), & V_{BD} \leq 0, \\ 0 & V_{BD} > 0. \end{cases}$$

The model parameters are given by

$$I_S = 10^{-15} \ A \qquad \text{and} \qquad V_{TH} = 25.85 \cdot 10^{-3} \ V.$$

The second nonlinear element, the current source I_{DS}, depends on the arguments V_{DS}, V_{GS}, V_{BS}, V_{GD}, and V_{BD}, i.e.

$$I_{DS} = I_{DS}(V_{DS}, V_{GS}, V_{BS}, V_{GD}, V_{BD}).$$

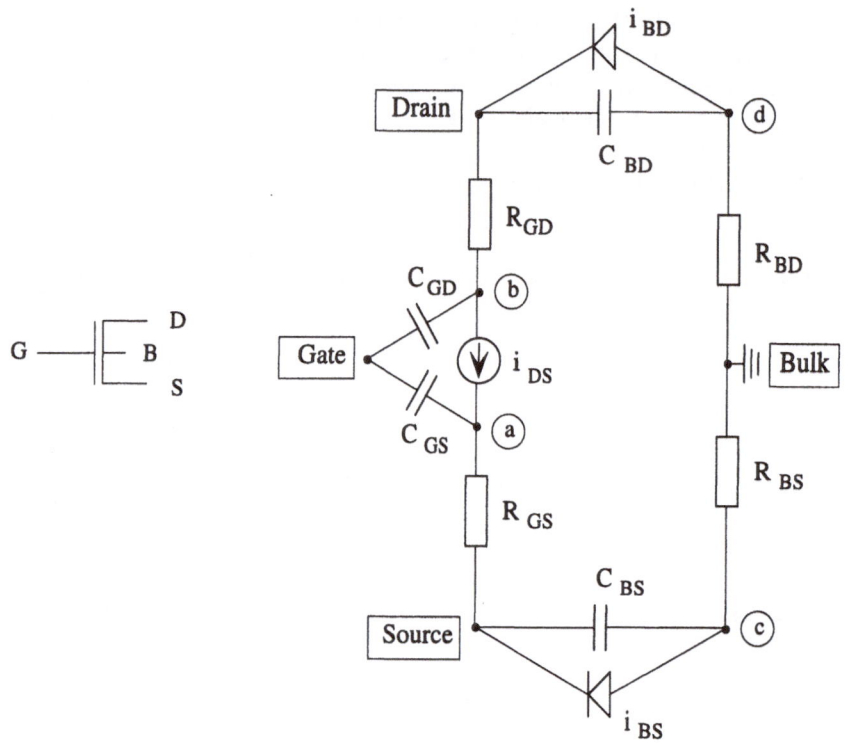

Figure 2: MOS-FET transistor: element symbol and equivalent circuit of Shichman and Hodges

Essential for its characteristic is the voltage V_{DS} between drain and source, which requires the distinction between two regions, namely:

Forward region $V_{DS} \geq 0$:

$$V_{TE} := V_{T0} + \gamma\{\sqrt{|\Phi - V_{BS}|} - \sqrt{\Phi}\},$$

$$I_{DS} = \begin{cases} 0, & V_{GS} - V_{TE} \leq 0 \\ -\beta(V_{GS} - V_{TE})^2 \cdot (1 + \lambda V_{DS}), & 0 < V_{GS} - V_{TE} \leq V_{DS} \\ \beta V_{DS} \cdot \{V_{DS} - 2(V_{GS} - V_{TE})\} \cdot (1 + \lambda V_{DS}), & 0 \leq V_{DS} < V_{GS} - V_{TE}. \end{cases}$$

Reverse region $V_{DS} < 0$:

$$V_{TE} := V_{T0} + \gamma \{ \sqrt{|\Phi - V_{BD}|} - \sqrt{\Phi} \},$$

$$I_{DS} = \begin{cases} 0, & V_{GD} - V_{TE} \leq 0 \\ \beta (V_{GD} - V_{TE})^2 \cdot (1 - \lambda V_{DS}), & 0 < V_{GD} - V_{TE} \leq |V_{DS}| \\ \beta V_{DS} \cdot [2(V_{GD} - V_{TE}) + V_{DS}] \cdot (\lambda V_{DS} - 1), & 0 \leq V_{DS} < V_{GD} - V_{TE}. \end{cases}$$

These drain–current relations depend on five model parameters V_{T0}, β, λ, γ, and Φ. Realistic values are:

$$V_{T0} = 1 \ V, \quad \beta = 8 \cdot 10^{-5} \ A/V^2, \quad \gamma = 0.37 \ V^{1/2}, \quad \lambda = 0.02 \ V^{-1}, \quad \Phi = 0.65 \ V.$$

The charge storage in the gate region is described by four capacitors C_{GS}, C_{GD}, C_{BD}, and C_{BS}. The last two capacitors are treated as voltage dependent:

$$(18) \qquad C_{BD}(V_{BD}) = \begin{cases} C_{BD0} \cdot (1 - \frac{V_{BD}}{\Phi_B})^{-1/2}, & V_{BD} \leq 0 \\ C_{BD0} \cdot (1 + \frac{V_{BD}}{2\Phi_B}), & V_{BD} > 0, \end{cases}$$

$$(19) \qquad C_{BS}(V_{BS}) = \begin{cases} C_{BS0} \cdot (1 - \frac{V_{BS}}{\Phi_B})^{-1/2}, & V_{BS} \leq 0 \\ C_{BS0} \cdot (1 + \frac{V_{BS}}{2\Phi_B}), & V_{BS} > 0. \end{cases}$$

The constants are:

$$C_{BS0} = C_{BD0} = 2.4 \cdot 10^{-14} \ F, \qquad \Phi_B = 0.87 \ V.$$

The remaining two capacitors C_{GS} and C_{GD} can be taken as constant ones:

$$C_{GS} = C_{GD} = 0.6 \cdot 10^{-13} \ F.$$

Remark: The expressions for C_{BD} and C_{BS} in (18) and (19) can complicate numerical integration, because they cause the resulting capacitance matrix to be voltage dependent. To simplify computation, here, both capacitors are assumed to be constant:

$$C_{BD} = C_{BS} = 2.4 \cdot 10^{-14} \ F.$$

This simplification makes the integration easier, because the capacitance matrix must be LU-decomposed only once; but it does not influence the index classification of the ring oscillator model.

The description of Shichman and Hodges' equivalent circuit is finished by the values of the remaining resistors:

$$R_{GD} = R_{GS} = 40 \ \Omega, \qquad R_{BD} = R_{BS} = 100 \ \Omega.$$

Let us now return to the circuit of the ring oscillator in figure 1. For its complete characterization only the quantities of the resistors, capacitors, and the voltage source are missing:

$$R = 10^4 \ \Omega, \qquad C = 0.2 \cdot 10^{-12} \ F, \qquad U_{OP} = 5 \ V.$$

To apply modified nodal voltage analysis to the ring oscillator, we introduce the nodes $1, 2, \ldots, N-1$, as in figure 1. Moreover, we must denote the internal nodes a, b, c, d in each MOS-FET equivalent (refer to figure 2). Altogether, we gain $5N - 1$ nodes, named: $T_{1a}, T_{1b}, T_{1c}, T_{1d}, T_{2a}, \ldots, T_{Nd}$, and $1, 2, \ldots, N-1$. Therefore, the differential equation has dimension $5N - 1$:

(20) $C \cdot \dot{x}(t) - f(x(t)) = 0,$

where C is the capacitance matrix and f is a highly nonlinear function. The components $x_j, j = 1, \ldots, 5N-1$ of the vector of voltages, x, are divided into the following subsections:

$$
\begin{aligned}
x_1, x_2, x_3, x_4 \quad &: \quad \text{voltages at the internal nodes } a, b, c, d \text{ of transistor } T_1, \\
x_{5j+i-6} \quad &: \quad \text{voltages at the internal nodes } a, b, c, d \text{ of transistor } T_j, \\
& \quad 2 \le j \le N, i = 1, 2, 3, 4 \text{ (related to the nodes } a, b, c, d), \\
x_{5j+4} \quad &: \quad \text{voltages at the nodes } j, 1 \le j \le N-1.
\end{aligned}
$$

Modified nodal analysis applied to transistor T_1 gives the equations:

(21) $C_{GS}\dot{x}_1 + R_{GS}^{-1}x_1 + I_{DS}(x_2 - x_1, -x_1, x_3, -x_2, x_4 - x_9) \ = \ 0,$

(22) $C_{GD}\dot{x}_2 + R_{GD}^{-1}(x_2 - x_9) - I_{DS}(x_2 - x_1, -x_1, x_3, -x_2, x_4 - x_9) \ = \ 0,$

(23) $C_{BS}\dot{x}_3 + R_{BS}^{-1}x_3 - I_{BS}(x_3) \ = \ 0,$

(24) $C_{BD}(\dot{x}_4 - \dot{x}_9) + R_{BS}^{-1}x_4 - I_{BD}(x_4 - x_9) \ = \ 0.$

Because of the reintroduced signal from transistor T_{N-1}, the equations for T_2 are more complex:

(25)
$$
\begin{aligned}
&C_{GS}(\dot{x}_5 - \dot{x}_{5N-1}) + R_{GS}^{-1}x_5 + \\
&+ I_{DS}(x_6 - x_5, x_{5N-1} - x_5, x_7, x_{5N-1} - x_6, x_8 - x_9) \quad = \quad 0,
\end{aligned}
$$

(26)
$$
\begin{aligned}
&C_{GD}(\dot{x}_6 - \dot{x}_{5N-1}) + R_{GD}^{-1}(x_6 - x_9) - \\
&- I_{DS}(x_6 - x_5, x_{5N-1} - x_5, x_7, x_{5N-1} - x_6, x_8 - x_9) \quad = \quad 0,
\end{aligned}
$$

(27) $C_{BS}\dot{x}_7 + R_{BS}^{-1}x_7 - I_{BS}(x_7) \quad = \quad 0,$

(28) $C_{BD}(\dot{x}_8 - \dot{x}_9) + R_{BD}^{-1}x_8 - I_{BD}(x_8 - x_9) \quad = \quad 0.$

The previous equations represent the voltages at the internal nodes a, b, c, and d of the inverters T_1 and T_2. At node 1 this yields:

(29)
$$
\begin{aligned}
&(C + C_{GS} + C_{GD})\dot{x}_9 - C_{GS}\dot{x}_{10} - C_{GD}\dot{x}_{11} - R_{GD}^{-1}(x_6 - x_9) + \\
&+ R_{BD}^{-1}x_8 + R^{-1}(x_9 - U_{OP}) - R_{GD}^{-1}(x_2 - x_9) + R_{BD}^{-1}x_4 \quad = \quad 0.
\end{aligned}
$$

For the remaining internal nodes the equations are structured regularly. Explicitly, we get for $j = 3, 4, \ldots, N$ and $k := 5j - 6$:

(30)
$$\begin{aligned} C_{GS}(\dot{x}_{k+1} - \dot{x}_k) + R_{GS}^{-1} x_{k+1} + \\ I_{DS}(x_{k+2} - x_{k+1}, x_k - x_{k+1}, x_{k+3}, x_k - x_{k+2}, x_{k+4} - x_{k+5}) \end{aligned} = 0,$$

(31)
$$\begin{aligned} C_{GD}(\dot{x}_{k+2} - \dot{x}_k) + R_{GD}^{-1}(x_{k+2} - x_{k+5}) - \\ I_{DS}(x_{k+2} - x_{k+1}, x_k - x_{k+1}, x_{k+3}, x_k - x_{k+2}, x_{k+4} - x_{k+5}) \end{aligned} = 0,$$

(32)
$$C_{BS}\dot{x}_{k+3} + R_{BS}^{-1} x_{k+3} - I_{BS}(x_{k+3}) = 0,$$

(33)
$$C_{BD}(\dot{x}_{k+4} - \dot{x}_{k+5}) + R_{BD}^{-1} x_{k+4} - I_{BD}(x_{k+4} - x_{k+5}) = 0.$$

The voltages at the nodes j, $2 \leq j \leq N - 2$, $k := 5j + 4$, are determined by:

(34)
$$\begin{aligned} (C + C_{GS} + C_{GD})\dot{x}_k - C_{GS}\dot{x}_{k+1} - C_{GD}\dot{x}_{k+2} - \\ -R_{GD}^{-1}(x_{k-3} - x_k) + R_{BD}^{-1} x_{k-1} + R^{-1}(x_k - U_{OP}) \end{aligned} = 0.$$

Finally, we receive for the output voltage at node $N - 1$ the expression:

(35)
$$\begin{aligned} -C_{GS}\dot{x}_5 - C_{GD}\dot{x}_6 + (C + C_{GS} + C_{GD})\dot{x}_{5N-1} - \\ -R_{GD}^{-1}(x_{5N-4} - x_{5N-1}) + R_{BD}^{-1} x_{5N-2} + R^{-1}(x_{5N-1} - U_{OP}) \end{aligned} = 0.$$

So far, the MOS-FET ring oscillator of figure 1 has been completely characterized. In the next section some numerical results are shown.

4. Numerical examples

In figure 3, the initial phase of a ring oscillator with four MOS-FET transistors is plotted. The curves show the time-voltage behaviour for the components $9, 14, 19$ of the vector x, i.e. the voltages at the nodes 1, 2, 3. The period of the limit cycle is $T = 15.16$ nsec. Figure 4 gives the voltages of this periodic solution at the same nodes as in figure 3. Within these calculations two versions of multiple shooting were compared: the standard implementation with numerical differentiation and the algorithm described in section 2. All computations were carried out on the CDC Cyber 995 vector machine. The coupling technique reduces computation time from 230 to 39 CPU seconds.

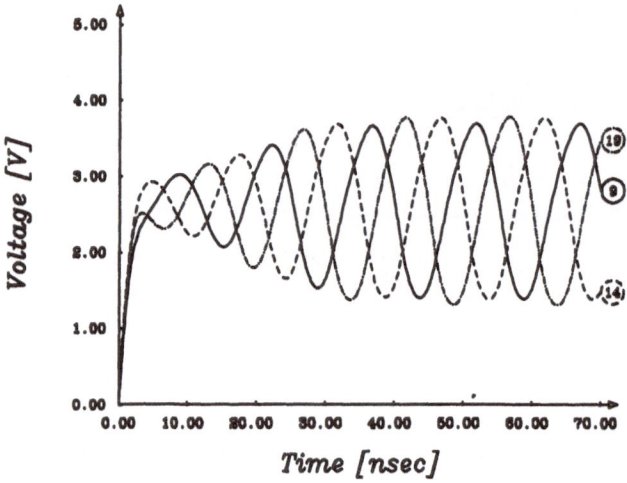

Figure 3: Initial phase of the ring oscillator with 4 inverters

Acknowledgements
I would like to thank the Numerical Analysis group of Prof. R. Bulirsch of the TU München for many helpful discussions.
My special thanks belong to Ms. Fiona Scanlon of the Institut für Englische Philologie, LMU München for her careful reading of the manuscript.
The author thanks the Bayerische Forschungsstiftung for the support of FORTWIHR: Bayerischer Forschungsverbund für Technisch-Wissenschaftliches Hochleistungsrechnen. This work is part of the project 4.4 "Numerical Simulation of Electric Circuits and Semiconductor Devices".

References

[1] R. Bulirsch, A. Gilg, K. Merten, K. Steger, *Numerische Simulation in der Halbleiterindustrie*, Informatik, Forsch. Entw. 5, (1990), pp. 42-56.

[2] P. Deuflhard, *Computation of Periodic Solutions of Nonlinear ODEs*, BIT 24, (1984), pp. 257-271.

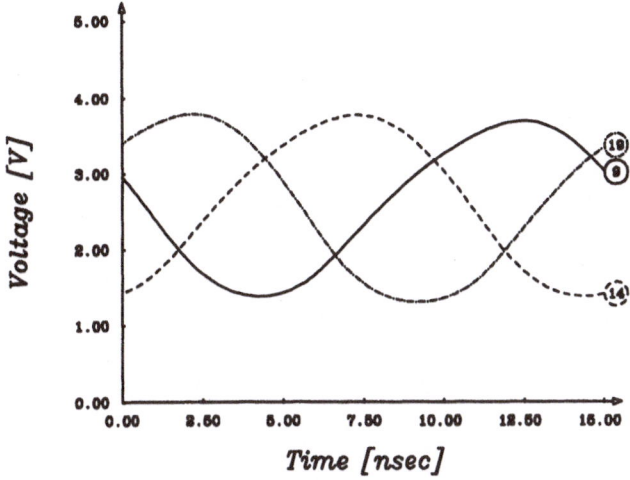

Figure 4: Periodic solution of the ring oscillator with 4 inverters

[3] M. Günther, P. Rentrop, *Multirate Row-Methods and Latency of Electric Circuits*, Mathematisches Institut, Technische Universität München, TUM-M9208 (1992)

[4] W. Kampowsky, P. Rentrop, W. Schmidt, *Classification and numerical simulation of electronic circuits*, Surv. Math. Ind. 2, (1992), pp. 23-65.

[5] H. Shichman, D. A. Hodges, *Insulated-gate field-effect transistors switching circuits*, IEEE J. Solid-State Circuits, SC-3, (1968), 285-289.

[6] R. Seydel, *Numerical computation of periodic orbits that bifurcate from stationary solutions of ODEs*, Appl. Math. Comp. 9 (1981), 257-271.

[7] J. Stoer, R. Bulirsch, *Introduction to numerical analysis*, Springer, Berlin Heidelberg New York (1979).

[8] Q. Zheng, M. Dellnitz, *Schwingungen eines Ringoszillators - eine numerische Behandlung unter Berücksichtigung der Symmetrie*, ZAMM, Z. angew. Mech. 70, (1990) 4, T135 - T138

International Series of Numerical Mathematics, Vol. 117, © 1994 Birkhäuser Verlag Basel

Timestep control for charge conserving integration in circuit simulation

E.-R. Sieber, TU Munich

U. Feldmann, Siemens AG Munich

R. Schultz, Siemens Nixdorf AG Munich

H. Wriedt, Uni Hamburg

Abstract. Charge oriented integration is necessary in general purpose circuit simulation for maintaining charge conservation. Generally, an approach derived by Ward and Dutton is implemented. Its drawbacks are higher expense for timestep control, loss of one integration order, and the fact that the user has no direct control of the variables which he really is interested in.

An error estimate is proposed which is closely related to error estimates given by Gear, Leimkuhler et al. for differential algebraic equations of index 2. A timestep control algorithm based on this estimate requires extra solutions of a linear system with the matrix of the Newton procedure. First results indicate that this additional overhead is compensated by a better adaption of the timesteps.

AMS subject classification (1980): 34A50, 65C20, 65D30, 65L05 and 94C99.

1   Introduction

For a wide class of integrated MOS circuits the distribution of charges is crucial. Therefore universal circuit analysis programs must assure charge conservation. This is not possible with the conventional capacitance oriented formulation of network equations, however it can be obtained with a charge oriented formulation together with numerical integration of charges rather than node potentials.

Straightforward implementations of charge oriented integration control accuracy of element charges or their time derivatives. These quantities however are of little interest for circuit designers. In this paper a timestep control algorithm is proposed, which is based on a prescribed accuracy of node potentials, although numerical integration of charges is performed. For this purpose charge errors are transformed into errors of node potentials by solving a linear system, whose matrix is just the Jacobian of Newton's procedure. A similar estimate was recommended by GEAR, LEIMKUHLER et al. [6, 9] for differential algebraic equations (DAEs) of index 2 (for an index classification of DAEs see e.g. [1]). It is well known [12] that such DAEs are difficult to solve, and even numerically stable methods like BDF require particular attention to timestep control in order to be successfull. Hence there is a second reason for the modified timestep control, because the network equations of many MOS circuits belong to this class of DAEs.

In Section 2 a motivation for charge oriented formulation of network equations is given. Section 3 describes the charge oriented formulation of WARD and DUTTON [18] in comparison with the conventional approach. In Section 4 the modified timestep control algorithm is presented, and in Section 5 first numerical results are discussed.

2   Why charge oriented formulation?

Charge conservation becomes essential for numerical analysis of electronic circuits in the time domain, if circuits are considered which exploit charge distribution effects. Many integrated circuits in MOS technology belong to this class, e.g. dynamic memories (DRAMs) and switched capacitor circuits. An example for the latter is the low-pass filter circuit of Fig. 1. Here Φ_A and Φ_B are nonoverlapping clocks. With each clock Φ_A some charge is transferred from the input node through an MOS transistor to the internal capacitor, and with clock Φ_B it is put to the output capacitor.

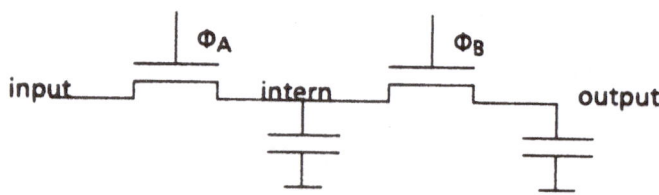

Figure 1: Switched capacitor low-pass filter

If this circuit is analyzed using the standard nonlinear capacitance model of MEYER [10] for describing the dynamic behaviour of MOS transistors (MOSFETs), then the results are wrong independently of tolerances for numerical integration. The problem has been investigated in several papers [18, 19, 14, 17, 2]. Already WARD and DUTTON noticed that the principal reason is nonconservation of charges for the MOSFETs. Further conclusions were:

- The intrinsic charges of MOSFETs are nonlinear functions of more than one controlling branch voltage with nonreciprocal derivatives.

- Such charges cannot be modelled with purely reciprocal elements like capacitors without violating the charge conservation law.

With other words: charge conservation is basically a modelling problem, not a problem of numerics. One should note that in such a situation even an attempt to put the charge conservation requirement into Lagrange multipliers (see e.g. [5]) when integrating network equations would fail.

As a consequence WARD and DUTTON suggested a charge oriented formulation of the network equations [18]. This is discussed in the following section.

3 Charge/flux versus conventional formulation

In this section the approach proposed by WARD and DUTTON for setting up network equations is shortly described. It includes not only electric charges but also magnetic fluxes as state variables. A comparison with the conventional formulation is given. We restrict ourselves to the widely used scheme of Modified Nodal Analysis (MNA) [7]. Basically three kinds of equations are available (see e.g. [16]):

- Kirchhoff's current law KCL: the sum of branch currents leaving a node is zero

- Kirchhoff's voltage law KVL: the sum of branch voltages along a closed network loop is zero

- the constitutive element equations, which describe the electrical characteristics of basic elements like resistors, capacitors, inductors, independent and controlled sources.

In MNA the following equations are explicitly used:

- KCL for each circuit node except ground

- the element equation for each voltage defining element, i.e. voltage source and inductor,

while KVL and the current defining relations of the other elements (resistors, capacitors, current sources,...) are inserted directly. The vector x of unknowns consists of

$$x = \begin{pmatrix} x_u \\ x_i \end{pmatrix} \in \mathbb{R}^n$$

with

x_u: vector of node potentials except ground
x_i: vector of branch currents through voltage sources and inductors.

Table I contains the major distinctions between capacitor/inductance and charge/flux oriented formulation and numerical solution.

Table I: Capacitance and charge oriented circuit analysis

	cap./induct. oriented	charge/flux oriented
element equations 	$i = C(u)\dot{u}$	$Q = Q(u)$ $i = \dot{Q}$
	$u = L(i)\dot{i}$	$\Phi = \Phi(i)$ $u = \dot{\Phi}$
network equations	$f(x, \dot{x}, t) = 0$ $x(0) = x_0$	$q = q(x)$ $f(x, \dot{q}, t) = 0$ $x(0) = x_0$ $q \in \mathbb{R}^m$: vector of terminal charges and branch fluxes
charge conservation	i.g. not guaranteed	for each charge storing element: the sum of all terminal charges is constant (is zero)
flux conservation	i.g. not guaranteed	for each flux storing element: the sum of branch fluxes along the loop of terminals is constant (is zero)
numerical integration with trapez or BDF	$\dot{x}(t_p) = \alpha x(t_p) + \beta$	$\dot{q}(t_p) = \alpha q(t_p) + \beta$
estimate for local truncation error LTE	of x; from difference $x_{corr} - x_{pred}$	of $\dot{q}$; from divided differences of q
Newton	$\left(\frac{\partial f}{\partial x} + \alpha \frac{\partial f}{\partial \dot{x}}\right)\Delta x = -f$	$\left(\frac{\partial f}{\partial x} + \alpha \frac{\partial f}{\partial \dot{q}} \frac{\partial q}{\partial x}\right)\Delta x = -f$
computational steps per timepoint	$x_{pred} \to \dot{x} \to \Delta x \to x$	$x_{pred} \to q(x) \to \dot{q} \to \Delta x \to x$

Some remarks follow, concerning charge/flux oriented formulation:

- Mathematically the vector of terminal charges/branch fluxes has to be included in the vector x of unknowns; however this is not explicitly done in practical implementations like SPICE [11], because the charges and fluxes are of almost no interest for the user and rather serve as intermediate variables.

- Numerical integration will not destroy charge conservation, if the multistep formula is applied for all but one terminal charges resp. branch fluxes of each element, and the missing derivative is set to the negative sum of all others.

- $\frac{\partial f}{\partial q}$ is a constant, usually rank deficient incidence matrix with entries 0, +1 and −1; the integrator however does not exploit this special structure.

- The index of the DAE system may be different, and we do not know any circuit where the charge oriented network equations have lower index than the conventional ones.

- Generally explicit integration is not possible, because x cannot be computed from q.

- In practice q has a higher dimension m than x: $m \approx 3n$; therefore the computational expense for numerical integration and timestep control is higher.

- Usually accuracy of $\dot{q}$ rather than of q is controlled, because q itself is no quantity of interest for the user; this implies the loss of one integration order (see formula in next section).

- The user has no direct control on the really interesting circuit variables, i.e. node potentials and voltage source currents.

In order to bypass some of the disadvantages associated with charge/flux oriented integration, DENK [3] used

$$\dot{q}(x(t)) = \frac{\partial q}{\partial x} \dot{x}(t),$$

which meant to

- assemble the terminal charges on circuit nodes

- perform classical integration on x rather than q

This method works well if Newton's procedure is started from a low order predictor. The problem however is that it requires the computation of the second derivatives

$$\frac{\partial^2 q}{\partial x^2},$$

which are hard to get in practice or do not even exist due to poor smoothness properties of transistor models. This was the reason to investigate the approach presented in the following section.

4   Error estimates and modified timestep control

The purpose of this section is to derive a timestep control algorithm, which is based on a prescribed accuracy for x rather than for q resp. $\dot{q}$. The basic idea is to transform an error ϵ_q for $\dot{q}$ into an estimate for the error $(x - x_p)$ of x.

By expanding $f(x, \dot{q}, t)$ at the actual timepoint t_p into a Taylor series around the approximate solution $(x_p, \dot{q}_p)$ and neglecting higher order terms, one gets:

$$f(x, \dot{q}, t_p) \approx f(x_p, \dot{q}_p, t_p) + \frac{\partial f}{\partial x}(x - x_p) + \frac{\partial f}{\partial \dot{q}}(\dot{q} - \dot{q}_p)$$

With the difference of exact and approximate value for $\dot{q}$

$$
\begin{aligned}
\dot{q}(x) - \dot{q}_p(x_p) &= \alpha q(x) + \beta + \epsilon_q - (\alpha q(x_p) + \beta) \\
&= \alpha(q(x) - q(x_p)) + \epsilon_q \\
&\approx \alpha \frac{\partial q}{\partial x}(x - x_p) + \epsilon_q
\end{aligned}
$$

follows:

$$f(x, \dot{q}, t_p) \approx f(x_p, \dot{q}_p, t_p) + \left(\frac{\partial f}{\partial x} + \alpha \frac{\partial f}{\partial \dot{q}}\frac{\partial q}{\partial x}\right)(x - x_p) + \frac{\partial f}{\partial \dot{q}}\epsilon_q$$

As f is zero for both the exact and the approximate solution, the desired error estimate $\epsilon_x = x - x_p$ can be computed from

$$\left(\frac{\partial f}{\partial x} + \alpha \frac{\partial f}{\partial \dot{q}}\frac{\partial q}{\partial x}\right)\epsilon_x = -\frac{\partial f}{\partial \dot{q}}\epsilon_q,$$

where ϵ_q is the LTE of $\dot{q}$.

Remarks:

- The matrix of this system is just the Jacobian of Newton's procedure (see Table I).

- ϵ_x can be interpreted as a linear perturbation of x, if f is perturbed with the LTE ϵ_q; this justifies to take ϵ_x as an error estimate for numerical integration.

- Weigthing the LTE via Newton's iteration matrix was already proposed by SACKS-DAVIS [13] for stiff ordinary differential equations (ODEs). GUPTA et al. [6] and LEIMKUHLER [9] recommended this kind of estimate for nonlinear DAEs of index 2 in order to prevent the well known failures of MILNE-type timestep control algorithms esp. in case of steep-gradient problems [12]. The key issue here is that the system behaviour is brought into account with Newton's matrix, which is more sensitive than local consideration of single variables.

Keeping in mind that many circuit models lead to DAEs of index 2, we get herewith a second motivation for using the modified error estimate, because this argument then should also be valid in our case.

The LTE of $\dot{q}$ can be computed as usual (see e.g. [11]) from:

$$\epsilon_{\dot{q}} \approx c_k h^k q^{(k+1)} \tag{1}$$

with

k: order of integration method
c_k: dependent of method and order and timestep history
$q^{(k+1)}$: $(k+1)$th derivative of q, estimated from divided differences

As a rule for timestep and order control follows with the modified error estimate:

* compute next timestep h from

$$\|\epsilon_x\| = \|x - x_p\| \quad = \quad \left\| -\left(\frac{\partial f}{\partial x} + \alpha \frac{\partial f}{\partial \dot{q}} \frac{\partial q}{\partial x}\right)^{-1} \frac{\partial f}{\partial \dot{q}} \epsilon_{\dot{q}} \right\| \tag{2}$$

$$= \quad h^k |c_k| \left\| \left(\frac{\partial f}{\partial x} + \alpha \frac{\partial f}{\partial \dot{q}} \frac{\partial q}{\partial x}\right)^{-1} \frac{\partial f}{\partial \dot{q}} q^{(k+1)} \right\| \tag{3}$$

$$\overset{!}{=} \quad \text{user given tolerance} \tag{4}$$

* check order k, $k-1$, $k+1$; choose order which gives maximal timestep

* in case of a posteriori test check (3), (4) with updated values of q <u>and</u> of the Jacobian

Remarks:

• There is a loss of one integration order in case of the conventional timestep control based on $\epsilon_{\dot{q}}$; this can be seen from (1).

• Because of

$$\alpha \sim \frac{1}{h}$$

the second term in Newton's iteration matrix may become dominant, if the timestep is sufficiently small. A sufficient, but not necessary condition is that $\frac{\partial f}{\partial \dot{q}} \frac{\partial q}{\partial x}$ has full rank. For highly dynamic circuits this dominance is likely to be met already for the size of timesteps used in practice. From (3) follows, that in such cases we can expect to get the lost integration order back with the modified timestep control.

- An extra computational effort is required per timestep:

  1 LU decomposition
  1 ... 3 forward backward substitutions, depending on order control.

  If the cost for an LU decomposition is high, one may accept some inaccuracy and take an already decomposed version of the matrix for the actual timestep.

- The a posteriori test is more rigorous than with the conventional strategy because of the inclusion of an updated iteration matrix. This leads in principle to a greater number of refused timesteps, which is not desirable and should be alleviated by more conservative a priori timesteps.

5   Results

The modified timestep and order control was implemented in Siemens' circuit simulator TITAN [4]. The computation of the LTE $\epsilon_{\dot{q}}$ for $\dot{q}$ was taken from the conventional method. Tolerances are in either case composed of absolute and relative errors. For testing the algorithm both artificial and real test circuits were used.

In a first step the code was verified with undamped and damped linear LC oscillator circuits. These circuits were chosen because their equations can be setup and solved analytically, and the timestep control is not affected by side effects like startup behaviour etc..

In a second step a 17 stage bipolar and a 65 stage MOS ringoscillator were used for adjusting tolerances and control parameters of the new algorithm. These circuits are characterized by a high degree of regularity and by a feedback loop, so that errors of numerical integration accumulate and are easy to observe.

With fixed tolerances several real life circuits were analyzed, and the results were compared with the conventional approach (see Table II). The Colpitts oscillator is investigated in [8]; the 2 bit adder is described in [15], and the word line booster is part of a DRAM which exploits dynamic circuit techniques to boost an internal signal beyond power supply voltage. TRBDF integration is described in [4].

Table II: Numerical results with benchmark circuits

circuit	Colpitts oscillator	2 bit adder	word line booster
# of transistors # of equations integration rule tolerance	1 bipolar 9 trapez standard/10	39 MOS 33 trapez standard	124 MOS 99 TRBDF standard

# of timesteps 	
  conventional method
    total/accepted/refused
  modified method
    total/accepted/refused
	
	4021/3700/321
	
	3250/2836/414
		
		514/508/6
		
		493/471/22
			
			410/407/3
			
			305/291/14

In no case there was a significant difference of output signal waveforms between conventional and modified timestep control. Nevertheless the total number of timesteps is decreased by 24 resp. 34 percent for the first and the third circuit. The second example is characterized by many breakpoints of the input signals, which cause side effects on timestep control.

The data of Table II also confirm that the number of refused timesteps is increased with the new timestep control. This effect should be accepted, because more pessimistic a priori timesteps would decrease the overall efficiency of the method.

As the word line booster works with dynamic circuit techniques, it was a first candidate for looking at the effective order of integration with the different methods. Taking Backward Euler as an integration rule, the circuit was again analyzed with tolerances varying by a factor of 10. For the conventional method one would expect from (1) a timestep relation of 10 : 1 (according to $order = 1$), while for the new method in the best case a timestep relation of $\sqrt{10} : 1$ is possible. The measured relations (see Table III) do not reflect this behaviour completely - probably because the timesteps are rather large - but there is a significant difference in favour of the modified timestep control.

Table III: Analysis of word line booster with different tolerances

	conventional method	modified method
tolerance	1 : 10	1 : 10
# of timesteps	3366 : 513	1178 : 280
relation timesteps	6.6 : 1	4.2 : 1

6 Conclusion

Charge/flux oriented formulation of the network equations is a convenient approach for assuring charge conservation, but is has some drawbacks concerning efficiency and reliability of numerical integration, if standard techniques for timestep control are applied. Improvements are possible by using a modified error estimate for numerical integration. It requires a higher effort per timestep and tends to increase the number of a posteriori refused timesteps; however our first numerical results indicate that this is at least compensated by a better timestep adaption to signal waveforms and by better overall efficiency.

Further investigations are necessary for getting experience esp. with large circuits and with circuits which lead to DAE systems of index 2. Another open question is an extension of the algorithm to multirate strategies.

Acknowledgment

The algorithm was implemented by the first author in Siemens' circuit simulator TITAN as a part of her diploma thesis at the Mathematische Institut, TU München. We would like to thank A. Gilg and P. Rentrop for enabling this work. Further we are indebted to P. Rentrop and to Q. Zheng for valuable discussions.

References

[1] Brenan, K.E.; Campbell, S.L.; Petzold, L.R. (1989) Numerical solution of initial value problems in differential algebraic equations. (Elsevier, New York)

[2] Cirit, M.A. (1989) The Meyer model revisited: why is charge not conserved? IEEE Trans. CAD 8, 1033-1037

[3] Denk, G. (1990) An improved numerical integration method in the circuit simulator SPICE2-S. Int. Ser. Numer. Math. 93 (Birkhäuser, Basel), 85-99

[4] Feldmann, U.; Wever, U.; Zheng, Q.; Schultz, R.; Wriedt, H. (1992) Algorithms for modern circuit simulation. AEÜ 46, 274-285

[5] Gear, C.W. (1986) Maintaining solution invariants in the numerical solution of ODEs. SIAM J. Sci. Stat. Comput. 7, 734-743

[6] Gupta, G.K.; Gear, C.W.; Leimkuhler, B.J. (1985) Implementing linear multistep formulas for solving DAEs. Rep. No. UIUCDCS-R-85-1205 (Univ. Illinois, Urbana)

[7] Ho, C.W.; Ruehli, A.E.; Brennan, P.A. (1975) The modified nodal approach to network analysis. IEEE Trans. CAS 22, 505-509

[8] Kampowsky, W.; Rentrop, P.; Schmidt, W. (1992) Classification and numerical simulation of electric circuits. Surv. Math. Ind. 2, 23-65

[9] Leimkuhler, B.J. (1986) Error estimates for differential-algebraic equations. Rep. No. UIUCDCS-R-86-1287 (Univ. Illinois, Urbana)

[10] Meyer, J.E. (1971) MOS models and circuit simulation. RCA Rev. 32, 42-63

[11] Nagel, L.W. (1975) SPICE2: A computer program to simulate semiconductor circuits. Tech. Rep. ERL M520 (Univ. California, Berkeley)

[12] Petzold, L.R. (1982) Differential/algebraic equations are not ODEs. SIAM J. Sci. Stat. Comput. 3, 367-384

[13] Sacks-Davis, R. (1972) Error estimates for a stiff differential equation procedure. Math. Comp. 31, 939-953

[14] Sakallah, K.A.; Yen, Y.T.; Greenberg, S.S. (1990) A first order charge conserving MOS capacitance model. IEEE Trans. CAD-9, 99-108

[15] Sieber, E.-R. (1992) Schrittweitensteuerung bei der numerischen Integration in der Schaltkreissimulation. Dipl. thesis (Math. Inst. Tech. Univ, Munich)

[16] Singhal, K.; Vlach, J. (1986) Formulation of circuit equations. In: Ruehli, A.E. (ed.) Circuit analysis, simulation and design. Part 1 (North Holland, Amsterdam), 45-70

[17] Turchetti, C.; Prioretti, P.; Masetti, G.; Profumo, E.; Vanzi, M. (1986) A Meyer-like approach for transient analysis of digital MOS ICs. IEEE Trans. CAD 5, 499-507

[18] Ward, D.E.; Dutton, R.W. (1978) A charge oriented model for MOS transistor capacitances. IEEE J. Solid-State Circuits SC-13, 703-708

[19] Yang, P. (1986) Capacitance modeling for MOSFETS. In: Ruehli, A.E. (ed.) Circuit analysis, simulation and design. Part 1 (North Holland, Amsterdam), 107-129

Address: U. Feldmann, Siemens AG, ZFE BT SE 43, 81739 Munich, Germany.

International Series of Numerical Mathematics, Vol. 117, © 1994 Birkhäuser Verlag Basel

Ein Zusammenhang zwischen Waveformrelaxation und Iterationsverfahren für nichtlinear gestörte Gleichungen

K. Taubert
Universität Hamburg
20146 Hamburg, Deutschland

Zusamenfassung. Mit Testabbildungen werden Aufgaben charakterisiert, bei denen die Gauß–Seidel Iteration und die Gauß–Seidel Waveformrelaxation konvergent ist. Die Testabbildungen liefern Abschätzungen für Störungen, welche die Konvergenz noch gewährleisten.

0. Einleitung.

Bekanntlich ist die Konvergenz vom Gauß–Seidel Verfahren für Gleichungssysteme

$$Ax = b$$

mit stark diagonaldominanten Matrizen gewährleistet. Diese Matrizen besitzen die sehr verallgemeinerungsfähige (§1) Akkretivitätseigenschaft

$$(Ax, \ell(x)) \geq \alpha ||x||^2, \quad \alpha > 0$$

mit dem euklidischen inneren Produkt $(\cdot, \cdot)$ und einer zur Maximum–Norm passenden Testabbildung $\ell : \mathbb{R}^n \to \mathbb{R}^n$.

Das Gauß–Seidel Verfahren (GSV) ist auch für die gestörte Gleichung

$$Ax = F(x) + b$$

konvergent (§2), wenn die Funktion $F : \mathbb{R}^n \to \mathbb{R}^n$ passend zur Konstanten α ist.

Das Transientenverhalten vieler digitaler MOS-Schaltungen führt auf Anfangswertaufgaben der Art

$$A\dot{y} = F(y), \quad y(0) = y_0.$$

Durch die Darstellung

$$A\dot{y}(t) = F(y_0 + \int_0^t \dot{y}(s)ds), \qquad t \in [0, T]$$

entsteht eine formal ähnliche Situation wie oben. Die Konstante α liefert aber jetzt eine Abschätzung für die Zeitfenster $[0, T_1] \subset [0, T]$ in denen die gleichmäßige Konvergenz der Gauß–Seidel Waveformrelaxation (GSWR) gegeben ist.

Geeignete Lipschitzbedingungen führen (§2) zu einer ähnlichen Aussage für die üblichen Aufgaben aus der MOS-Technologie

$$A(t, y)\dot{y} = F(t, y), \quad y(0) = y_0.$$

Beispiele im §2 und die zugehörigen Bilder belegen die Bedeutung einer „richtigen" Wahl der Zeitfenster bei der Waveformrelaxation.

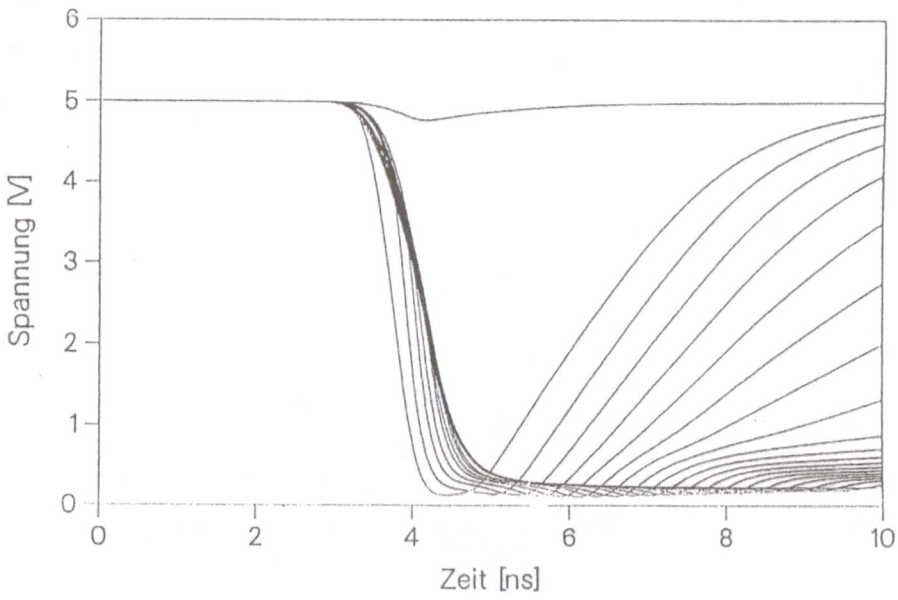

Bild 1.   2Bit-Volladdierer
Waveformrelaxation, Iterationen 1 – 20, Potential Knoten 14

Die Testabbildung und die zugehörigen Systeme können (§1) auf der elektrischen Ebene interpretiert werden. Einige Bemerkungen über die Wahl anderer iterativer Verfahren und Testabbildungen ℓ schließen die Arbeit ab.

1. Auf die Maximum–Norm im $\mathbb{R}^n$ oder $C[0,T]$ bezogene Testabbildungen. Anwendungen.

Es sei $(\cdot,\cdot)$ das euklidische innere Produkt im n-dimensionalen linearen Raum $\mathbb{R}^n$ der Vektoren $u = (u_1, u_2, \ldots, u_n)$ und $A = (a_{ij})$ eine $n \times n$ Matrix mit reellen Einträgen und $a_{ii} > 0$.

Für $c = (c_1, c_2, \ldots, c_n) \in \mathbb{R}^n$ mit positiven Komponenten sei

$$\|u\|_c = \max_{i=1,2,\ldots,n} c_i |u_i|.$$

Die Norm $\|u\|_c$ führt auf eine „Testabbildung" $\ell_c : \mathbb{R}^n \to \mathbb{R}^n$:

Ist u_k die erste Komponente von $u \in \mathbb{R}^n$ mit $c_k |u_k| = \|u\|_c$, dann sei

$$\ell_c(u) = (0, \ldots, 0, u_k, 0, \ldots, 0).$$

Mit diesen von c abhängigen Testabbildungen können bekannte Klassen von Matrizen A charakterisiert und Aussagen über die Lösungen spezieller Gleichungssysteme $Ax = b$ gemacht werden. Außerdem charakterisieren die Testabbildungen auch das Verhalten bestimmter elektrischer Netzwerke.

Bemerkung 1
Die Bedingung $(\alpha > 0)$

$$(Au, \ell_c(u)) \geq \alpha \|u\|_c^2 \quad \text{für alle} \quad u \in \mathbb{R}^n$$

ist äquivalent dazu, daß die skalierte Matrix

$$\begin{pmatrix} c_1 & \ldots & 0 \\ \vdots & \ddots & \vdots \\ 0 & \ldots & c_n \end{pmatrix} A \begin{pmatrix} c_1 & \ldots & 0 \\ \vdots & \ddots & \vdots \\ 0 & \ldots & c_n \end{pmatrix}^{-1}$$

das starke Zeilensummenkriterium erfüllt. Im Fall $c^1 = (1, 1, \ldots, 1)$ gilt also $a_{ii} > \sum_{j \neq i} |a_{ij}|$ für alle i.

Bemerkung 2
Die Eigenschaft einer Matrix

$$a_{ii} \geq \sum_{j \neq i} |a_{ij}| \quad \text{für alle } i$$

ist äquivalent zu

$$(Au, \ell_{c^1}(u)) \geq 0 \quad \text{für alle} \quad u \in \mathbb{R}^n.$$

Ist die Matrix A außerdem nicht zerfallend und gibt es ein k mit

$$a_{kk} > \sum_{j \neq k} |a_{kj}|,$$

dann gibt es stets eine Testabbildung ℓ_c und ein $\alpha > 0$ mit

$$(Au, \ell_c(u)) \geq \alpha \|u\|_c^2 \quad \text{für alle} \quad u \in \mathbb{R}^n.$$

Es wurde bereits erwähnt, daß mit den Testabbildungen Aussagen über die Lösungen bestimmter Gleichungssysteme $Ax = b$ erzielt werden können. Dazu:

Bemerkung 3

Es sei $x^* \neq 0$ eine Lösung von $Ax = b$ und die i-te Komponente b_i von b gleich Null. Genügt die Matrix A der Bedingung

$$(Au, \ell_c(u)) > 0 \quad \text{für alle} \quad u \neq 0,$$

dann gilt

$$c_i |x_i^*| < \max_{j \neq i} c_j |x_j^*|.$$

Mit den Testabbildungen können auch bestimmte Klassen von linearen elektrischen Netzwerken charakterisiert werden:

Bemerkung 4

Der Einfachheit halber betrachten wir zunächst nur das Beispiel

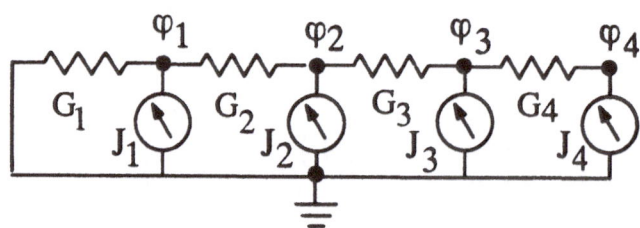

Bild 2.

Die reine Knotenspannungsanalyse führt auf ein Gleichungssystem $A\varphi = J$ der Form

$$
\begin{pmatrix}
G_1 + G_2 & -G_2 & 0 & 0 \\
-G_2 & G_2 + G_3 & -G_3 & 0 \\
0 & -G_3 & G_3 + G_4 & -G_4 \\
0 & 0 & -G_4 & G_4
\end{pmatrix}
\begin{pmatrix}
\varphi_1 \\
\varphi_2 \\
\varphi_3 \\
\varphi_4
\end{pmatrix}
=
\begin{pmatrix}
J_1 \\
J_2 \\
J_3 \\
J_4
\end{pmatrix}.
$$

Die Matrix A erfüllt die Bedingung

$$(Au, \ell_{c^1}(u)) \geq 0 \quad \text{für alle} \quad u \in \mathbb{R}^4.$$

Sind die Komponenten von $J = (J_1, J_2, J_3, J_4)$ bis auf die i-te Komponente J_i alle gleich Null, dann ergibt die genannte Bedingung, daß φ_i das betragsgrößte Potential sein muß, oder

$$|\varphi_i| \geq |\varphi_j| \quad \text{für alle} \quad j \neq i.$$

Dieses Verhalten überträgt sich auf alle elektrischen Netzwerke, die auf lineare Gleichungssysteme führen und das schwache Zeilensummenkriterium erfüllen. Aber auch umgekehrt: Wird ein lineares elektrisches Netzwerk $A\varphi = J$ mit den Einheitsvektoren $J = e_i$, $i = 1, 2, \ldots, n$ „getestet" und ergibt sich stets $|\varphi_i| \geq |\varphi_j|$ für alle $j \neq i$, dann muß die Matrix A das schwache Zeilensummenkriterium erfüllen.

Die bisher angegebenen Begriffe und gemachten Bemerkungen können auch auf Funktionsräume übertragen werden.

Es sei $C[0,T]$ der n-dimensionale Raum der stetigen Funktionen $z = (z_1, z_2, \ldots, z_n)$ mit

$$z_i : [0,T] \to \mathbb{R}.$$

Es sei $BV[0,T]$ der n-dimensionale Raum der Funktionen von beschränkter Variation $g = (g_1, g_2, \ldots, g_n)$ mit

$$g_i : [0,T] \to \mathbb{R}$$

und $g_i(0) = 0$.

Für $z \in C[0,T]$ und $g \in BV[0,T]$ sei

$$[z, g] = \sum_{i=1}^{n} \int_0^T z_i(t) dg_i(t).$$

Für ein $c = (c_1, c_2, \ldots, c_n) \in \mathbb{R}^n$, mit $c_i > 0$ für alle i, sei jetzt

$$\|z\|_c = \max_i (c_i \max_{t \in [0,T]} |z_i(t)|).$$

Auch hier induziert die Norm $\| \cdot \|_c$ eine „Testabbildung" $\ell_c : C[0,T] \to BV[0,T]$:

Ist t^* der erste Zeitpunkt und z_k die erste zugehörige Komponente von z mit

$$\|z\|_c = c_k |z_k(t^*)|,$$

dann sei

$$\ell_c(z) = (0, \ldots, 0, g_k, 0, \ldots, 0).$$

Dabei ist g_k jene Funktion von beschränkter Variation für die gilt

$$g_k(t) = \begin{cases} 0 & \text{für} \quad 0 \leq t \leq t^* \\ c_k & \text{für} \quad t^* < t \leq T. \end{cases}$$

Die für $\mathbb{R}^n$ gemachten Bemerkungen und Beispiele können nun auf zeitabhängige Matrizen $A = (a_{ij}(t))$ mit stetigen Komponenten

$$a_{ij} : [0, T] \to \mathbb{R}$$

übertragen werden.

Bemerkung 5
Aus der Bedingung $(\alpha > 0)$

$$[A(t)z, \ell_c(z)] \geq \alpha |z|_c^2 \quad \text{für alle} \quad z \in C[0, T]$$

folgt, daß die durch c skalierte Matrix $A(t)$ (gleichmäßig) für alle t das starke Zeilensummenkriterium erfüllen muß und umgekehrt.

Bemerkung 6
Es sei z^* eine nichttriviale Lösung von $A(t)z^* = b(t)$ auf $[0, T]$ und $b \in C[0, T]$. Ist die i-te Komponente von b gleich Null für alle t und genügt die Matrix $A(\cdot)$ der Bedingung

$$[Az, \ell_c(z)] > 0 \quad \text{für alle} z \neq 0,$$

dann gilt die nützliche Ungleichung

$$c_i \max_{t \in [0,T]} |z_i^*(t)| < \max_{j \neq i} \left(c_j \max_{t \in [0,T]} |z_j^*(t)| \right).$$

2. Konvergenzbedingungen für das Gauß–Seidel Verfahren und die Gauß–Seidel Waveformrelaxation.

Gegeben sei das nichtlinear gestörte Gleichungssystem

$$Ax = F(x) + b \qquad (*)$$

mit einer reellen $n \times n$ Matrix A, einem reellen Vektor $b = (b_1, b_2, \ldots, b_n)$ und einer Funktion $F = (f_1, f_2, \ldots, f_n) : \mathbb{R}^n \to \mathbb{R}^n$.

Beim Gauß–Seidel Verfahren ergibt sich die i-te Komponente x_i^{k+1} der $(k+1)$-ten Iterierten x^{k+1} aus der Gleichung

$$a_{i1}x_1^{k+1} + \ldots + a_{ii}x_i^{k+1} + a_{i,i+1}x_{i+1}^k + \ldots + a_{in}x_n^k - f_i(x_1^{k+1}, \ldots, x_i^{k+1}, x_{i+1}^k, \ldots, x_n^k) - b_i = 0.$$

Es gilt nun der

Satz 1
Es gelte für alle $u, v \in \mathbb{R}^n$

1. $(Au, \ell_c(u)) \geq \alpha \|u\|_c^2$
2. $|(F(u) - F(v), \ell_c(u - v))| \leq \beta \|u - v\|_c^2$ und
3. $\beta < \alpha$.

Dann ist das GSV für die Gleichung $(*)$ für jeden Startvektor $x^0 \in \mathbb{R}^n$ durchführbar und konvergent.

Beweis

Wegen der Bedingungen 1–3 ist zunächst einmal die Existenz einer (eindeutigen) Lösung x^* von $(*)$ und die Durchführbarkeit des GSV gegeben (Fixpunktsatz für kontrahierende Abbildungen).

Für die $\imath$-te Komponente der $(k+1)$-ten Iterierten beim GSV gilt

$$a_{\imath 1}x_1^{k+1} + \ldots + a_{\imath\imath}x_\imath^{k+1} + a_{\imath,\imath+1}x_{\imath+1}^k + \ldots + a_{\imath n}x_{\imath n}^k - f_\imath(x_1^{k+1},\ldots,x_\imath^{k+1},x_{\imath+1}^k,\ldots,x_n^k) - b_\imath = 0.$$

Gleichzeitig gilt auch für die Lösung x^* des nichtlinearen Gleichungssystems

$$a_{\imath 1}x_1^* + \ldots + a_{\imath\imath}x_\imath^* + a_{\imath,\imath+1}x_{\imath+1}^* + \ldots + a_{\imath n}x_{\imath n}^* - f_\imath(x_1^*,\ldots,x_\imath^*,x_{\imath+1}^*,\ldots,x_n^*) - b_\imath = 0.$$

Für die Differenz gilt also

$$\sum_{j=1}^\imath a_{\imath j}(x_j^{k+1} - x_j^*) + \sum_{j=\imath+1}^n a_{\imath j}(x_j^k - x_j^*) -$$

$$-f_\imath(x_1^{k+1},\ldots,x_k^{k+1},x_{\imath+1}^k,\ldots,x_n^k) + f_\imath(x_1^*,\ldots,x_\imath^*,x_{\imath+1}^*,\ldots,x_n^*) = 0.$$

Nach Voraussetzung ist

$$(Ax - Ax^* - (F(x) - F(x^*)), \ell_c(x - x^*)) \geq \alpha\|x - x^*\|_c^2 - \beta\|x - x^*\|_c^2 > 0$$

für alle $x - x^* \neq 0$.

Ähnlich wie in der Bemerkung 3 ergibt sich daraus

$$c_\imath|x_\imath^{k+1} - x_\imath^*| < \max\left(\max_{j<\imath} c_j|x_j^{k+1} - x_j^*|, \max_{j>\imath} c_j|x_j^k - x_j^*|\right).$$

Da die letzte Ungleichung für alle $\imath = 1, 2, \ldots, n$ richtig ist, folgt

$$\|x^{k+1} - x^*\|_c < \|c^k - x^*\|_c,$$

wenn nicht schon $x^k = x^*$ ist.

Wegen der Existenz einer (eindeutigen) Lösung x^* ist auch die Konvergenz $x^{k+1} \to x^*$ gegeben. ∎

Es wurde bereits erwähnt, daß das Transientenverhalten vieler MOS–Schaltungen auf Gleichungen der Form ($t \in [0, T]$)

$$A(t,y)\dot{y} = F(t,y), \quad y(0) = y_0 \tag{$**$}$$

führt.

Ausgehend von $(y_1^k, y_2^k, \ldots, y_n^k)$ mit $y_i^k : [0,T] \to \mathbb{R}$ ergibt sich bei der GSWR die i–te Komponente y_i^{k+1} der $(k+1)$–ten Iterierten aus der Gleichung

$$a_{i1}(t, y^{k+1,i})\dot{y}_1^{k+1} + \ldots + a_{ii}(t, y^{k+1,i})\dot{y}_i^{k+1} + a_{i,i+1}(t, y^{k+1,i})\dot{y}_{i+1}^k + \ldots +$$

$$+ a_{in}(t, y^{k+1,i})\dot{y}_n^k = F(t, y^{k+1,i}).$$

Mit $y^{k+1,i}$ wird dabei jeweils der Vektor $(y_1^{k+1}, \ldots, y_i^{k+1}, y_{i+1}^k, \ldots, y_n^k)$ bezeichnet.

Wird jede auftretende Funktion y_j durch das Integral $y_j(t) = y_{0,j} + \int_0^t \dot{y}_j(s)ds$ dargestellt, dann kann die GSWR als Einzelschrittverfahren in den Ableitungen aufgefaßt werden.

Die Modelle in der MOS-Technologie führen überwiegend dazu, daß die Voraussetzungen des folgenden Satzes erfüllt sind.

Satz 2
Es sei

1. $F : [0,T] \times \mathbb{R}^n \to \mathbb{R}^n$ eine stetige Funktion mit

$$|(F(t,u) - F(t,v), \ell_c(u-v))| \leq K\|u-v\|_c^2$$

   für alle t, u, v und

2. $A(\cdot, \cdot)$ eine von $[0,T] \times \mathbb{R}^n$ abhängige stetige Matrix mit den Eigenschaften:

   (a) Es gibt eine Konstante $\alpha > 0$, so daß für alle $(t,u,v) \in [0,T] \times \mathbb{R}^n \times \mathbb{R}^n$ gilt

$$(A(t,v)u, \ell_c(u)) \geq \alpha\|u\|_c^2.$$

   (b) Für jedes j gilt

$$\sum_{k=1}^n \frac{c_j}{c_k}|a_{jk}(t,u) - a_{jk}(t,v)| \leq L\|u-v\|_c$$

   mit einer von t, u, v unabhängigen Konstanten L.

Ist y^* die (eindeutige) Lösung von $(**)$, dann konvergiert die von der GSWR erzeugte Folge von Näherungen $(\dot{y}^k)_{k\in\mathbb{N}}$ gleichmäßig gegen $\dot{y}^*$ in jedem Intervall $[0,T]$ mit $\alpha - KT - LT|\dot{y}^*|_c > 0$.

Beweis
Wegen den Voraussetzungen hat die Differentialgleichung $(**)$ eine eindeutige Lösung y^* für ein beliebiges Intervall $[0,T]$. Auch die Waveformrelaxation ist in jedem Intervall $[0,T]$ durchführbar.

Für jede stetig differenzierbare Funktion $y : [0, T] \to \mathbb{R}^n$ mit $y(0) = y^*(0)$ betrachte den Ausdruck

$$B(y, y^*) = A(t, y)\dot{y} - A(t, y^*)\dot{y}^* - F(t, y) + F(t, y^*).$$

Dann gilt

$$[B(y, y^*), \ell_c(\dot{y} - \dot{y}^*)] \geq (\alpha - KT - LT|\dot{y}^*|_c)|\dot{y} - \dot{y}^*|_c^2.$$

Aus der Bedingung $\alpha - KT - LT|\dot{y}^*|_c > 0$ und der Bemerkung 6 folgt dann (wie im Beweis von Satz 1), daß die Folge $(\dot{y}^k)_{k \in \mathbb{N}}$ die Ungleichung

$$|\dot{y}^{k+1} - \dot{y}^*|_c < |\dot{y}^k - \dot{y}^*|_c$$

erfüllen muß, sofern nicht schon $|\dot{y}^k - \dot{y}^*|_c \equiv 0$ ist.

Da die Folge $(\dot{y}^k)_{k \in \mathbb{N}}$ auch gleichgradig stetig und gleichmäßig beschränkt ist, und eine Lösung $\dot{y}^*$ existiert, liegt die gleichmäßige Konvergenz von $\dot{y}^k \to \dot{y}^*$ auf $[0, T]$ vor. ∎

Bemerkungen

Im nichtlinearen Fall wird die Konstante K wegen der MOS–Modelle „groß" sein. Dieses bedeutet, daß eine einheitliche Fensterwahl für alle Komponenten nicht angemessen ist.

Die Konvergenz der GSWR kann unter den vorliegenden Bedingungen für ein beliebiges Intervall $[0, T]$ nachgewiesen werden. Es liegt dann gleichmäßige Konvergenz bezüglich einer exponentiell gewichteten Norm vor. Diese Gewichtung kann numerisch ungünstig und dramatisch sichtbar sein.

Die angegebenen Testabbildungen sind Dualitätsabbildungen aus der Theorie der monotonen Operatoren. Auf den ersten Blick ist nun denkbar, daß das GSV und die GSWR auch für Aufgaben mit Operatoren A konvergent ist, wenn die Operatoren A eine (starke) Akkretivitätsbedingung bezüglich irgend einer Dualitätsabbildung erfüllen. Einfache Beispiele zeigen, daß dieses i.a. nicht richtig ist.

Die Sätze 1 und 2 bleiben richtig für das Gesamtschrittverfahren und die SOR Varianten.

Beispiele

Die Waveformrelaxation wurde zur

1. Transientenanalyse eines 2Bit Volladdierers in der N-MOS-Technologie und zur

2. Berechnung der Sprungantwort eines RCGL-Gliedes verwendet.

Bild 1 zeigt die Näherungslösungen nach der 1.–20. Iteration beim 2Bit Volladdierer. Das System umfaßt 19 Differentialgleichungen mit 19 Unbekannten. Dargestellt werden die Näherungen zu einem dieser Knotenpotentiale. Die Startnäherung war $y(t) \equiv 5$. Die exakte Lösung liegt im stark geschwärzten Bereich. Man erkennt zum einen wieder, daß die Näherungslösungen um so besser sind, je kleiner t und je größer die Iterationszahl ist, zum anderen, daß die steile Flanke der Lösungskurve Probleme bereitet.

Das Bild 3 zeigt die von der GSWR gelieferten Näherungslösungen nach der 1., 2., 5. und 8. Iteration beim RCGL-Glied. Die exakte Lösung ist gepunktet dargestellt. Die Startnäherung war $y \equiv 0$.

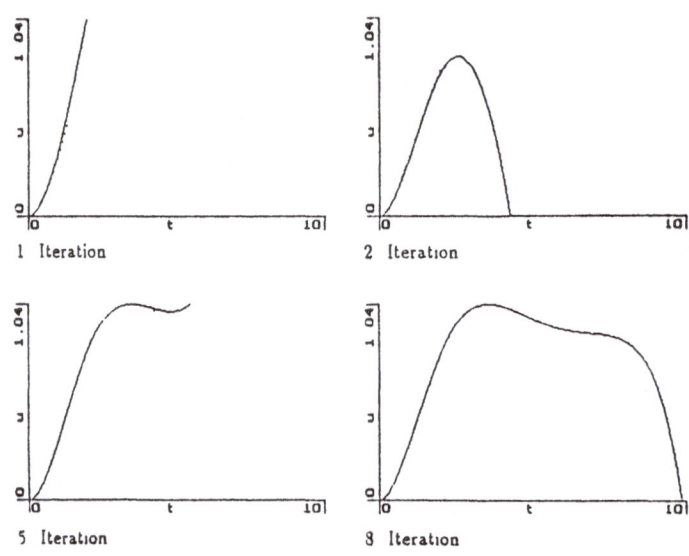

Bild 3.

Literatur

[1] Crandall, M.G.: Ligget, T.: Generation of semigroups of nonlinear transformations on general Banach spaces. Amer.J.Math. 93, 265–298. 1974.

[2] Morè, J.J.: Nonlinear generalizations of matrix diagonal dominance with applications to Gauß–Seidel iterations. SIAM J. Numer. Anal. 9, 357–378, 1972.

[3] Taubert, K.: Accretive Operators with applications to numerical integration of ordinary differential equations. In: Ròzsa, P. (ed.): Numerical Methods, Colloquia Mathematica Societatis János Bolyai 50, North-Holland Publishing Company, 111–125, 1986.

[4] Taubert, K.; Wiedl, W.: Waveformrelaxation und ein Störungssatz für akkretive Operatoren in der Maximumnorm. ZAMM, Z.angew.Math.Mech. 73 T925–T927, 1993.

[5] White, J.K.; The multirate integration properties of waveform relaxation with applications to circuit simulation and parallel computation. Thesis, Memorandum Nr. VCB/ERL 85/90, Elektronics Research Laboratory, University of California, Berkeley 1985.

Prof. Dr. Klaus Taubert
Institut für Angewandte Mathematik
der Universität Hamburg
Bundesstraße 55
20146 Hamburg
Deutschland

International Series of Numerical Mathematics, Vol. 117, © 1994 Birkhäuser Verlag Basel

MULTILEVEL–NEWTON–VERFAHREN
IN DER TRANSIENTENANALYSE ELEKTRISCHER NETZWERKE

Wolfgang Wiedl

Regionales Rechenzentrum der Universität Hamburg

Mit zunehmender Integrationsdichte und zunehmender Chipfläche werden die Anforderungen der Schaltkreis–Simulation an die Rechner immer größer, so daß heute häufig Vektorrechner für diese Aufgabe eingesetzt werden. Als Alternative dazu bieten sich jetzt mehr und mehr Cluster relativ preiswerter Workstations an. Inzwischen wurden schon eine Reihe von Tools entwickelt, die das Parallele Rechnen auf solchen Clustern unterstützen, als Beispiel sei PVM erwähnt.

Die Nutzung einer solchen Umgebung für die Transientenanalyse elektrischer Netzwerke setzt voraus, daß das zu analysierende Problem in geeigneter Weise partitioniert wird. Bei der numerischen Lösung kann die Partitionierung auf verschiedenen Ebenen berücksichtigt werden. Bei der 'Standardmethode' zur Lösung der Netzwerkgleichungen wird das vorliegende Anfangswertproblem:

$$f(z, z', t) = 0 \; ; \quad z(0) = z_0$$

zunächst mit Hilfe einer BD–Formel:

$$z'_{n+1} = -\frac{1}{h} \sum_{i=0}^{k} a_i \cdot z_{n+1-i}$$

auf ein algebraisches, in der Regel nicht–lineares Gleichungssystem:

$$f(z_{n+1}, z_n, z_{n-1}, \ldots\ldots\ldots, z_{n+1-k}, t_{n+1}) = 0$$

mit der Unbekannten z_{n+1} zurückgeführt. Dieses wird dann mittels Newton– oder inexaktem Newton–Verfahren und Sparse–Matrix–Techniken gelöst. Die Partitionierung kann man nun bei der Lösung der linearen algebraischen Gleichungen berücksichtigen, man kann sie aber mit wenig mehr Aufwand schon bei der Lösung der nicht–linearen algebraischen Gleichungen oder bereits bei der Lösung der Differentialgleichungssysteme berücksichtigen.

Am einfachsten wird die Numerik, wenn man nur die linearen Systeme verteilt rechnet, andererseits lassen sich viele spezielle Eigenschaften partitionierter Systeme auch an anderen Stellen des Lösungsprozesses ausnutzen. Beim Newton–Verfahren kann man z.B. den unterschiedlichen

Grad der Nichtlinearität, bei der Schrittweitensteuerung den unterschiedlichen Grad der Aktivität in den einzelnen Teilsystemen ausnutzen. Ersteres bedeutet, daß man bei der Lösung der nicht–linearen Gleichungssysteme die Newton–Iteration bei den einzelnen Teilsystemen nach unterschiedlich vielen Iterationsschritten abbricht, z.B. ein lineares Teilsystem nur im ersten Iterationsschritt berücksichtigt, nicht–lineare Teilsysteme dagegen in mehreren. Letzteres berücksichtigt, daß sich, insbesondere bei digitalen Netzwerken, die Aktivitäten immer nur auf relativ wenige Bereiche des Netzwerkes konzentrieren, wobei diese Bereiche sich im Verlaufe der simulierten Zeit allerdings ändern. Die beiden Latenzkriterien ergänzen sich insofern, als Teilsysteme, die bei letzterem irrtümlich als aktiv eingeordnet wurden, bei der Newton–Iteration meist als latent erkannt werden. Es gibt schon eine Reihe von Arbeiten, die die Ausnutzung dieser zeitlichen, räumlichen und nicht–linearen Latenz bei der Simulation elektrischer Netzwerke beschreiben und auch entsprechende Implementationen [1], [2]. Details dazu findet man auch in der inzwischen recht umfangreichen Literatur zur Multirate–Integration. Die Zielrichtung ist dabei immer die Reduzierung der Rechenarbeit. Erreicht wird dies dadurch, daß man die als latent eingeschätzten Teilsysteme in einer Reihe von Integrationsschritten nicht neu berechnet, sondern extrapolierte oder interpolierte Werte zu ihrer Beschreibung verwendet. Dadurch wird gleichzeitig eine Entkopplung der Teilsysteme voneinander erreicht. In unserem Zusammenhang ist die damit verbundene Reduzierung der Prozessor-Kommunikation mindestens ebenso interessant wie die Reduzierung der Rechenarbeit, denn dieser Aufwand betrifft einen der wesentlichen Engpässe beim Verteilten Rechnen.

Bei der Partitionierung sind unterschiedliche Gesichtspunkte zu berücksichten bzw. gegeneinander abzuwägen, da diese nicht immer miteinander verträglich sind. Zu berücksichtigen sind z.B. die Anzahl der Prozessoren und ihre Leistungsfähigkeit, der Umfang der lokalen Haupt– und Cache–Speicher und wie schon erwähnt, der Aufwand für die Kommunikation der Prozessoren untereinander. Im Hinblick auf die verfügbaren Speicher wird man eine Zerlegung in allzu große Teilprobleme vermeiden, im Hinblick auf die Kommunikation dagegen eine Zerlegung in allzu kleine. Die Anzahl der verfügbaren Prozessoren ist bei einem Workstation–Cluster in der Regel recht begrenzt. Wir gehen davon aus, daß wir mehr parallel zu bearbeitende Teilprobleme als Prozessoren haben und daß die Teilprobleme über eine Warteschlange an die Prozessoren verteilt werden.

Natürlich wird man ganz wesentlich eine modulare Struktur des Netzwerkes berücksichtigen, das sich in der Regel aus einzelnen Funktionseinheiten zusammensetzt. Innerhalb dieser Funktionseinheiten sind die Netzvariablen häufig stark und vielfältig gekoppelt, Kopplungen mit Netzvariablen in anderen Funktionseinheiten sind dagegen viel seltener. Tatsächlich ist es so, daß wir von vorgegebenen Partitionierungen der Netzwerke ausgehen, die immer an der modularen Struktur des Netzwerkes ausgerichtet sind.

Die bekanntesten Techniken zur Partitionierung elektrischer Netzwerke sind Branch– und Node–Tearing. Ersteres läuft numerisch auf eine Anwendung der Formel von Sherman–Morrison–Woodbury hinaus, letzteres führt zu einem hierarchisch gegliedertem System von Teilnetzen und läuft numerisch auf die Anwendung des Schur–Komplementes hinaus. Die einzelnen Teilsysteme kann man bei Bedarf wiederum zerlegen, so daß man zu mehreren Hierarchie–Ebenen kommt. Wir haben uns bisher immer auf zwei bis drei Hierachie–Ebenen beschränkt, bei mehr als zwei Hierachie–Ebenen nimmt der Aufwand für die Programmierung erheblich zu.

Eine programmgesteuerte Partitionierung, die die genannten und weitere hier nicht erwähnte Gesichtspunkte berücksichtigt, wäre von großem Vorteil. Wir haben mit einigen Ansätzen dazu experimentiert [3] und bei speziellen Beispielen auch gute Ergebnisse erhalten. Allerdings erweist sich die Problemklasse, für die diese automatischen Partitionierer anwendbar sind, bisher als viel zu eng.

Zur Lösung der auftretenden nicht–linearen algebraischen, partitionierten Gleichungssysteme benötigt man nun einen Algorithmus, der es in einfacher Weise gestattet, die als latent eingeschätzten Teilsysteme bei den weiteren Rechenschritten auszulassen. Wichtig dabei ist auch, daß

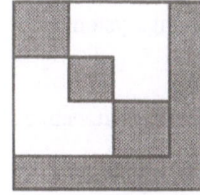

Bild 1

man Teilsysteme, die irrtümlich als latent eingestuft wurden, ohne allzu großen Aufwand nachträglich mit in die Rechnung einbeziehen kann. Wir benutzen beim Partitionieren eine Variante des Node–Tearing mit 'Verdoppelung der Randknoten', was dazu führt, daß die Jacobi–Matrizen bei der Newton–Iteration BBD–Struktur (bordered block diagonal) aufweisen (Bild 1).

In Anlehnung an [4] verwenden wir zur Lösung der auftretenden Gleichungssysteme ein Multi-Level–Newton–Verfahren, das hier skizziert werden soll. Da die formale Beschreibung eines MLN–Verfahrens schnell unübersichtlich wird, beschränken wir uns dabei, wie auch bei der graphischen Darstellung der BBD–Struktur in Bild 1, auf zwei–stufige Verfahren, bzw. auf die oberste Ebene eines mehrstufigen Verfahrens.

x bezeichne den Vektor der Variablen der Teilsysteme, etwa die betreffenden Knotenpotentiale. y bezeichne den Vektor der Variablen des Verbindungsnetzes.

Wünschenswert ist es, wenn die Anzahl m der Komponenten von y wesentlich kleiner ist als die Anzahl n der Komponenten von x. Das zu lösende Gleichungssystem ist dann von der Form:

$$\begin{aligned} H(x, y) &= 0 \\ F(x, y) &= 0 \end{aligned} \tag{1}$$

wobei H die Teilsysteme und F das Verbindungsnetzwerk beschreibt. Entsprechend der Anzahl k der Teilsysteme zerfällt H in entsprechend viele Blöcke H_i. dasselbe gilt für den Vektor x. Wir haben also statt $H(x, y) = 0$ eigentlich:

$$H_1 (x_1, y) = 0$$
$$\cdots\cdots\cdots\cdots\cdots\cdots\cdots$$
$$H_k (x_k, y) = 0$$

Für die Parallelisierung ist das von Bedeutung, für die Beschreibung des ML–Newton–Verfahrens reicht Darstellung (1) aber aus.

Physikalisch gesehen richten sich die Werte der Variablen x nach den Werten der Variablen y, denn letztere beschreiben die Anregungen der einzelnen Teilsysteme. Betrachten wir also x als einen von y abhängigen Vektor:

$$x = \psi(y)$$

Anstelle von (1) müssen wir dann lösen:

(2) $$F(\psi(y) , y) = 0$$

wobei wir voraussetzen müssen, daß zu (1) eine Lösung $(x^*, y^*) \in \mathbb{R}^{n+m}$, Umgebungen $U(y^*)$, $U(x^*)$ und eine Abbildung $\psi: U(y^*) \to U(x^*)$ mit der Eigenschaft:

(3) $$H (x , y) = 0 \wedge y \in U(y^*) \Rightarrow x = \psi(y)$$

auch existieren. Das Newton–Verfahren für (2) führt dann auf lineare Gleichungssysteme:

(4) $$[F_x \cdot \psi_y + F_y] \cdot \Delta y = - F(\psi(y) , y)$$

in denen allerdings $\psi_y(y)$ und $\psi(y)$ selbst noch unbekannt sind. Um $\psi(y)$ zu bestimmen, betrachten wir eine innere Newton–Iteration zur Lösung des Gleichungssystems:

$$H (x , y) = 0$$

bei festgehaltenem y. Dies führt auf die Lösung eines linearen Gleichungssystems:

(5) $$H_x \cdot \Delta x = - H (x , y)$$

Um $\psi_y(y)$ zu bestimmen, gehen wir von (3) aus. Es gilt für $y \in \quad U(y^*)$:

$$H (\psi(y) , y) \equiv 0$$

und damit:

$$H_x \cdot \psi_y + H_y \equiv 0$$

also:

(6) $$\psi_y = - H_x^{-1} \cdot H_y$$

Die linearen Gleichungssysteme der äußeren Iteration (4) lauten damit

(7) $$[- F_x \cdot H_x^{-1} \cdot H_y + F_y] \cdot \Delta y = - F(\psi(y) , y)$$

Man sieht, daß die Berechnung von Δx in (5) und ψ_y in (6) parallel durchgeführt werden kann, denn H_x ist eine Blockdiagonal–Matrix (die linke, obere Teilmatrix in Bild 1). An die verfügbaren Prozessoren verteilen wir als Aufgaben also die Lösung von Folgen linearer Gleichungssysteme, die entweder zu einer inneren Newton–Iteration oder zum Aufbau der Iterationsmatrix für das äußere Newton–Verfahren dienen. Es ist offensichtlich kein Problem, latente Teilsysteme in diesem Prozess zu überspringen. Offensichtlich ist auch, daß die Ausnutzung der Latenz insbesondere dann von Vorteil ist, wenn weniger Prozessoren zur Verfügung stehen als Teilprobleme zu berechnen sind, was wir bei Workstation–Clustern und vielen Parallelrechnern, sofern es sich nicht um massiv parallele Systeme handelt, aber annehmen können. Problematisch ist es aber, mit diesem Verteilungsschema allein zu einer ausgewogenen Auslastung der verfügbaren Prozessoren zu kommen, wenn eine ungünstige Partitionierung vorgegeben wird. Das ist z.B. dann der Fall, wenn der Arbeitsaufwand zur Lösung eines einzelnen Teilsystems den für die anderen dominiert. Problematisch sind auch größere Verbindungssysteme, denn während der Lösung von (7) können keine Teilsysteme bearbeitet werden. Einen Ausweg aus dieser Problematik bietet die verteilte Lösung der linearen Gleichungssysteme, bei größeren Gleichungssystemen ist dies durchaus sinnvoll. Dabei stehen direkte und iterative Verfahren zur Verfügung, denn es ist nicht festgelegt, wie man innerhalb eines MLN–Schemas die einzelnen linearen Gleichungssysteme zu lösen hat.

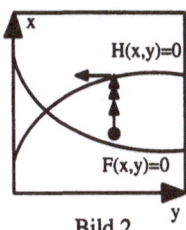

Bild 2

Wenn wir die innere Iteration bis zur Konvergenz durchführen (Bild 2), ist das vorgestellte Verfahren nichts anderes als ein gewöhnliches Newton–Verfahren für (2), zu vorgegebenem y wird jeweils x mit $H(x,y)=0$ berechnet und ein Newton–Schritt in y–Richtung ausgeführt. MLN–Verfahren unterscheiden sich demgegenüber dadurch, daß die innere Iteration nicht bis zur Konvergenz durchgeführt wird, wodurch sie numerisch überhaupt erst interessant werden. In [4] wird gezeigt, daß die quadratische Konvergenz des Newton–Verfahrens sich auf das MLN–Verfahren überträgt, wenn man als Abbruchkriterium für die innere Iteration:

$$\|\Delta x\| \leq \tau^{i+1} = min \ (\ \tau^0, \ \|\Delta y\|^{\gamma} \ )$$

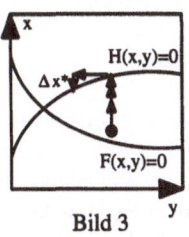

Bild 3

mit $\gamma = 2$ verwendet. Für $\gamma = 1$ kann man paarweise quadratische Konvergenz nachweisen, also quadratische Konvergenz, wenn man jeweils zwei aufeinanderfolgende Schritte zusammenfaßt [5].

Praktisch macht sich der Einfluß von γ aber wenig bemerkbar, wenn die Abbruchkriterien in beiden Iterationen 'zusammenpassen', wenn also z.B. als Abbruchkriterium in der äußeren Iteration $\|\Delta y\| \leq \eta = 10^{-8}$ und für die innere Iteration $\tau^0 = 10^{-4}$ gewählt wird. Das liegt auch daran, daß häufig gar nicht solange iteriert wird, bis spezielle Konvergenzeigenschaften zum Tragen kommen.

Eine naheliegende Variante zum MLN–Verfahren besteht darin, die äußere Iteration nicht in y–Richtung sondern im Tangentialraum von ψ durchzuführen (Bild 3). Das läuft darauf hinaus, daß man nach einer Δy–Korrektur noch eine Korrektur $\Delta x^* = \psi_y \cdot \Delta y$ anbringt. Da ψ_y und Δy

zu diesem Zeitpunkt bekannt sind, ist dies mit wenig Aufwand zu bewerkstelligen. Diese Korrektur hat keine Auswirkungen auf die Konvergenzeigenschaften, führt aber zu weniger inneren Iterationen. Da diese aber parallel ausgeführt werden, ist der Nutzen nicht allzu groß.

In [6] werden verwandte Verfahren auf ihre Konvergenz hin untersucht, die innere Iteration wird dort jeweils auf einen Schritt beschränkt. Eventuell vorhandene nichtlineare Latenz kann daher nicht berücksichtigt werden, Latenz auf der Ebene der Differentialgleichungen aber sehr wohl. Da wir bei der inneren Iteration aber meist nur 1 – 2 Iterationen beobachtet haben (das hängt von der Schrittweitensteuerung ab), dürfte dies kein sonderlicher Nachteil sein. Für dieses Verfahren wird, unter entsprechenden Regularitätsvoraussetzungen, eine paarweise Q–Ordnung 2 angegeben. Für dasselbe Verfahren, ergänzt um einen Korrekturschritt Δx^* ergibt sich eine (1–Schritt) Q–Ordnung 2.

Die Ausnutzung der Latenz bietet im Zusammenspiel mit einem MLN–Verfahren ein beträchtliches Potential zur Parallelisiereung und zur Einsparung von Rechenarbeit, ist jedoch auch mit einem beträchtlichen Aufwand im Programm verbunden, so daß wir erst ab Systemen mit 100 Variablen Vorteile gegenüber einem Standardalgorithmus beobachten konnten. Problematisch ist die Abhängigkeit von einer geeignet vorgegebenen Partitionierung des Netzwerkes, was es notwendig macht, zusätzliche Strategien einzubauen um zu vernünftigen Lastverteilungen zu kommen. Brauchbare Partitionierungsalgorithmen für elektrische Netzwerke wären hier sehr hilfreich.

Literatur:

[1] Sakallah, K.A. und Director, S.W. (1985) An Event Driven VLSI Circuit Simulator. IEEE Trans. CAD–4, No. 4, pp. 668 – 684

[2] Schultz, R. (1990) Local Timestep Control for Simulating Electrical Circuits. in: Bank, R.E., Burlirsch, R., Merten, K.: Mathematical modeling and simulation of electrical circuits and semiconductor devices.

[3] Wallat, O. (1991) Experimente mit dem Multilevel–Newtonverfahren, insbesondere bei der Netzwerksimulation, Diplomarbeit, Institut für Angew. Mathematik, Universität Hamburg.

[4] Rabbat, N.B.G., Sangiovanni–Vincentelli, A.L., Hsieh (1979) A Multilevel Newton Algorithm with Macromodelling and Latency for the Analysis of Large–Scale Nonlinear Circuits in the Time Domain, IEEE Transactions Vol–CAS 26, pp. 733 – 740.

[5] Lin, P.M. und Nordsieck, A.W. (1982) A Study of the Convergence Rate of a Multilevel Newton Algorithm, School of Electrical Engineering, Purdue University.

[6] Hoyer, W. und Schmidt, J. W. (1984) Newton–Type Decomposition Methods for Equations Arising in Network Analysis, ZAMM 64, pp. 397 – 405

Wolfgang Wiedl, Regionales Rechenzentrum der Universität Hamburg, Schlüterstr. 70, D–20146 Hamburg

International Series of Numerical Mathematics, Vol. 117, © 1994 Birkhäuser Verlag Basel

TRANSIENTENSIMULATION ELEKTRISCHER NETZWERKE MIT TRBDF

Hartmut Wriedt

Regionales Rechenzentrum der Universität Hamburg

Zusammenfassung: In der Transientensimulation elektrischer Netzwerke hat man ein Algebro-Differential-Gleichungssystem mit einem
Index, welcher in der Regel nicht größer als Zwei ist, numerisch
zu lösen. Die in der Schaltkreissimulation hierfür bevorzugten
Verfahren sind die 'Backward Differentiation Formulas' und die
Trapezregel. Jedes Verfahren hat für sich genommen seine speziellen Vorzüge, damit verbunden aber auch gewisse Nachteile. Durch
eine geeignete Kombination beider Verfahren lassen sich die jeweiligen Vorteile nutzen, und die entsprechenden Nachteile ab-
schwächen. Diese Strategie wurde im Schaltkreissimulator TITAN
unter dem Namen TRBDF implementiert.

1. Einleitung

Eine wichtige Aufgabe innerhalb der Schaltkreissimulation ist die
Simulation im Zeitbereich. Dazu wird der Schaltkreis üblicher-
weise durch ein Algebro-Differential-Gleichungssystem (ADG-
System) der Art

$$(1) \qquad G(\ dX(t)/dt,\ X(t),\ t\ ) = 0\ , \qquad\qquad X(t_0) = X_0$$

beschrieben. Der Index von (1) ist meist Null, Eins oder Zwei.
Der hier benutzte Begriff des Index entspricht dem des in BRENAN
et.al [2] definiertem (Def. 2.2.2). Gebräuchliche Methoden zum
Aufstellen der Netzwerkgleichungen sind die Sparse-Tableau-
Formulierung sowie die modifizierte Knotenanalyse (MNA). Letztere
läßt sich aus der Sparse-Tableau-Formulierung durch Elimination

der Ströme aller stromdefinierenden Elemente, sowie aller Zweig-
spannungen, herleiten.

Im Schaltkreissimulator TITAN werden die Netzwerkgleichungen
etwas anders aufgestellt. Hierüber berichtete bereits DENK [3]
und SCHULTZ [6]:

(2) $F(\ dQ(X(t))/dt,\ X(t),\ t) = 0\ ,$ $X(t_0) = X_0$

     X Vektor mit den Größen V, U, I1, I2
     V Knotenpotentiale
     U Zweigspannungen
     I1 Ströme spannungsdefinierender Zweige
     I2 Ströme stromdefinierender Zweige
     Q Ladungen, Flüsse

Der Grund hierfür ist, daß bei einer Diskretisierung von dX/dt in
(1) die sogenannte Ladungserhaltung (numerisch) nicht gewährleistet ist. Daher wird in (2) nach dQ/dt diskretisiert. Desweiteren
wird in TITAN eine Kombination zwischen der Sparse-Tableau-
Formulierung und der MNA benutzt: es werden grundsätzlich alle
Variablen berechnet, wie dies auch bei der Sparse-Tableau-
Formulierung geschieht. Die Organisation der Gleichungen entspricht jedoch mehr der MNA.

(3a) | $F0(\ dQ(Y)/dt,\ Y,\ t) = 0$
(3b) | U $= F1*V$
(3c) (2) <=> | Q $= F2(Y)$
(3d) | $dQ(Y)/dt$ $= F3(Q,\ Q(alt),\ dQ/dt(alt))$
(3e) | $I2$ $= F4(U,\ dQ/dt)$

     Y Vektor mit den Größen V, I1
     F0 Gleichungen gemäß der MNA
     F1 Knoten-Kanten-Incidencematrix
     F2, F4 gemäß den Zweigrelationen
     F3 entspricht dem gewählten Diskretisierungsverfahren

Die Aufgabe (3) wird wie folgt berechnet: es wird zunächst das
ADG-System (3a) gemäß der MNA aufgestellt, mit Hilfe von (3d)
diskretisiert und unter Verwendung von (3c) nach Y, beispiels-
weise mit dem Newtonverfahren, gelöst:

(4)   FO(dQ(Y(t))/dt, Y(t), t) = 0

(5)   FO(dQ(Y_n)/dt, Y_n, t_n) = 0

         dQ(Y_n)/dt := α*Q(Y_n)+β

(6)   FO(α*Q(Y_n)+β, Y_n, t_n) = 0   ----> Lösen nach Y_n

Mit Y_n und den bereits aus den vorherigen Integrationsschritten
bekannten Werten für Q und dQ/dt lassen sich die restlichen
Variablen ohne Lösen eines weiteren Gleichungssystems berechnen.

Der Index des ADG-Systems (2) ist bei MOS-Schaltungen, abhängig
vom verwendeten Transistormodell, meist Zwei. Hingegen ist der
Index des ADG-Systems (3a) (MNA) meist nur Eins. Der Grund für
den Index Zwei des Systems (2) sind die in der Transistor-
ersatzschaltung enthaltenen Schleifen, die nur aus Kapazitäten
bestehen, näheres hierzu in SINCOVEC et. al. [7].

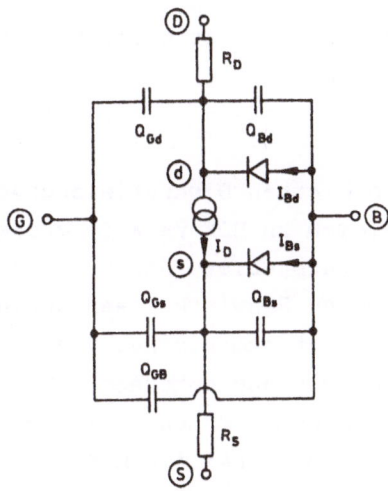

Bild 1: NMOS-Ersatzschaltung (Hoefer [5])

Die zu diesen Kapazitäten gehörenden Zweigströme besitzen ein
Index-Zwei Verhalten. Diese Ströme sind jedoch nicht Bestandteil

der MNA-Gleichung (3a), weshalb der Index von (3a) meist Eins
ist.

2. Integration der Netzwerkgleichungen

Die in der Schaltkreissimulation meist eingesetzten Verfahren
sind die 'Backward Differentiation Formulas'

$$(7) \qquad dQ/dt_n = \frac{1}{h} [\sum_{i=0}^{k} \alpha_{n,i} * Q_{n-i}] \qquad\qquad (BDF-k)$$

sowie die Trapezregel

$$(8) \qquad dQ/dt_n = 2 * [Q_n - Q_{n-1}]/h - dQ/dt_{n-1} \qquad\qquad (TR)$$

In der Praxis sind die BDF-k nur für k=1,2 und seltener 3 in
Gebrauch.
Jedes der Verfahren (7) und (8) hat seine Vorzüge, damit ver-
bunden aber auch Nachteile:

Trapezregel:
  + Hohe Effizienz durch kleinen Diskretisierungsfehler. Dies
    bewirkt die im Vergleich zu BDF relativ großen Schrittweiten.
  + Symmetrisches Stabilitätsgebiet, daher ist die TR gut für die
    Simulation von autonomen Schwingkreisen geeignet.
  - Nicht L-stabil, daher ist numerisches Oszillieren,
    insbesondere bei den oben angesprochenen Strömen, möglich.
  - Eine weitere Folge des numerischen Oszillierens ist das
    Versagen der Latenzdiagnose bei der Methode der lokalen Zeit-
    schrittsteuerung. Dies bewirkt einen unnötig hohen Aufwand bei
    den einzelnen Newtoniterationen.

BDF-k:
  + L-Stabilität, numerisches oszillieren der Lösung ist bei den

betrachteten Problemen nicht moglich;

+ damit verbunden. eine sichere Diagnose von Latenz ist bei der
  lokalen Zeitschrittsteuerung moglich. Dies bewirkt einen
  geringeren Aufwand bei den einzelnen Newtoniterationen.

- Der lokale Diskretisierungsfehler ist bei den BDF relativ
  groß, so daß nur kleinere Schrittweiten moglich sind: Unter
  der Annahme konstanter Schrittweiten erlaubt die TR ca. 40%
  großere Schrittweiten als BDF-2.

- Die hohe Stabilitat der Verfahren bewirkt eine starke
  Dampfung der Losung, so daß autonome Schwingkreise nicht
  korrekt mit BDF gerechnet werden konnen.

Das folgende Bild illustriert die Stabilitatsprobleme der TR:

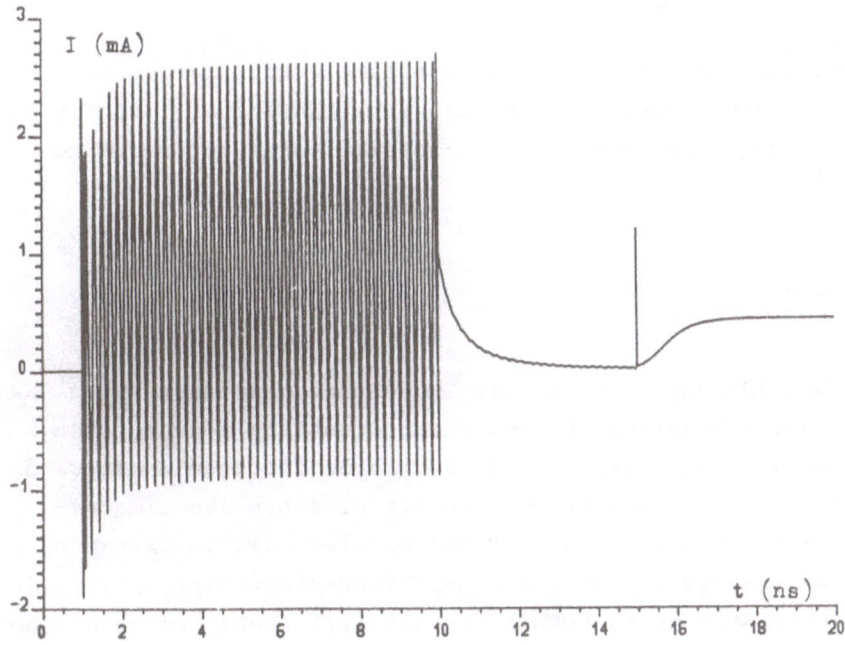

Bild 2: Stromkurve einer analogen Schaltung, gerechnet mit der TR

Dieses numerische Oszillieren der Strome hat aber keinen nachtei-
ligen Einfluß auf die in TITAN verwendete Schrittweitenkontrolle.

Hingegen wirken sich die numerischen Oszillationen der TR nach-
teilig bei der Methode der lokalen Zeitschrittsteuerung aus. Die
Methode der lokalen Zeitschrittsteuerung ist in SCHULTZ [6]
beschrieben. Es ergibt sich, daß bei gewissen Schaltungen BDF-2
eine höhere Effizienz als die TR besitzt. Dies ist zunächst
unerwartet, da die Anzahl der benötigten Newtoniterationen bei
BDF-2 (bedingt durch die größere Anzahl der Integrationsschritte)
deutlich höher ist. Die Ursache ist, daß bei der TR die in der
Schaltung vorhandene Latenz nicht ausreichend diagnostiert werden
kann. Dies bewirkt einen deutlich höheren CPU-Zeitverbrauch pro
Newtoniteration bei der TR gegenüber BDF-2.

3. TRBDF

Es liegt nahe, durch eine geeignete Kombination beider Verfahren
die jeweiligen Vorteile zu nutzen. Von BANK et. al. [1] ist das
TRBDF-Verfahren bekannt, wo dies bereits vorgeschlagen wird:

```
t_{n-2}      t_{n-1}    t_n
 :            :          :            mit        h_{n-1} = √2*h_n
 :---TR---:              :
 :----------BDF-2-:
```

Nach einem TR-Schritt folgt ein BDF-2 Integrationsschritt. Das
vorgegebene Schrittweitenverhältnis garantiert den gleichen
führenden Koeffizienten, so daß für beide Integrationsschritte
ein vereinfachtes Newtonverfahren mit gleicher Jacobimatrix
möglich ist. Wird statt der TR ihr one-leg Zwilling benutzt, so
entspricht das resultierende Einschrittverfahren $t_{n-2} \rightarrow t_n$ dem
in der Literatur (z.B. BRENAN et. al. [2]) wohbekannten Diagonal-
impliziten Runge-Kutta Verfahren:

(9)
$$\begin{array}{c|cc}
\mu & \mu & 0 \\
1 & 1-\mu & \mu \\
\hline
 & 1-\mu & \mu
\end{array}$$
mit $\mu = 1 - \sqrt{2}/2$

TRBDF ist ein L-stabiles Integrationsverfahren, besitzt aber eine
nicht so starke numerische Dampfung wie BDF-2. Hinsichtlich der
moglichen Schrittweitengroße ist TRBDF effizienter als BDF-2.
Die in TITAN unter dem Namen TRBDF implementierte Strategie
weicht vom obigen Ansatz ab. Sie hat zum Ziel, so oft wie moglich
die TR einzusetzen (Effizienz) und nur, wenn es fur die Stabilitat des Verfahrens erforderlich ist, die BDF-k.

(10)
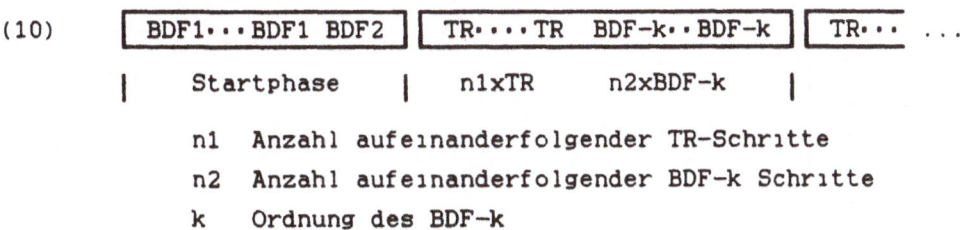

| | Startphase | | n1xTR | n2xBDF-k | |

        n1   Anzahl aufeinanderfolgender TR-Schritte
        n2   Anzahl aufeinanderfolgender BDF-k Schritte
        k   Ordnung des BDF-k

Nach einer Startphase wird die TRBDF-Strategie aktiv. Nach n1
aufeinanderfolgenden Integrationsschritten mit der TR folgen n2
weitere Schritte mit BDF-k. Die jeweiligen Schrittweiten werden
mit der herkommlichen Schrittweitenkontrolle festgelegt. Aufgrund
des BDF-Schlußschrittes ist die Sequenz [TR···BDF-k] auf jeden
Fall L-stabil, numerische Oszillationen werden ausgedampft. Zur
Steuerung der Parameter n1, n2 und k gibt es drei Moglichkeiten:

- Die Parameter werden dem Simulator von außen als feste Großen
  mitgeteilt. Typische Werte sind etwa n1=8, n2=1, und k=2.

- In einer Laborversion von TITAN wurde ein Stabilitatsmonitor
  eingebaut. Dieser untersucht die Werte dQ/dt auf numerische
  Oszillationen. Es wird solange mit der TR gerechnet, bis der
  Monitor einen BDF-Schritt vorschlagt.

– Nähere Untersuchungen haben gezeigt, daß das Auftreten von
  numerischen Oszillationen in dQ/dt bei der Trapezregel an
  bestimmte Ereignisse gekoppelt ist. Einige dieser Ereignisse
  lassen sich ohne Aufwand diagnostizieren bzw. vorhersagen. Zu
  diesen Zeitpunkten wird prophylaktisch mit BDF gerechnet.

Die TRBDF-Strategie erfüllt die in sie gesetzten Erwartungen:
– Numerische Oszillationen treten nur noch selten auf, und dann
  auch nur über wenige Zeitschritte hinweg.

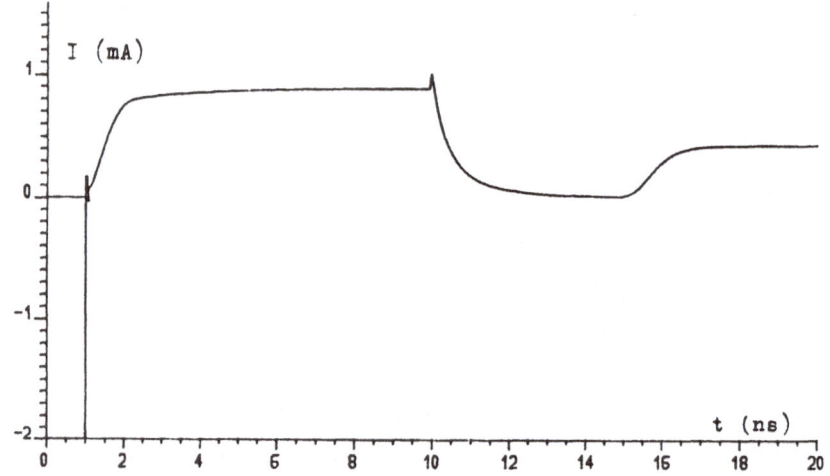

Bild 3: Schaltung wie in Bild 2, mit TRBDF gerechnet. Der
Ausschlag bei t=1[ns] ist kein numerischer Effekt.

– Hinsichtlich der Effizienz ist die TRBDF-Strategie mit
  der TR vergleichbar.
– TRBDF ist in der Lage, autonome Schwingkreise zu berechnen.
– Besonders erfreulich ist das Verhalten von TRBDF, wenn die
  Methode der lokalen Zeitschrittsteuerung aktiv ist:

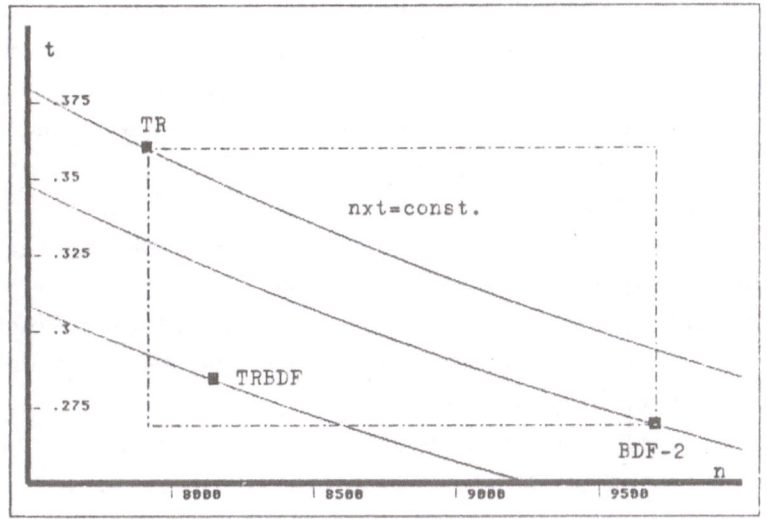

Bild 4: Effizienz verschiedener Integrationsverfahren bei
Gebrauch der lokalen Zeitschrittsteuerung. (Gerechnet wurde ein
Schaltkreis mit 2005 Transistoren auf einer Siemens 7590).
Aufgetragen ist die CPU-Zeit t für LOAD pro Newtoniteration
gegen die Anzahl der Newtoniterationen.

Danksagung: Der Autor dankt den Herren Dr. A. Gilg,
Dr. U. Feldmann, Dr. Q. Zheng und Dr. Dr. R. Schultz von der
Firma SIEMENS AG, ZFE BT SE 4, für ihre herzliche Unterstützung.

Literatur:

[1] Bank, R. E., Coughran, W. M., Fichtner, W., Grosse, E. H.,
    Rose, D. J., Smith, R. K.: Transient simulation of silicon
    devices and circuits, IEEE Transactions on computer-aided
    design, Vol. CAD-4, Oct. 1985
[2] Brenan, K. E., Campbell, S. L., Petzold, L. R.: Numerical
    Solution of Initial-Value Problems in Differential-Algebraic
    Equations, North-Holland, New York-Amsterdam-London 1989

[3] Denk, G.: An improved numerical integration method in the
    circuit simulator SPICE2-S, International Series of Numerical
    Mathematics, Vol. 93, Birkhäuser Verlag Basel 1990

[4] Ho, C. W., Ruehli, A. E., Brennan, P. A.: The modified nodal
    approach to network analysis, IEEE Trans. Circuits Syst.,
    Vol. CAS-22, June 1975, pp. 504-509

[5] Hoefer, E. E. E., Nielinger, H.: SPICE, Springer Verlag,
    Berlin-Heidelberg-New York-Tokyo 1985

[6] Schultz, R.: Local timestep control for simulating electrical
    circuits, International Series of Numerical Mathematics, Vol.
    93, Birkhäuser Verlag Basel 1990

[7] Sincovec, R. F., Dembart, B., Epton, M. A.,Erisman, A. M.,
    Manke, J. W., Yip, E. L.: Solvability of large scale
    descriptor systems, Technical Report, Boeing Computer
    Services Company, Seattle, WA, 1979

Hartmut Wriedt
Regionales Rechenzentrum der Universität Hamburg
Schlüterstr. 70
D-20146 Hamburg

International Series of Numerical Mathematics, Vol. 117, © 1994 Birkhäuser Verlag Basel

The Transient Behavior of an Oscillator

Qinghua Zheng

Siemens AG, Corporate Research and Development

January 18, 1994

Abstract. In this paper the nonlinear autonomous equation

$$\dot{z} = g(z),$$

$g : \mathbb{R}^n \to \mathbb{R}^n$ which describes an electrical oscillator, is treated. For the numerical simulation and the computation of the limit cycle, the analysis and the approximation of the transient behaviors are the main constraints. Here, some typical phenomena of electrical oscillators are very important. The periodic linearization technique can be introduced on this basis. Using this technique, some properties of the transient behaior are shown and an algorithm for the numerical approximation of this behavior is also presented. Finally, two numerical examples demonstrate the results.

AMS subject classification (1980): 34A45, 34A50, 34C15, 34C27, 35C35 and 65L05.

Introduction

Theoretical interest in the dynamic behavior of nonlinear autonomous differential equations has a long tradition. Their technical application to dynamic systems is becoming more and more important. Limit cycles play an important role in the study of dynamic behavior as well as in the technical applications. Numerous theorems about the existence and stability of periodic solutions have been developed during this century (an overview can be foud in Amann 1983, Reithmeier 1991 [1, 6]). The numerical computation of the periodic solutions and determining the solution near the limit cycle has been a very active research area in resent years (see e. g. Kundert et. al. 1988, Petzold 1981 [4, 5]). It is also considered in this conference by Dr. G. Denk (A New Discretization for the Efficient Integration of Highly Oscillatory ODEs) and Dr. W. Schmidt (Limit Cycle Computation of Oscillating Electrical Circuits).

This paper discusses the dynamic behavior of a nonlinear autonomous differential equation which describes an oscillator on the way to a steady state. This approach is based on some typical phenomena of oscillators. They are given in section 1. In section 2 this behavior is treated analytically by a periodic linearization technique. The analytical results can be used to build a numerical algorithm. A possible algorithm and two examples are shown in section 3.

1   Motivation

We start the investigation with a typical oscillator circuit in the the domain of microelectronics, namely a quartz oscillator. Fig. 1 shows a simplified circuit. Its dynamic behavior is also shown in Fig. 2. It takes approximately 40 000 oscillations to come to a steady state (very near the limit cycle) in ca. 2.5 milliseconds, because of the enormously different time constants.

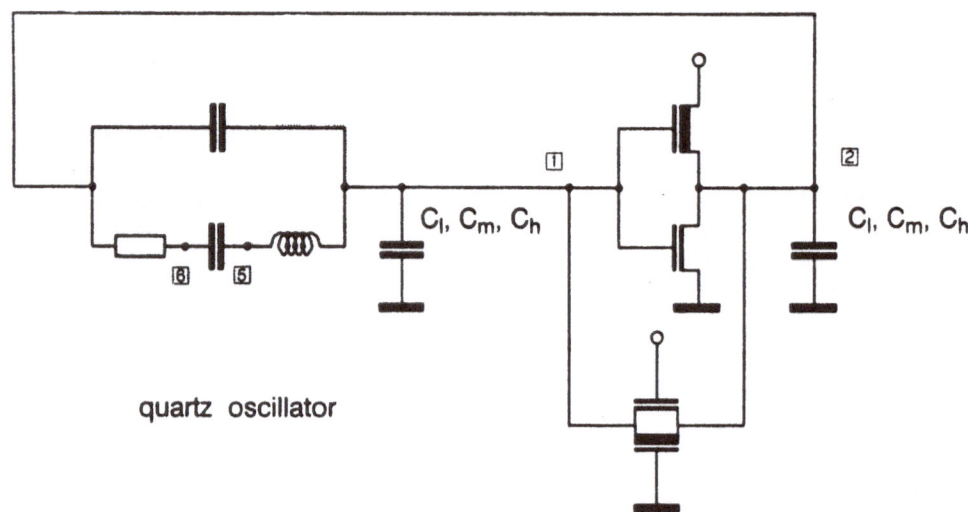

quartz oscillator

Figure 1: A simple quartz oscillator circuit with switched capacitor technique.

Three important properties are easy to perceive:

a) It takes a very long time to come to a steady state;

b) All trajectories are almost closed in every time;

c) All trajectories are shaped like an ellipse.

These are the basis phenomena for the following discussion.

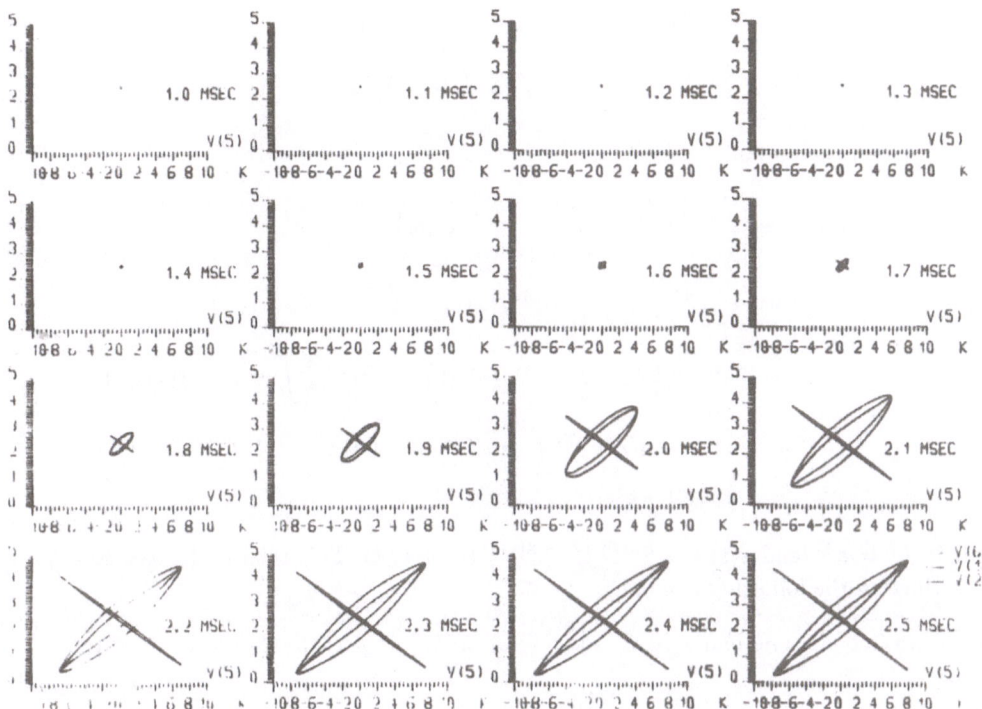

Figure 2: The transient behavior of the oscillator is shown in the phase space (x-axis: the voltage on node 5, y-axis: the voltage on nodes 1, 2 and 6; from top/left to bottom/right: the oscillations near the time points 1.0, 1.1, 1.2, ... 2.5 milliseconds after starting).

2 Periodic Linearization

In this section we will assume that the positive number T is periodic. Before starting the analysis of the nonlinear autonomous differential equation

$$\dot{z} = g(z), \tag{1}$$

$g : \mathbb{R}^n \to \mathbb{R}^n$, $n \in \mathbb{N}$, which describes the oscillator, the linear T-periodic system is treated. For the dynamic system

$$\begin{cases} \dot{x} = Ax + f(t) \\ x(0) = x_0 \end{cases},$$

where $A \in \mathbb{R}^{n \times n}$ and $f : \mathbb{R}^n \to \mathbb{R}^n$ are T-periodic, we know the solution exactly:

$$x(t) = e^{At}x_0 + e^{At}\int_0^t e^{-As}f(s)\,ds$$

$$= e^{At}\left(x_0 + \int_0^t e^{-As}f(s)\,ds\right).$$

By reason of the periodicity, we have

$$x(t+T) = e^{A(t+T)}\left(x_0 + \int_0^{t+T} e^{-As}f(s)\,ds\right)$$

$$= e^{At}e^{AT}\left(x_0 + \int_0^t e^{-As}f(s)\,ds + \int_t^{t+T} e^{-As}f(s)\,ds\right)$$

$$= e^{AT}e^{At}\left(x_0 + \int_0^t e^{-As}f(s)\,ds\right) + e^{At}e^{AT}\left(\int_t^{t+T} e^{-As}f(s)\,ds\right)$$

$$= e^{AT}x(t) + e^{A(t+T)}\int_0^T e^{-A(s+t)}f(s+t)\,ds$$

$$= M\,x(t) + q(t),$$

where $M = e^{AT}$ and $q(t) = e^{A(t+T)}\int_0^T e^{-A(s+t)}f(s+t)\,ds$. This relation for the solution is described in the following lemma:

Lemma 2.1 *Assume that x is the solution of the linear periodic differential equation:*

$$\begin{cases} \dot{x} = Ax + f(t) \\ x(0) = x_0 \end{cases} \qquad f(t+T) = f(t) \quad \forall t \in \mathbb{R}.$$

We have the relation:

$$x(t+T) = Mx(t) + q(t),$$

where q and M are independent of x_0 and q is also a T-periodic function.

M is known as the monodromy matrix (see Collatz 1990 [3] p. 294) and $\mathcal{L} : x \to Mx + q$ as the monodromy operator (see [1] p. 309).

For the more general case:

$$\begin{cases} \dot{x} = A(t)x + f(t) \\ x(0) = x_0 \end{cases},$$

where $A(t) \in \mathbb{R}^{n \times n}$ and $f(t) \in \mathbb{R}^n$ are T-periodic functions, we can use Floquet's theorem (see [1] pp. 307 ff.). According to that, there is a nonsingular T-periodic matrix $Q(t)$ such that for the transformation $y(t) := Q^{-1}(t)x(t)$ is:

$$\dot{y} = \tilde{A}y + \tilde{f}(t),$$

where $\tilde{f}(t+T) = \tilde{f}(t)\forall t \in \mathbb{R}$ and the matrix $\tilde{A} \in \mathbb{R}^{n \times n}$ is constant. By reason of Lemma 2.1 we also have:

$$y(t+T) = \tilde{M}y(t) + \tilde{q}(t)$$

or

$$\begin{aligned}
x(t+T) &= Q(t+T)y(t+T) \\
&= Q(t)y(t+T) \\
&= Q(t)(\tilde{M}y(t) + \tilde{q}(t)) \\
&= Q(t)\tilde{M}y(t) + Q(t)\tilde{q}(t) \\
&= Q(t)\tilde{M}Q^{-1}(t)Q(t)y(t) + Q(t)\tilde{q}(t) \\
&= M(t)x(t) + q(t),
\end{aligned}$$

where $M(t) = Q(t)\tilde{M}Q^{-1}(t)$ and $q(t) = Q(t)\tilde{q}(t)$. We now have a generalization of Lemma 2.1:

Lemma 2.2 *Assume that x is the solution of the linear periodic differential equation:*

$$\begin{cases} \dot{x} = A(t)x + f(t) \\ x(0) = x_0 \end{cases} \qquad A(t+T) = A(t), f(t+T) = f(t) \quad \forall t \in \mathbb{R}.$$

We have the relation:

$$x(t+T) = M(t)x(t) + q(t),$$

where q and M are T-periodic functions and independent of x_0.

Some properties and consequences which are useful for studying of transient behavior are given in the following remarks:

Remark 2.1 *The monodromy matrix in Floquet's theorem is given by*

$$M(t) = Q(t)\tilde{M}Q^{-1}(t).$$

It follows that:

$$\sigma(M(t)) = \sigma(\tilde{M}) \quad \forall t \in \mathbb{R},$$

where $\sigma(.)$ means the spectrum of a matrix.

Remark 2.2 *Let $x^*(t)$ be a fixed point of the monodromy operation:*

$$\begin{aligned}
x(t) &= \mathcal{L}(x)(t) \\
&= M(t)x(t) + q(t).
\end{aligned}$$

Then we have:

$$\begin{aligned}
x^*(t) - x(t+nT) &= M(t)x^*(t) + q(t) - x(t+nT) \\
&= M(t)x^*(t) + q(t) - [M(t)x(t+(n-1)T) - q(t)] \\
&= M(t)[x^*(t) - x(t+(n-1)T)] \\
&\quad \vdots \\
&= M^n(t)[x^*(t) - x(t)],
\end{aligned}$$

for all $n \in \mathbb{N}$.

Remark 2.3 *In the case of $f(t) \equiv 0$, also $q(t) \equiv 0$. Using Lemma 2.2 we have:*

$$x(t + T) = M(t)x(t) \quad \text{for} \quad \dot{x} = A(t)x$$

or:

$$X(t + T) = M(t)X(t) \quad \Leftrightarrow \quad M(t) = X(t + T)X^{-1}(t),$$

where $X(t) \in \mathbb{R}^{n \times n}$ is the solution of the matrix equation:

$$\begin{cases} \dot{X} = AX \\ X(0) = E \end{cases}$$

On the other hand we have:

$$q(t) = x(t + T) - M(t)x(t),$$

where x is a solution of $\dot{x} = A(t)x + f(t)$.

Form the last remark follows a principle way to compute $M(t)$ and $q(t)$ numerically.

After the preparation we can start the analysis of the nonlinear autonomous differential equation (1) with the typical phenomena described in section 1. According to the properties a) and b) in section 1, the solution z can be represented as a periodic function $\bar{x}$ and a very small correction x for a large time window. We note that

$$z = \bar{x} + x.$$

Therefore we have the approximation:

$$\begin{aligned} \dot{x} + \dot{\bar{x}} &= \dot{z} \\ &= g(z) \\ &= g(\bar{x} + x) \\ &\approx g(\bar{x}) + Dg(\bar{x})x, \end{aligned}$$

because $\|x(t)\|$ is very small for a large time window. We now define

$$f(t) \stackrel{\text{def}}{=} g(\bar{x}(t)) - \dot{\bar{x}}(t) \qquad A(t) \stackrel{\text{def}}{=} Dg(\bar{x}(t)).$$

In this way, the approximation is transformed into the linear periodical system whose solution properties have already been discussed:

$$\begin{aligned} \dot{x} &= Dg(\bar{x}(t))x + g(\bar{x}) - \dot{\bar{x}} \\ &= A(t)x + f(t), \end{aligned}$$

This procedure will be called *periodic linearization*.

Remark 2.4 *The matrix $A(t) = Dg(\bar{x}(t))$ describes the sensitivity of the autonomous equation $\dot{z} = g(z)$ with respect to the periodic approximation $\bar{x}$. The asymptotic behavior near the curve is given by this matrix.*

The dynamics of a periodic linear system near a periodic orbit are characterized in Lemma 2.2 and Remark 2.2. Using the periodic linearization technique, the transient behavior of an oscillator is also given approximately in a time window $[t_a, t_a + mT]$, if $\|x(t)\| \ll 1$ $\forall t \in [t_0, t_0 + mT]$. In most applications, we have exactly this case because of property a) in section 1. Furthermore, the stranger this property the greater the time window can be (i.e. larger m). The following remark points out the relation between the periodic linearization and the theorem of Poincaré:

Remark 2.5 *Let $\bar{x} = z$ be a periodic solution of $\dot{z} = g(z)$. In this case we have $f(t) \equiv 0$. Furthermore, let 1 be an algebraic simple eigenvalue of $M(t)$ (cf. remark 2.1). Then we have:*

- *$M(t_a)|_H$ is the linearization of a Poincaré mapping on the fixed point $\bar{x}(t)$, where*

$$H := \bigoplus_{\lambda \in \sigma(M(t_a)) \backslash \{1\}} \ker([\lambda E - M(t_a)]^{k_\lambda})$$

   (k_λ is the algebraic multiple of eigenvalue λ),

- *Assume $|(\sigma(M(t_a))\backslash\{1\})| < 1$. Then we have the asymptotic stability of the periodical solution $\bar{x} = z$,*

for all $t_a \in \mathbb{R}$.

3   An Algorithm for Periodic Linearization and Examples

The monodromy matrix $M(t)$ and the periodic function $q(t)$ can be computed by means of remark 2.3. For a numerical implementation of the periodic linearization technique, the periodic approximation $\bar{x}$ for the solution z must be treated. By reason of property c) in section 1 we will assume that:

$$\bar{x}(t) = a \cos \omega t + b \sin \omega t + c$$

for the approximation in the time window $[t_a, t_a + T]$, where $\omega = 2\pi/T$, $a, b, c \in \mathbb{R}^n$. Furthermore, we will assume that the approximation via periodic linearization is exactly enough in the time window $[t_a, t_a + mT]$ and we now need a new periodic approximation

$$\bar{x}'(t) = a' \cos \omega' t + b' \sin \omega' t + c'$$

in order of accuracy. We define:

$$\Delta a = a' - a, \quad \Delta b = b' - b, \quad \Delta c = c' - c, \quad \Delta \omega = \omega' - \omega$$

and $t_k = t_a + mT + k(\pi/2\omega)$, $k = 0, 1, 2, 3$. According to the linearization we have:

$$(\bar{x}'(t_k) - \bar{x}(t_k))_{k=0,1,2,3}$$

$$\approx \left(\frac{\partial \bar{x}(t_k)}{\partial \omega} \Delta\omega + \frac{\partial \bar{x}(t_k)}{\partial a} \Delta a + \frac{\partial \bar{x}(t_k)}{\partial b} \Delta b + \frac{\partial \bar{x}(t_k)}{\partial c} \Delta c \right)_{k=0,1,2,3}$$

$$= \left(\left. \frac{\partial \bar{x}(t_k)}{\partial \omega} \right|_{k=0,1,2,3} \quad \left. \frac{\partial \bar{x}(t_k)}{\partial a} \right|_{k=0,1,2,3} \quad \left. \frac{\partial \bar{x}(t_k)}{\partial b} \right|_{k=0,1,2,3} \quad \left. \frac{\partial \bar{x}(t_k)}{\partial c} \right|_{k=0,1,2,3} \right) \begin{pmatrix} \Delta\omega \\ \Delta a \\ \Delta b \\ \Delta c \end{pmatrix}$$

$$= \begin{pmatrix} 0 & E & 0 & E \\ a & 0 & E & E \\ 2b & -E & 0 & E \\ -3a & 0 & -E & E \end{pmatrix} \begin{pmatrix} \alpha \\ \Delta a \\ \Delta b \\ \Delta c \end{pmatrix},$$

where $\alpha = -\pi \Delta\omega/2\omega$.

$\bar{x}'$ is the periodic approximation of $z = \bar{x} + x$ for $t \in [t_a + mT, t_a + (m+1)T]$. Therefor we have the following least-square problem to determine the parameters $\Delta a, \Delta b, \Delta c$ and α for $\Delta\omega$:

$$\begin{pmatrix} 0 & E & 0 & E \\ a & 0 & E & E \\ 2b & -E & 0 & E \\ -3a & 0 & -E & E \end{pmatrix} \begin{pmatrix} \alpha \\ \Delta a \\ \Delta b \\ \Delta c \end{pmatrix} \overset{!}{=} \begin{pmatrix} x^0 \\ x^1 \\ x^2 \\ x^3 \end{pmatrix} \overset{\text{def}}{=} \begin{pmatrix} x(t_0) \\ x(t_1) \\ x(t_2) \\ x(t_3) \end{pmatrix}.$$

The optimum solution of this problem is obtained by solving the following normal equation (see Stoer Bulirsch 1980 [2] pp. 201–202):

$$\begin{pmatrix} 10a^T a + 4b^T b & -2b^T & 4a^T & 2(b-a)^T \\ -2b & 2E & 0 & 0 \\ 4a & 0 & 2E & 0 \\ 2(b-a) & 0 & 0 & 4E \end{pmatrix} \begin{pmatrix} \alpha \\ \Delta a \\ \Delta b \\ \Delta c \end{pmatrix} = \begin{pmatrix} a^T x^1 + 2b^T x^2 - 3a^T x^3 \\ x^0 - x^2 \\ x^1 - x^3 \\ x^0 + x^1 + x^2 + x^3 \end{pmatrix},$$

The matrix is nonsingular if and only if

$$\|a\| + \|b\| \neq 0.$$

This means that the approximation $\bar{x}$ is not trivial (e.g. $\bar{x}(t) \not\equiv const.$). In this case the equation has a unique solution. Using Schur's complement principle, it is given by:

$$\begin{cases} \begin{pmatrix} \Delta a \\ \Delta b \\ \Delta c \end{pmatrix} = X_1 - \alpha X_2 \\ \\ \alpha = \dfrac{(a+b)^T \left(x^0 - x^1 + x^2 - x^3 \right)}{2 (a+b)^T (a+b)} \end{cases},$$

where

$$X_1 = \frac{1}{4} \begin{pmatrix} 2(x^0 - x^2) \\ 2(x^1 - x^3) \\ x^0 + x^1 + x^2 + x^3 \end{pmatrix}, \quad X_2 = \frac{1}{2} \begin{pmatrix} -2b \\ 4a \\ b - a \end{pmatrix},$$

which are the solutions of:

$$\begin{pmatrix} 2E & 0 & 0 \\ 0 & 2E & 0 \\ 0 & 0 & 4E \end{pmatrix} X_1 = \begin{pmatrix} x^0 - x^2 \\ x^1 - x^3 \\ x^0 + x^1 + x^2 + x^3 \end{pmatrix},$$

and

$$\begin{pmatrix} 2E & 0 & 0 \\ 0 & 2E & 0 \\ 0 & 0 & 4E \end{pmatrix} X_2 = \begin{pmatrix} -2b \\ 4a \\ 2(b - a) \end{pmatrix}.$$

With $t'_a \stackrel{\text{def}}{=} t_a + mT$ and $\bar{x}'$ we can repeat the periodic linearization. This algorithm is shown

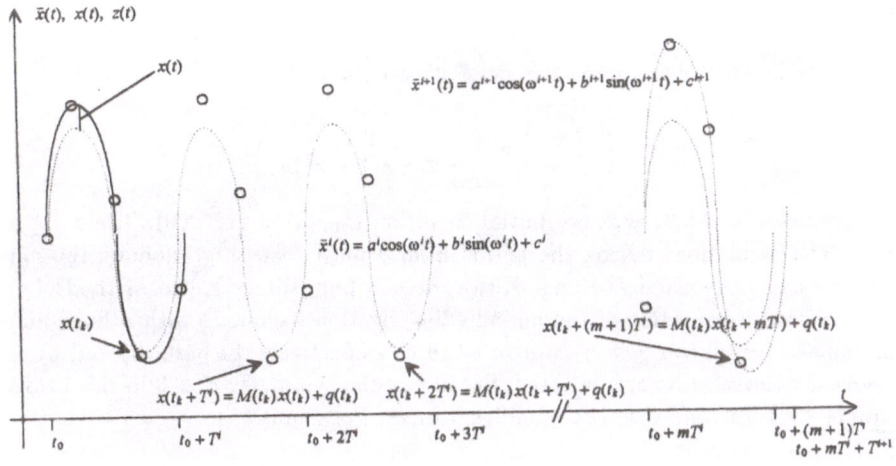

Figure 3: An algorithm based on the periodic linearization technique shown schematically in the time domain.

schematically in Fig. 3 in the time domain and in Fig. 4 in the phase space. This method is demonstrated below:

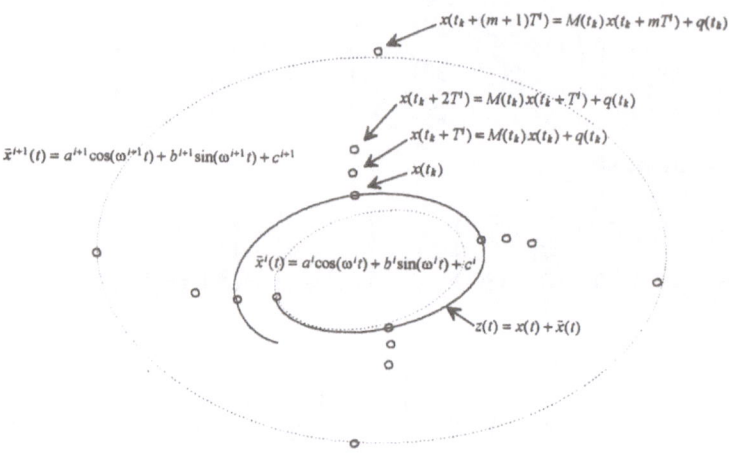

Figure 4: An algorithm based on the periodic linearization technique shown schematically in the phase space.

Example 1: The first equation is Van der Pol's oscillator:

$$\frac{d}{dt}\begin{pmatrix} x \\ y \end{pmatrix} = \begin{pmatrix} y \\ -x + \lambda(1 - x^2)y \end{pmatrix}.$$

The parameter $\lambda = 0.02$ and the initial condition $(x_0, y_0) = (10^{-5}, 0)$. There are approximately 250 oscillations during the period from 0 until 1500. Fig. 5 shows the numerical results by using the classical Runge-Kutta method (e.g. Runge-Kutte method of order 4) a) and the implementation of the periodic linearization technique with a fixed number of approximation cycles $m = 5$ b) and $m = 10$ c) respectively. In parts b) and c) only the envelopes of the solution are depicted. Because of the symmetry $c^i \equiv 0$ in this exsample.

Example 2: The second case is a modified Van der Pol's equation:

$$\frac{d}{dt}\begin{pmatrix} x \\ y \\ z \end{pmatrix} = \begin{pmatrix} y - z \\ -x + \lambda(1 - x^2)(y - z) \\ \alpha(2 - z)z \end{pmatrix}.$$

The parameters are $\alpha = 0.035$, $\lambda = 0.01$ and the initial condition is $(x_0, y_0, z_0) = (1.3 \cdot 10^{-5}, 0, 10^{-5})$. In this case, the parameters c^i's in the periodic approximations z^i's are not constant. The numerical results are shown in Fig. 6. The same methods are used as in example 1. The time window $[0, 3000]$ comprises approximately 500 oscillations.

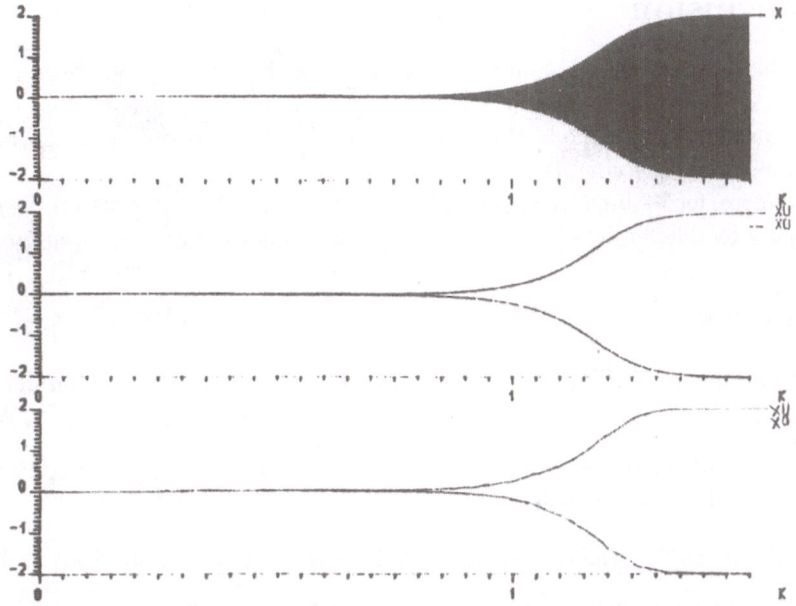

Figure 5: Computation results for example 1 of Von der Pol's equation a) by using the classical Runge-Kutta method and the periodic linearization technique $m = 5$ b) and $m = 10$ c) respectively

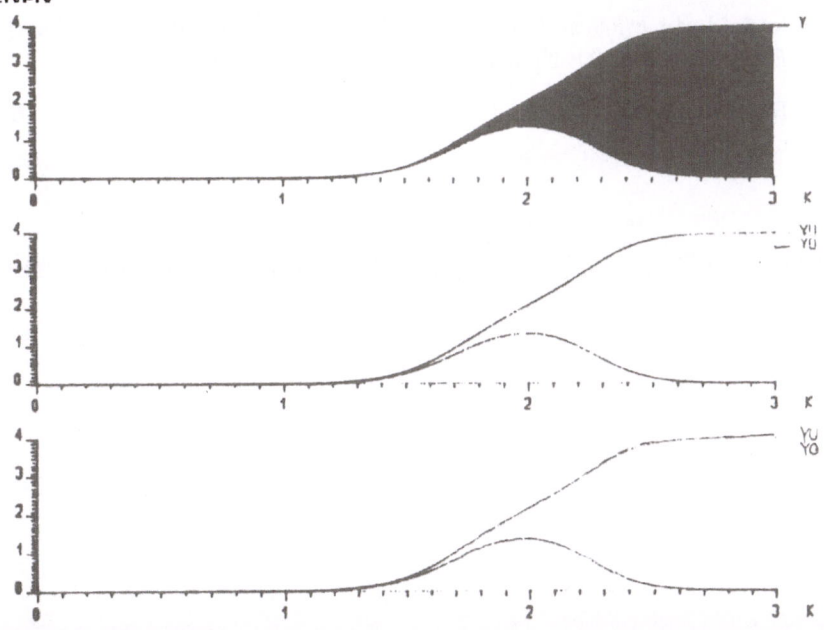

Figure 6: Computation results for the second example, the modified Von der Pol's equation. The same algorithms as in example 1 are used in a), b) and c).

4 Conclusion

The periodic linearization technique may be interpreted as an explicit Euler method in Banach space. The idea of periodic linearization can also be applied to other techniques such as linear multistep methods, Runge-Kutta methods etc.. It will be of great interest for the analysis of oscillator circuits.

Furthermore, for its broader application in industrial practice, it is useful to generalize this technique for differential algebraic equations and to improve the periodic approximation.

References

[1] **H. Amann**: *Gewöhnliche Differentialgleichungen.* Walter de Gruyter, Berlin New York 1983.

[2] **J. Stoer and R. Bulirsch** *Introduction to Numerical Analysis.* Springer-Verlag, New York Heidelberg Berlin 1980.

[3] **L. Collatz**: *Differentialgleichungen.* 7. Auflage B. G. Teubner, Stuttgart 1990.

[4] **K. Kundert, J. White, A. Sangiovanni-Vincentelli**: *An Envelope-Following Method for the Efficient Transient Simulation of Switching Power and Filter Circuits.* pp. 446–449 IEEE 1988.

[5] **L. Petzold**: *An Efficient Numerical Method for Highly Oscillatory Ordinary Differential Equations.* SIAM. J. Numer. Anal. Vol. 18, No. 3 1981.

[6] **E. Reithmeier**: *Periodic Solutions of Nonlinear Dynamical Systems* LNM 1483, Springer-Verlag 1991.

Address: Qinghua Zheng   Siemens AG, Dept.ZFE BT SE 43, 81730 Munich, Germany.

Device Simulation

International Series of Numerical Mathematics, Vol. 117, © 1994 Birkhäuser Verlag Basel

NUMERICAL SIMULATION OF THE CARRIER TRANSPORT IN SEMICONDUCTOR DEVICES ON THE BASE OF AN ENERGY MODEL

GÜNTER ALBINUS

The energy model is one of the phenomenological models which allow the simulation of carrier transport in semiconductor devices. It can be considered as an extension of the widely accepted drift-diffusion model which consists in the van Roosbroeck system of partial differential equations for the electrostatic potential and for the densities of electrons and holes in a semiconductor device. Solving this system under some prescribed boundary conditions and in the transient case under some initial conditions enables one to simulate the function or to analyse the operating schema of this element. In the drift-diffusion model the temperature is a constant parameter. An electric current, however, transforms electric energy into heat. In many applications the amount of this energy is negligible small, but in some situations it may be desirable to consider the temperature as an dynamic variable too. Thermoelectric effects, in particular, can not be described by means of the drift-diffusion model. H. Gajewski and co-workers [8] implemented a two-dimensional version of the drift-diffusion model in the program package TOSCA. This package has been successfully applied (cf. [4], [10],[15]) such that we have decided to extend it by including an energy model too.

In this paper this extension is introduced. Roughly speaking, the extension concerns three points:

(1) An additional equation is necessary to determine the additional dynamic variable, i.e. the energy balance equation is applied to calculate the temperature.
(2) Carrier currents are caused not only by gradients of their electrochemical potential, but by the gradients of the temperature too.
(3) In general the temperature will be considered in a larger and more heterogeneous domain as in the channel region. Boundary conditions of the third kind are chosen for the temperature.

The inclusion of conductive regions, however, will not be discussed in this paper, but it remains open for a time when some experience with the inclusion of the energy equation is available. As a phenomenological model the energy model is based on the band theory and effective mass approximation of the semiconductor and on the assumption of a local partial equilibrium. That means, in particular, the existence of thermodynamic state variables as functions of the coordinates and of time as well as the validity of some thermodynamic relations. H. Gajewski and K. Gröger [6] used thermodynamic concepts both in the mathematical analysis as well as in the numerical treatement of the drift-diffusion model. Thus we also spend some attention to thermodynamic aspects (cf. [1]).

Energy models are also used to describe hot electrons. In that case the electrons and the holes are considered as thermodynamic subsystems which have a temperature of their own, which may differ from the lattice temperature. Such models differ from that one described here mainly in the energy balance equation. The formal treatement seems to be quite similar, but difficulties may arise from the fact that the elliptic operator in the heat equation for the electron or hole gas may degenerate.

1. The Energy model

The system of equations under consideration is

$$
\begin{aligned}
-\nabla \cdot (\epsilon \nabla U) &= D + p - n \,, \\
\frac{\partial}{\partial t} n + \nabla \cdot j_n &= -R \,, \\
\frac{\partial}{\partial t} p + \nabla \cdot j_p &= -R \,, \\
\frac{\partial}{\partial t} w + \nabla \cdot j_w &= S \,.
\end{aligned}
$$

The first equation therein is the Poisson equation with a given fixed doping profile D and the dielectric permittivity ϵ of the semiconductor which may depend on the coordinates. The other three equations are continuity equations for the electrons, holes and for the energy, respectively. Thus n, p denote the carrier densities, j_n, j_p the corresponding carrier current densities, and R means the netto recombination rate. We fix a reference temperature T^* in Kelvin , e.g. $300K$, and measure the temperature in T^*. Energy quantities are measured in $k_B T^*$ where k_B denotes the Boltzmann constant. The electrostatic potential U is the usual potential ψ measured in V, but scaled according to $U = -q \cdot \psi / k_B \cdot T^*$ (q is the negative charge of the electron).

Of course, the continuity equations must be completed by state equations and current equations. The band theory and the effective mass approximation provide the state equations for the carrier densities

$$
n = c_n^* T^{3/2} \exp \frac{F_n + U - E_c}{T} \quad , \quad p = c_p^* T^{3/2} \exp(-\frac{F_p + U - E_v}{T})
$$

as a relation between the temperature, the electrostatic potential and the electrochemical potentials or quasi-Fermi levels F_n and F_p. The band edges E_c and E_v and the coefficients c_o^* which represent the power $3/2$ of the effective masses m_o (always $o = n$ or p) are considered as known material properties.

Carrier currents are caused by gradients of the electrochemical potentials and by the gradient of the temperature.

$$
\begin{aligned}
j_n &= -\mu_n n[\nabla F_n + (R_n - \frac{F_n + U - E_c}{T})\nabla T]\\
j_p &= -\mu_p p[-\nabla F_p + (R_p + \frac{F_p + U - E_v}{T})\nabla T]
\end{aligned}
$$

The carrier mobilities $\mu_o = \mu_o(., T)$ and the other "energy transport" coefficients R_o are also considered as given material laws. Such laws and the structure of the current equations, in particular the coefficient of the thermoelectric power, will be discussed below.

In the literature different energy balance equations are considered. In connection with our topic the dielectric permittivity is usually assumed to be independent of the temperature. Furthermore it is assumed that the heat capacity c_L of the lattice is independent of the electric field. Thus we consider the total energy of our system as the sum of its interior energy and of its electrostatic energy. Since the balance of the electrostatic energy is guaranteed by the Poisson equation we consider only the balance of the interior energy. Thus the energy density of the system consisting of the lattice and the carriers is given by

$$
w = \int^T c_L(y)dy + \frac{3}{2}nT + nE_c + \frac{3}{2}pT - pE_v \ .
$$

The energy current density is split up into a conductive part with coefficients of thermal conductivity and a convective part

$$
j_w = -[D_L + T(nD_n + pD_p)]\nabla T + (R_nT + E_c)j_n + (R_pT - E_v)j_p \ .
$$

The structure of the current densities reflects the Onsager symmetry relations as we will see below. The thermal conductivity D_L of the lattice dominates the heat conduction of the whole system. This is just the point mentioned above where the heat equation may degenerate if one considers an analogeous balance equation for hot electrons. The energy supply S is the Joule heating $S = (j_n - j_p) \cdot \nabla U$. In semiconductor materials the current must not have the direction of the electric field as in the Ohmic law. Thus S must not be nonnegative. Therefore the use of the name Joule heating is not generally accepted.

We do not apply the system of equations in this form, since material laws on the intrinsic carrier density n_i or the band gap $E_g = E_c - E_v$ are more usual. The densities $n = n_-(T)v$ and $p = n_+(T)w$ are factorized by

$$
n_- = n_i(c_n^*/c_p^*)^{1/2} = c_-(T)\exp(-\frac{E_g}{2T}) \ ,
$$

$$
n_+ = n_i/(c_n^*/c_p^*)^{1/2} = c_+(T)\exp(-\frac{E_g}{2T})
$$

with $c_l(T) = c_l^* T^{3/2}$ $(l = +, -)$ in such a way that

$$j_n = -n_- \mu_n(T) T \{ \nabla v + v \left[(R_n + E_g/2T) \nabla T - \nabla U \right]/T \},$$

$$j_p = -n_+ \mu_p(T) T \{ \nabla w + w \left[(R_p + E_g/2T) \nabla T + \nabla U \right]/T \}.$$

The energy balance equation is actually implemented in the following version

$$c_L \frac{\partial T}{\partial t} + \frac{3}{2} \frac{\partial}{\partial t}[(n+p)T] + \nabla \cdot \{ -[D_L(T) + T(nD_n + pD_p)]\nabla T + T(R_n j_n + R_p j_p) \}$$

$$= RE_g + (j_n - j_p)\nabla U - (j_n + j_p)\nabla E_g/2.$$

Both versions of the system of equations, that one with electrochemical potentials and the band edges introduced at first and that one with functions v, w and E_g which is actually implemented, are equivalent only if the sum $E_c + E_v$ is a constant. Otherwise the electrostatic potentials in both systems are not quite the same, but the difference can be compensated by modifying the doping profile, which is a given fixed function of the coordinates.

The band gap as well as the effective masses may depend on the temperature. If the band gap depends on the temperature one should expect that the sum of the band edges also depends on the temperature. Up to now, however, we neglect such a dependence on the sum of band edges.

As a result of the band theory, of the effective mass approximation and of further simplifications which are usually made in this field the electrons and holes are described as an ideal gas, more or less, maybe of fermions. If the effective masses depend on the temperature, then not all relations which are familiar to us from the ideal gas can be applied. For instance, if we take the state equation

$$n = c_-(T) \exp(\frac{F_n + U - E_g/2}{T})$$

and choose the expression

$$f(n,T) = nT \log [n/c_-(T)] - nT$$

as the free energy of the electrons, then its interior energy becomes

$$w_n = f - T\partial_T f = nT^2 \partial_T \log c_-(T),$$

which is just $w_n = \frac{3}{2}nT$ if $c_-(T) = c_-^* T^{3/2}$. A factor $a(T)$ of the state density can at least formally be interpreted as a modification of the band gap. Thus some arbitrariness may appear in any formulation.

The boundary conditions for n, p and U will be chosen in the same way as in the drift-diffusion model, but for the temperature we will choose boundary condition of third kind

$$\nu \cdot j_w = -D_{bd}(T_{bd} - T)/d_{bd}$$

with heat conductivity D_{bd} of the boundary region of the width d_{bd} and a temperature T_{bd} on its outside. With regard to the gate contacts and the Dirichlet contacts and still other features of the device the boundary is divided into some parts for each of which individual values of D_{bd}, d_{bd} and T_{bd} can be prescribed. The thermal boundary condition suggested for the temperature neglects the convective contribution of the carrier current to the energy current in the boundary region at an electrical Dirichlet contact. This question should be investigated in connection with including conductive regions.

The system of partial differential equations which represents the energy model seems to be decisively more complicate than the drift-diffusion model. The coefficients in the continuity equations depend essentially nonlinearly on the temperature. The single equations are coupled by terms which contain not only the unknown functions, but its derivatives too. Furthermore, the lower order terms contain the gradients of the sought functions quadraticly such that the existence of solutions is a mathematical question for which there is no ready-made answer in the theory of partial differential equations.

2. Transport coefficients

The transport coefficients are considered as given material laws. They may depend on the coordinates and on the state of the system, mainly on the temperature. Usual mobility models like the Caughey-Thomas formula are used, but for the beginning we consider only mobilities which do not depend on the electric field. In the simplest case the mobilities are potential laws

$$\mu_o(T) = \mu_{o\beta}T^\beta \quad (\text{e.g. } \beta = -1/2).$$

Such potential laws arise from modeling just one carrier-lattice scattering process using further simplifying arguments. In the case of Boltzmann statistics one obtains on that way

$$R_{o\beta} = 2.5 + \beta \quad \text{and} \quad D_{o\beta} = [(3.5 + \beta)(2.5 + \beta) - (2.5 + \beta)^2]\mu_o = (2.5 + \beta)\mu_o .$$

More realistic mobility models are linear superpositions of several potential laws with different exponents β_i or superpositions of potential laws according to the Mathiessen rule. In such cases we suggest weighted averages of the $2.5 + \beta_i$'s or $(2.5 + \beta_i)\mu_o^i$'s for R_o or D_o, respectively. For widely accepted mobility models, however, rather unrealistic coefficients R_o or D_o arise, if the averages are taken over all contributions. Usually only two exponents α and β dominate the total mobility in a realistic temperature region. In such a case $\mu_o(T) = \mu_{o\alpha}T^\alpha + \mu_{o\beta}T^\beta + ...$ we take for the other transport coefficients weighted averages

$$R_o = 2.5 + \frac{\alpha \mu_{o\alpha} T^\alpha + \beta \mu_{o\beta} T^\beta}{\mu_o(T)} \; ,$$

$$D_o = R_{o\alpha} \mu_{o\alpha} T^\alpha + R_{o\beta} \mu_{o\beta} T^\beta + (R_{o\alpha} - R_{o\beta})^2 \mu_{o\alpha} T^\alpha \mu_{o\beta} T^\beta / \mu_o(T) \; .$$

Of course, we are not specialists for such physical questions. Thus we are fairly open for more qualified suggestions.

With regard to theoretical results in the case of potential laws we generally expect that the mobilities μ_o and the coefficients D_o of heat conduction heavily depend on the temperature, meanwhile the transport coefficients R_o only gently depend on the temperature. If material laws for the transport coefficients are derived theoretically from special solutions of the Boltzmann equation , in particular, if one takes the Fermi-Dirac statistics instead of the Boltzmann statistics, then the transport coefficients depend not only on the temperature but also on the other state variables in the special combination $(F_n + U - E_c)/T$ or $(F_p + U - E_v)/T$. We do not know, whether such a remarkable behavior is realistic or whether it only reflects the special ansatz made in the investigations.

3. NUMERICAL TREATMENT

There are at least two reasons for dividing each iteration step of the solving process for the energy model into two steps. In a 1st step the Poisson equation and the continuity equations for the carrier densities are solved with the old temperature, but in a 2nd step the energy balance equation is solved with the potential and the densities from the first step. One reason is the existence of the comfortable and approved frame of the program package TOSCA. The other reason is that the substantial contributions of the lattice to the total heat capacity and to the total heat conduction suggest the convergence of the two-step iteration process. Another aspect is that one will not always know in advance, whether the energy model should be used or whether the drift-diffusion model would do it too. In the proposed strategy it is easy to switch out or in the energy balance equation. The results of both the steps are compared by integral quantities

$$d_T(n, p, U, n_0, p_0, U_0) =$$

$$\int \left\{ \tfrac{1}{2}\epsilon[\nabla(U - U_0)]^2 + T\left[(n - n_0)\log(n/n_0) + (p - p_0)\log(p/p_0) \right] \right\}$$

and

$$d_{n,p,U}\,(T, T_0) = \int [\, c_L + \tfrac{3}{2}(n + p)\,](T - T_0)\log(T/T_0).$$

These expressions are derived from the free energy of the system, and they are applied to control the iteration process.

In the same way as in the nonlinear Poisson equation the dependence of the densities n and p upon the sought potential U is used to improve the iteration process by applying a

Newton-like improvement in each step, one could also use state equations in the energy balance equation and try to improve the iteration process by Newton-like modifications. In this case, however, it is not evident at all, whether such Newton-like modifications really improve the iteration process. One should also be aware that such a modification reflects some physical process. In the linear Poisson equation, e.g. , the electrostatic potential is determined for a fixed density, meanwhile in the nonlinear Poisson equation the electrostatic potential is determined for a fixed electrochemical potential.

In the numerical treatment of the drift-diffusion model the Scharfetter-Gummel difference approximation is a powerful tool for the spatial discretization of the continuity equations in two dimensions. We remember briefly to it. On the sides of the elements an initial-value problem for the onedimensional current equation $I_n = -\bar{\mu}_n(v' - v\psi')$ with a constant averaged mobility is solved for a given potential ψ. The value I_n of the constant current determines the value v_f of v in the endpoint of the side. Finally the current is expressed by the values v_i, v_f and $\Delta\psi = \psi_f - \psi_i$ by means of the Bernoulli function $b(x) = \frac{x}{\exp(x)-1}$:

$$l I_n = \bar{\mu}_n \left(v_i b(\Delta\psi) - v_f b(-\Delta\psi) \right) \qquad (l \text{ denotes the length of the side}).$$

For the energy model only a modified version of this approximaton schema is possible. The onedimensional current equation

$$I_n = -n_-(T)\, \mu_n(T)\, T \left\{ v' + v[\ (R_n + E_g/2T)T' - U'\ ]/T \right\}$$

is considered as an equation

$$I_n = -M_n(s)(v' + Q_n v)$$

with a constant

$$Q_n = \frac{1}{\bar{T}} \left[(\bar{R}_n + \frac{\bar{E}_g}{2\bar{T}}) \frac{\Delta T}{l} - \frac{\Delta U}{l} \right]$$

and a linear function $M_n(s)$ which is obtained by interpolating $n_-[s, T(s)]\, \mu_n[s, T(s)]\, T(s)$ linearly. Then we have

$$l I_n = \frac{\Delta M_n}{\log(M_{nf}/M_{ni})} \left\{ v_i\, b\left[-\frac{l Q_n}{\Delta M_n} \log\left(\frac{M_{nf}}{M_{ni}}\right) \right] - v_f\, b\left[+\frac{l Q_n}{\Delta M_n} \log\left(\frac{M_{nf}}{M_{ni}}\right) \right] \right\}$$

quite analogeously. For constant temperature the modification agrees with the Scharfetter-Gummel difference approximation. Tha approximation of the coefficients in the differential equation is justified to a certain degree by the assumption that R_n only gently varies with the temperature and by the finite-element philosophy according to which the gradients of U and T are constant on the elements.

The modification is applied once more to the spatial discretization of the energy balance equation in which the temperature T_0 of the preceding iteration step is partially used. In the onedimensional energy current equation

$$j_w = -\left[D_L(T_0) + T_0(nD_n(T_0) + pD_p(T_0))\right] T' + (R_n j_n + R_p j_p) T$$

the coefficient $D_L[T_0(s)] + T_0(s)[\, n(s)D_n(T_0(s)) + p(s)D_p(T_0(s))\,]$ of T' is substituted by its linear interpolation formula $M(s)$, meanwhile the coefficient $Q = (R_n I_n + R_p I_p) l$ of T is constant in a natural way. As a result we get also

$$lI_w = \frac{\Delta M}{\log(M_f/M_i)} \left\{ T_i\, b\left[\, \frac{lQ}{\Delta M}\log\left(\frac{M_f}{M_i}\right)\right] - T_f\, b\left[-\frac{lQ}{\Delta M}\log\left(\frac{M_f}{M_i}\right)\right]\right\}\,.$$

4. FREE ENERGY

In the drift-diffusion model the free energy has turned out to be a very useful quantity. Gajewski and Gröger applied it in the analysis of the transient initial-boundary value problem and Gajewski also used it to control the step width in the time discretization. As a functional of the carrier densities the free energy is a convex functional and a thermodynamic potential. Therefore it is not surprising that the free energy is a very nice quantity. On the other hand, as an integral quantity it is not too sensitive to local deviations of either the carrier densities or the electric field.

In the case of variable temperature we also expect the free energy to be a competent quantity for comparing different approximative solutions. We omit here some technical details which arise from the fact that U, n and p are not independent of another. The technical details consist in describing the free energy as a functional and in defining partial derivatives of functionals. Let us describe the free energy formally by its density

$$f(.,n,p,T) = f_L(.,T) + nT\log(n/n_-) - nT + pT\,log(p/n_+) - pT$$

$$+\ electrostatic\ energy$$

depending on n, p and T. This free energy has still the essential property that the first partial derivatives in a funcional sense are physical quantities, e. g.

$$\partial_n f = T\frac{F_n + U}{T} + nT\frac{1}{n} - T - U = F_n.$$

The last summand $-U$ comes from the electrostatic energy, and it is just here where we have to take regard to both the self-consistency of the potential and the functional character of the derivative. In the same way $\partial_p f = -F_p$ (the sign of F_p depends only on our notation; cf. also the next section). The negative value of the partial derivative

$$\partial_T f = \partial_T f_L + n \log(n/n_-) - nT \, \partial_T n_-/n_- - n$$
$$+ \, p \log(p/n_+) - pT \, \partial_T n_+/n_+ - p$$

with

$$T \frac{\partial_T \, n_l(T)}{n_l(T)} = \frac{3}{2} + \frac{E_g}{2T} - \frac{1}{2}\partial_T E_g$$

should be the entropy density. The terms like $n \log(n/n_-) - \frac{5}{2}n$ are quite familiar, but the terms containing the band gap are due to the modifications mentioned in section 1.

We have not defined yet the free energy of the lattice. We could take e.g. the Debye interpolating formula. We remember, however, that the interior energy is the Legendre transform with respect to the temperature, i.e.

$$w_L = u_L(s) = f_L(T) + Ts = f_L(T) - T \, \partial_T f_L(T) \, .$$

With regard to $du_L = c_L dT$ one obtains

$$-T\partial_T^2 f_L(T) = c_L \, .$$

The matrix of the second-order partial derivatives of the free energy (as a function of n, p and T) is then

$$d^2 f(n,p,T) = \begin{pmatrix} \frac{T}{n} & 0 & \log\frac{n}{n_-} - \frac{5}{2} - \frac{E_g}{2T} - \frac{1}{2}\partial_T E_g \\ 0 & \frac{T}{p} & \log\frac{p}{n_+} - \frac{5}{2} - \frac{E_g}{2T} - \frac{1}{2}\partial_T E_g \\ \ldots & \ldots & -\frac{1}{T}[c_L + \frac{3}{2}(n+p)] - \frac{n+p}{T}[\frac{E_g}{2T}(1-\frac{1}{T}) - \partial_T E_g] \end{pmatrix} .$$

It seems to be realistic to assume that the term c_L/T in $\partial_T^2 f$ dominates the indefinite terms, i.e. that $\partial_T^2 f < 0$. Then it follows from the Routh-Hurwitz criteria that the matrix $d^2 f$ is indefinite. Nevertheless the upper left 2×2 matrix and the lower right corner are definite. They provide the distances

$$d_T(n,p,U,n_0,p_0,U_0) =$$
$$\int \{ \, \frac{1}{2}\epsilon[\nabla(U - U_0)]^2 + T \, [\, (n - n_0)\log(n/n_0) + (p - p_0)\log(p/p_0) \,] \, \}$$

and

$$d_{n,p,U} \, (T,T_0) = \int [\, c_L + \frac{3}{2}(n+p) \,](T - T_0)\log(T/T_0)$$

which we use to measure the distance between states and to control the iteration process.

5. AN ALTERNATIVE SYSTEM

Since the free energy is not a convex functional the question arises, whether another thermodynamic potential has this property. Let us discuss the topic for the version of the energy model given in the beginning. Furthermore, let us consider the electrostatic potential as the potential of an exterior field, i.e. let us neglect the self-consistent coupling of U, n and p. The last assumption corresponds to the spirit of the Gummel iteration process. A consequence of this assumption is that in the following densities of quantities like the free energy and partial derivatives of them with respect to other functions like n or T can be taken in the usual sense instead of quantities which are functionals and their partial derivatives in a funcional-analytic meaning, respectively. Then the system has the free energy

$$f_U(n,p,T) \; = \; f_L(T) + \; nT \log(n/c_n T^{3/2}) \; - nT \; + \; n(E_c - U)$$

$$+ \; pT \log(p/c_p T^{3/2}) \; - pT \; - \; p(E_v - U)$$

where the coefficients c_o are assumed to be independent of T. The partial derivatives of the function f_U with respect to n or p are the electrochemical potentials F_n and $-F_p$ and the partial derivative with respect to the temperature is the negative entropy,

$$\partial_T f_U \; = \; \partial_T f_L + \; n \log(n/c_n T^{3/2}) \; - \; \frac{5}{2}n$$

$$+ \; p \log(p/c_p T^{3/2}) \; - \; \frac{5}{2}p \; = \; -s \;.$$

As in the preceding section the matrix of second-order partial derivatives of the free energy f_U is indefinite. The Legendre transform of the function f_U with respect to the densities n and p is given by

$$g_U(F_n, F_p, T) = \; f_U \; - \; nF_n \; + \; pF_p \; = \; f_L(T) \; - \; nT \; - \; pT$$

$$= \; f_L(T) \; - \; c_n T^{3/2} \exp \frac{F_n + U - E_c}{T} \; - \; c_p T^{3/2} \exp(-\frac{F_p + U - E_v}{T}) \;.$$

With the notation

$$z_n = \frac{F_n + U - E_c}{T} \;,\; z_p = \frac{F_p + U - E_v}{T} \;,\; X = z_n - \frac{3}{2} \;,\; Y = z_p - \frac{3}{2}$$

the matrix of the second-order partial derivatives of the function g_U is

$$d^2 g_U(F_n, F_p, T) = \begin{pmatrix} -\frac{n}{T} & 0 & \frac{n}{T}(z_n - \frac{3}{2}) \\ 0 & -\frac{p}{T} & \frac{p}{T}(z_p - \frac{3}{2}) \\ \cdots & \cdots & -\frac{1}{T}\left[c_L + n(\frac{3}{2} + X^2) + p(\frac{3}{2} + Y^2)\right] \end{pmatrix}.$$

Applying the Routh–Hurwitz theorem to the characteristic polynomial of this matrix we see that the matrix is definite in semiconductor regions where the densities n and p do not vanish. The matrix is degenerated, of course, where at least one of the densities n or p vanishes, in particular, in the oxid region. For a self-consistently coupled electrostatic potential, however, the matrix is not definite. Nevertheless the corresponding thermo-dynamic potential G is quite a nice quantity. Its conjugate functional is the energy as a function of the densities of the carriers and of the entropy. Thus we are lead to ask for the continuity equation of the entropy. This continuity equation arises, if we differentiate the expression for the entropy density s given above with respect to the time and if we use the other continuity equations of the energy model. The result is

$$\dot{s} + \nabla j_s = -\frac{1}{T}(j_n \cdot \nabla F_n - j_p \cdot \nabla F_p + j_s \cdot \nabla T)$$
$$+ R(z_n - z_p)$$

with the entropy currence density

$$j_s = -(D_L/T + nD_n + pD_p)\nabla T +$$
$$+ (R_n - z_n)j_n + (R_p + z_p)j_p .$$

The energy balance equation in the energy model can be substituted by the entropy balance equation. On this way we obtain an euivalent description of the energy model. The equivalent system consists in the Poisson equation for the potential U and in the system of three balance equations for n, p and s . The three balance equations have a very nice symmetric form, in particular, if we write them in matrix form:

$$\begin{pmatrix} \dot{n} + \nabla \cdot j_n \\ \dot{p} + \nabla \cdot j_p \\ \dot{s} + \nabla \cdot j_s \end{pmatrix} = \begin{pmatrix} - & R \\ - & R \\ & R(z_n - z_p) + \sigma \end{pmatrix}.$$

With the abbreviation

$$E = n\mu_n(R_n - z_n)^2 + p\mu_p(R_p + z_p)^2$$

the corresponding current vector is written

$$
\begin{pmatrix} j_n \\ j_p \\ j_s \end{pmatrix} = - \begin{pmatrix} n\mu_n & 0 & n\mu_n(R_n - z_n) \\ 0 & p\mu_p & p\mu_p(R_p + z_p) \\ \dots & \dots & \frac{D_L}{T} + nD_n + pD_p + E \end{pmatrix} \cdot \begin{pmatrix} \nabla F_n \\ -\nabla F_p \\ \nabla T \end{pmatrix}
$$

with a symmetric matrix. The entropy production σ has the form

$$
T\sigma = -(\, j_n \cdot \nabla F_n - j_p \cdot \nabla F_p + j_s \cdot \nabla T \,)
$$

$$
= (\, \nabla F_n \,,\, -\nabla F_p \,,\, \nabla T \,) \times
$$

$$
\begin{pmatrix} n\mu_n & 0 & n\mu_n(R_n - z_n) \\ 0 & p\mu_p & p\mu_p(R_p + z_p) \\ \dots & \dots & \frac{D_L}{T} + nD_n + pD_p + E \end{pmatrix} \cdot \begin{pmatrix} \nabla F_n \\ -\nabla F_p \\ \nabla T \end{pmatrix}
$$

$$
= \frac{j_n^2}{\mu_n n} + \frac{j_p^2}{\mu_p p} + \frac{(D\nabla T)^2}{D}
$$

with $D = \frac{D_L}{T} + nD_n + pD_p$. The system current equations reveals the symmetry of Onsager's relations. The right-hand side of the entropy balance equation can be naturally interpreted as entropy production. In the matrix form of the three continuity equations the divergence term is a definite symmetric operator. This model suggests an iteration process, which is similiar to Gummel's iteration process for the drift-diffusion model and which differs from that one described above. Thus we consider it worthwhile to study also this version of the energy model.

The assumption that the potenntial U is independent of n and p influences only the interpretation of f_U as free energy density, but the equivalence of the system of euations derived in this section with that one introduced at the beginning does not depend on this assumption. Furthermore, the equivalence remains also true if we substitute the state equations for n, p and the interior energy $\frac{3}{2}T(n + p)$ by the corresponding state equations which hold in the case of Fermi-Dirac statistics for electrons and holes. In the case of Fermi-Dirac statistics, however, the definiteness or indefiniteness of $d^2 g_U$ or $d^2 f_U$, respectively, depend on some relations between Fermi integrals of different degrees, and we do not know, whether such relations are proved or not.

REFERENCES

1. Albinus, G.: *Thermodynamic Aspects of the Energy model of hot carrier Transport in semiconductor Devices*. In eds. W. Fichtner, D. Aemmer: SISDEP (Zurich 1991). Vol. 4, 493–498.
2. Bass, F.G., Ju. G. Gurevič: *Gorjačie Élektrony i sil'nye élektromagnitnye volny v plazme poluprovodnikov i gazovogo razrjada*. Moskva 1975.

3. Birjukova, L. Ju., V.A. Nikolaeva, V.I. Ryžij, V.N. Četveruškin: *Algoritmy kvazigidrodinamičeskoj modeli dlja rasčeta processov v élektronnoj plazme submikronnych poluprovodnikovych struktur* Matem. Modelirovanie **1** (1989) 5, 11–22.

4. Erlebach, A., A. Hürrich, R. Stephan, U. Todt: *Einsatz der Simulatoren DIOS und TOSCA im VLSI-Bereich.* In eds H Gajewski, P Deuflhard, P. A. Markowich. NUMSIM '91 Konrad-Zuse-Zentrum für Informationstechnik Berlin Technical Report TR 91-8, 96–105

5. Gajewski, H.: *Analysis und Numerik des Ladungsträgertransports in Halbleitern* To appear in GAMM Mitt. 1993.

6. Gajewski, H , K. Gröger: *Semiconductor Equations for variable Mobilities Based on BOLTZMANN Statistics or FERMI-DIRAC Statistics.* Math. Nachr. **140** (1989), 7–36

7. Gajewski, H., K. Gärtner *On the Iterative Solution of van Roosbroeck's Equations* Z angew Math. Mech. **72** (1992), 19–28.

8. Gajewski, H., u.a. *TOSCA Handbuch (TwO-dimensional SemiConductor Analysis package)* Berlin 1987.

9. Gnudi, A., F. Odeh: *An efficient discretization scheme for the energy-continuity equation in semiconductors.* In eds. G. Baccarani, M. Rudan: SISDEP (Bologna 1988) Vol 3, 387–9

10. Heinemann, B., R. Richter· *TOSCA-Simulations of Silicon Devices.* In eds H Gajewski, P Deuflhard, P. A Markowich· NUMSIM '91 Konrad-Zuse-Zentrum für Informationstechnik Berlin Technical Report TR 91-8, 90–96

11. Landau, L.D., E.M. Lifschitz: *Lehrbuch der theoretischen Physik* Bd V , Berlin 1979

12. McAndrew, C C., E.L. Heasell, K Singhal. *A comprhensive transport model for semiconductor device simulation* Semicond. Sci. Technol **2** (1987) 10, 643–8 and *Modelling thermal effects in energy models of carrier transport in semiconductors.* Semicond Sci Technol. **3** (1988) 3, 758–765

13. Selberherr, S.: *Analysis and Simulation of Semiconductor Devices* Wien 1984.

14. Wachutka, G.: *Rigorous thermodynamic treatment of heat generation and conduction in semiconductor device modelling* In eds. G Baccarani, M. Rudan: SISDEP (Bologna 1988). Vol. 3, 83–95

15. Wünsche, H.-J., H. Wenzel, U. Bandelow et al.: *2D Modelling of Distributed Feedback Semiconductor Lasers.* In eds. W. Fichtner, D Aemmer· SISDEP (Zurich 1991) Vol 4, 65–69.

INSTITUT FÜR ANGEWANDTE ANALYSIS UND STOCHASTIK, HAUSVOGTEIPLATZ 5-7, D-O-1086 BERLIN, GERMANY

E-mail address: albinus@iaas-berlin dbp.de

International Series of Numerical Mathematics, Vol. 117, © 1994 Birkhäuser Verlag Basel

On uniqueness of solutions to the drift–diffusion–model of semiconductor devices

H. Gajewski

Institut für Angewandte Analysis und Stochastik, Mohrenstrasse 39, O–1086 Berlin, Germany

Abstract. We prove a uniqueness result for the drift– diffusion–model of semiconductor devices under weak regularity assumptions. Our proof rests on the convexity of the free energy functional and uses a new concavity argument.

Key words: drift–diffusion–model, device equations, uniqueness of transient solutions, energy functional, Lyapunov function.

Introduction

Since the drift–diffusion–model has been established by van Roosbroeck in 1950 [17] it has proven to be of fundamental significance for the mathematical describing and numerical simulation of carrier transport in semiconductor devices. The drift–diffusion–model is formed by a coupled system of a Poisson equation for the electrostatic potential and continuity equations for positive and negative carriers (holes and electrons). The existence of solutions to these device equations has been proved under natural assumptions (comp. [13, 4]). With respect to the uniqueness of solutions the situation is more complex. Whereas steady state solutions to the device equations in general cannot exspected to be unique by physical reasons (comp. [12, 11]), uniqueness of transient solutions should hold in principle, but has been proved only under unpleasant restrictions up to now.

The first uniqueness result for the transient device equations was published by Mock [12] under strong regularity assumptions, excluding nonsmooth domains as well as mixed boundary conditions. More recently Gajewski & Gröger [3] have shown weak solutions to be unique, provided the semiconductor obeys Boltzmann statistics. Concerning the physically more realistic Fermi–Dirac statistics, these authors [4] proved uniqueness of solutions having bounded gradients. In a forthcoming paper Gröger & Rehberg [9] could relax this regularity condition essentially in the case of two space dimensions.

The main result of this paper is a uniqueness result which rests on a new concavity argument involving density and conductivity as functions of the chemical potentials of electrons and holes. This argument allows to remove all regularity assumptions except for a mild L_q–condition with respect to the gradient of the electrostatic potential. Our idea of proof is based on the convexity of the free energy functional, which induces a natural metric in the space of solutions to the device equations. This approach has been discussed in [1] in a more abstract way.

The plan of the paper is as follows: First we formulate the complete initial boundary value problem for the drift–diffusion–model. In Section 2 suitable function spaces, definitions and assumptions are introduced. The energy functional is discussed in Section 3. Section 4 is devoted to the proof of the uniqueness result. Finally, the assumptions are verified in Section 5.

1   The initial boundary value problem

Let be: $S = (0, T)$ a bounded time interval, $\Omega \subset \mathbb{R}^N, 1 \leq N \leq 3$, a bounded Lipschitzian domain and $Q = S \times \Omega$. We suppose that $\partial\Omega = \Gamma_D \cup \Gamma \cup \Gamma_N$, where the subsets $\Gamma_D, \Gamma, \Gamma_N$ are pairwise disjoint and Γ_D is closed and possesses positive surface measure. Moreover, $\nu(x_0)$ denotes the outer unit normal at $x_0 \in \partial\Omega$.

The transient carrier transport in a semiconductor occupying the domain Ω can be described by the following system of partial differential equations:

$$-\nabla(\cdot\varepsilon\nabla v_0) = D + u_p - u_n, \qquad (1.1)$$

$$\frac{\partial u_0}{\partial t} + e_i \nabla \cdot J_i + r_i = 0, \quad i = n, p, \quad e_n = -1, e_p = +1. \qquad (1.2)$$

Here and in what follows the subscript i stands for electrons $i = n$ (negative) resp. holes $i = p$ (positive). The physical meaning of the other quantities is:

v_0 – electrostatic potential,

v_i – chemical potential,

$u_i = f_i(v_i)$ – carrier density,

$J_i = J_i(x, v_i, z_i), z_i = \nabla(v_0 + e_i v_i)$ – current density,

ε – dielectric permittivity,

D – density of impurities,

$r_i = r_i(x, v), v = (v_l), l = 0, n, p$ – recombination/generation rate.

From the functional analytic point of view it turns out to be advantageous to replace the Poisson equation (1.1) by the current conservation equation

$$\nabla \cdot J + r_0 = 0, \quad J = -\varepsilon \nabla \frac{\partial v_0}{\partial t} + J_n + J_p, \quad r_0 = r_p - r_n - D_t, \tag{1.3}$$

which results from (1.1) after differentiating with respect to time and eliminating u_{it} from (1.2).

We complete (1.2)–(1.3) by initial conditions

$$v_l(0, \cdot) = v_{l0} \quad on \ \Omega \tag{1.4}$$

and boundary conditions

$$
\begin{aligned}
v_l &= v_{l\Gamma} \quad on \ S \times \Gamma_D, \\
\nu \cdot J &= k_0 u_{0t} + \alpha, \quad v_i = v_{i\Gamma} \quad on \ S \times \Gamma, \\
-\nu \cdot \nabla v_{0t} &= k_0 u_{0t}, \quad \nu \cdot J_i = 0 \quad on \ S \times \Gamma_N,
\end{aligned}
\tag{1.5}
$$

where the functions $v_{l\Gamma}$ represent given boundary values and

$$
\begin{aligned}
u_0 &= v_0 - v_{0\Gamma}, \\
\alpha &= exp(-k_1 t)(k_2 + k_3 \int_0^t exp(k_1 s) u_0(s) \, ds).
\end{aligned}
$$

Remark 1.1 *From the physical point of view Γ_D models Ohmic contacts. The condition on Γ describes interaction of the semiconductor with an outer electric circuit formed by inductivity ($L = \frac{1}{k_3}$), resistance ($R = \frac{k_1}{k_3}$) and parallely switched capacity ($C = k_0$). Finally, Γ_N can be interpreted as interface between semiconductor and an isolator.*

2 Definitions and assumptions

We denote by $L_q, (\| \cdot \|_q, \| \cdot \| = \| \cdot \|_2), 1 \le q \le \infty, H^1, H^{-1}$ the usual spaces of functions defined on Ω (comp. [5, 6, 18]). Additionally we introduce

$$
\begin{aligned}
V_0 &= \{h \in H^1, h = 0 \ on \ \Gamma_D\}, \quad V = \{h \in V_0, \ h = 0 \ on \ \Gamma\}, \\
H &= \{v = (v_0, v_i), \ v_0 \in V_0, \ v_i \in L_2\}, \quad \|v\|_H^2 = \||\nabla v_0|\|^2 + \|v_n\|^2 + \|v_p\|^2.
\end{aligned}
$$

If X is a Banach space, $L_q(S; X)$ is the space of Bochner measurable functions $t \to h(t) \in X$ such that

$$\int_0^T \|h(t)\|_X^q \, dt < \infty.$$

Definition 2.1 *A function vector* $v = (v_l), l = 0, n, p,$ *is called (weak) solution of* (1.2)–(1.5), *if:*

(D1) $v_l \in L_\infty(Q) \cap L_2(S; H^1)$, $v_{0t} \in L_2(S; H^1)$, $v_{it} \in L_2(S; H^{-1})$;

(D2) $v_l(0) = v_{l0}$;

(D3) $v_0 = v_{0\Gamma}$ *on* $S \times \Gamma_D$, $v_i = v_{i\Gamma}$,   *on* $S \times (\Gamma_D \cup \Gamma)$;

(D4) *for almost every* $t \in S$ *it holds*

$$\int (J \cdot \nabla h + r_0 h)\, d\Omega = \int (k_0 u_{0t} + \alpha) h\, d\Gamma + \int k u_{0t} h\, d\Gamma_N, \quad \forall h \in V_0,$$

$$\int [(u_{it} + r_i) h - e_i J_i \cdot \nabla h]\, d\Omega = 0, \quad \forall h \in V.$$

Remark 2.1 *The integral involving* u_{it} *has to be understood in the sense of distributions, that means as dual pairing between* H^{-1} *and* V.

Definition 2.2 $\beta \in (\Omega \times I\!\!R^N \to I\!\!R^N)$ *is called Carathéodory function if:*

(*i*) *the function* $x \to \beta(x, z)$ *is measurable for every* $z \in I\!\!R^N$;

(*ii*) *the function* $z \to \beta(x, z)$ *is continuous in* $I\!\!R^N$ *for almost every* $x \in \Omega$;

(*iii*) $|\beta(x, z)| \leq C(\beta_0(x) + |z|)$, $\beta_0 \in L_2$

We suppose the current densities J_i to have the representation

$$J_i(x, s, z) = -(a_i(s)z + b_i(s)\beta_i(x, z) + f_i'(s)\gamma_i(x, z)). \tag{2.1}$$

Moreover, we suppose constants $K, \kappa > 0$ to exist such that for a. e. $x \in \Omega$, $\forall s, \forall s_j \in I = [-K, K], \ 0 < |s_1 - s_2| \leq \kappa, \ \forall z_j \in I\!\!R^N, \ j = 1, 2,$ the following assumptions are satisfied:

(A1) $f_i \in C^1(I)$, f_i' is Lipschitzian , $f_i'(s) > 0$;

(A2) $g_i = f_i' \circ f_i^{-1}$ is uniformly concave on I such that

$$G_i \geq c|f_i(s_1) - f_i(s_2)|^2, \ c > 0,$$

where

$$G_i = 2 - g_{i1} - g_{i2}, \quad g_{ij} = \frac{g_i(f_i(s_j))}{g_{im}}, \quad g_{im} = g_i\left(\frac{f_i(s_1) + f_i(s_2)}{2}\right);$$

(A3) $a_i \in C(I)$, $a_i(s) > 0$;

(A4)

$$\rho_i(s_1, s_2) = \frac{(g_{i1}a_{i2} - g_{i2}a_{i1})^2}{8G_i a_{i1} a_{i2}} < 1, \quad a_{ij} = a_i(s_j);$$

(A5) b_i is Lipschitzian on I, $b_i(s) \geq 0$;

(A6) β_i is Carathéodory function such that

$$|\beta_i(x, z)| \leq c_\beta, \quad \beta_i(x, 0) = 0, \quad (\beta_i(x, z_1) - \beta_i(x, z_2)) \cdot (z_1 - z_2) \geq 0;$$

(A7) γ_i is Carathéodory function such that

$$\gamma_i(x, 0) = 0, \quad (\gamma_i(x, z_1) - \gamma_i(x, z_2)) \cdot (z_1 - z_2) \geq m_\gamma |z_1 - z_2|^2, \quad m_\gamma > 0;$$

(A8) $r_l(x, v)$, $l = 0, n, p$, is Carathéodory function, locally Lipschitzian with respect to v;

(A9) $\varepsilon \in L_\infty$, $\varepsilon(x) \geq \varepsilon_0 > 0$, $0 \leq k_0 \in L_\infty(\Gamma_N)$, $0 \leq k_1, k_2, k_3 \in L_\infty(\Gamma)$;

(A10) $v_{l0} \in L_\infty(Q)$, $v_{l\Gamma} \in L_\infty(Q) \cap L_2(S; H^1)$, $v_{0\Gamma t} \in L_2(S; H^1)$.

Remark 2.2 *The coefficients a_i, b_i and f'_i may be interpreted as electrical conductivities.*

Remark 2.3 *As a consequence of* (A1), (A2) *it holds*

$$G_i \geq c|s_1 - s_2|^2 \geq c|g_{i1} - g_{i2}|^2, \quad \forall s_1, \forall s_2 \in \mathbb{R}, \ |s_j| \leq K, \ 0 < |s_1 - s_2| \leq \kappa, \qquad (2.2)$$

where G_i and g_{ij} are given by (A2).

3   Energy functional

For $u = (u_l) \in L_2(S; H)$ we define the energy functional

$$F(u) = \int [\frac{\varepsilon}{2}|\nabla u_0|^2 + \sum_{i=n}^{p} (\int_0^{u_i} f_i^{-1}(s)\,ds - u_i v_{i\Gamma})]\,d\Omega$$
$$+ \int [\frac{k_0}{2}u_0^2 + \frac{\alpha^2}{2k_3}]\,d\Gamma + \int \frac{k_0}{2}u_0^2\,d\Gamma_N.$$

For a solution v to (1.2)-(1.5) the function

$$t \to \phi(t) = F(Ev(t)), \quad Ev = u = (u_0, u_n, u_p),$$

is absolutely continuous. Moreover, for a. e. $t \in S$ we find by (D4)

$$
\begin{aligned}
\phi'(t) &= \int [\varepsilon \nabla v_{0t} \cdot \nabla u_0 + \sum_{i=n}^{p} u_{it}(v_i - v_{i\Gamma})] \, d\Omega \\
&+ \int [(k_0 u_{0t} + \alpha) u_0 - \frac{k_1}{k_3} \alpha^2] \, d\Gamma + \int k_0 u_{0t} u_0 \, d\Gamma_N \\
&= \int [-J \cdot \nabla u_0 + \sum_{i=n}^{p} (J_i \cdot \nabla u_0 + u_{it}(v_i - v_{i\Gamma}))] \, d\Omega \qquad (3.1) \\
&+ \int [(k_0 u_{0t} + \alpha) u_0 - \frac{k_1}{k_3} \alpha^2] \, d\Gamma + \int k_0 u_{0t} u_0 \, d\Gamma_N \\
&= \int [r_0 u_0 + \sum_{i=n}^{p} (J_i \cdot (z_i - z_{i\Gamma}) - r_i(v_i - v_{i\Gamma}))] \, d\Omega - \int \frac{k_2}{k_3} \alpha^2 \, d\Gamma,
\end{aligned}
$$

where $z_i = \nabla(v_0 + e_i v_i)$, $z_{i\Gamma} = \nabla(v_{0\Gamma} + e_i v_{i\Gamma})$.

Remark 3.1 *From (3.1) it is clear, that ϕ can be looked at as Lyapunov function of (1.2)-(1.3) in some situations. Indeed, let for example $v_{0\Gamma} + e_i v_{i\Gamma} = r_0 = 0, r_i = exp(v_n + v_p) - 1$. Then, we infer from (3.1), using (2.1),(A3),(A5)–(A7),*

$$
\phi'(t) = \int [\sum_{i=n}^{p} J_i \cdot z_i - (exp(v_n + v_p) - 1)(v_n + v_p)] \, d\Omega \le 0.
$$

4 Uniqueness

In this section the uniqueness result will be proved. To this end let us suppose that we are given two solutions $v_j = (v_{lj})$, $j = 1, 2$, of (1.2)-(1.5) such that

$$
\|v_{ij}\|_{L_\infty(Q)} \le K, \quad \|v_{i1} - v_{i2}\|_{L_\infty(Q)} \le \kappa.
$$

We introduce

$$
d(t) = d(v_1(t), v_2(t)) = [F(Ev_1) + F(Ev_2) - 2F\left(\frac{Ev_1 + Ev_2}{2}\right)](t)
$$

as a kind of distance between v_1 and v_2. Our aim is to show that

$$
c_1 \|v(t)\|_H^2 \le d(t) \le c_2 \int_0^t \sigma(s) \|v(s)\|_H^2 \, ds, \quad t \in S, \quad v = v_1 - v_2,
$$

where $\sigma \in L_1(S)$.

Lemma 4.1 *There exists a constant $c > 0$ such that*

$$c\|v(t)\|_H^2 \leq d(t).$$

Proof. Because of (A1) the function $\psi(y) = \int_0^y f_i^{-1}(s)\, ds$, $y \in \mathbb{R}$, is convex. Moreover, setting

$$c_0 = min\{f_i'(s), |s| \leq K\}, \quad c_1 = max\{f_i'(s), |s| \leq K\}, \tag{4.1}$$

we find by elementary calculations that for $y_j = f_i(s_j)$, $|s_j| \leq K$,

$$\psi(y_1) + \psi(y_2) - 2\psi\left(\frac{y_1 + y_2}{2}\right) \geq \frac{1}{4c_1}|y_1 - y_2|^2 \geq \frac{c_0}{4c_1}|s_1 - s_2|^2.$$

This estimate along with the uniform convexity of the energy functional F with respect to u_0 proves the lemma ∎

Lemma 4.2 *Suppose in addition to (D1) that the solutions v_i satisfy the regularity condition*

$$|\nabla v_{0j}| \in L_{2q/(q-N)}(S; L_q), \ j = 1, 2, \ q > N. \tag{4.2}$$

Then there exists a constant c such that

$$d(t) \leq c \int_0^t \sigma(s)\|v(s)\|_H^2 \, ds, \quad \sigma(t) = 1 + \sum_{j=1}^2 \|y_j(t)\|_q^{2q/(q-N)}, \quad y_j = \nabla v_{0j}. \tag{4.3}$$

Proof. Setting

$$
\begin{aligned}
v_l &= v_{l1} - v_{l2}, \quad v_{0m} = \frac{v_{01} + v_{02}}{2}, \quad v_{im} = f_i^{-1}\left(\frac{u_{i1} + u_{i2}}{2}\right), \quad u_{ij} = f_i(v_{ij}), \\
y_j &= \nabla v_{0j}, \quad z_{ij} = \nabla(v_{0j} + e_i v_{ij}), \quad z_{im} = \nabla(v_{0m} + e_i v_{im}), \\
\alpha &= \alpha_1 - \alpha_2, \quad J = J_1 - J_2, \quad r_l = r_{l1} - r_{l2},
\end{aligned}
$$

we infer from (3.1) and (A8)

$$
\begin{aligned}
d'(t) &= \int [\frac{1}{2} r_0 v_0 + \sum_{i,j=1}^2 (J_{ij} \cdot (z_{ij} - z_{im}) - r_{ij}(v_{ij} - v_{im}))]\, d\Omega - \frac{1}{2}\int \frac{k_2}{k_3}\alpha^2 \, d\Gamma \\
&\leq \int \sum_{i,j=1}^2 J_{ij} \cdot (z_{ij} - z_{im})\, d\Omega + c\|v(t)\|_H^2.
\end{aligned}
\tag{4.4}
$$

In view of (A2) we have

$$\nabla v_{im} = \frac{1}{2}\sum_{j=1}^{2} g_{ij}\nabla v_{ij}$$

and hence

$$z_{im} = \frac{1}{2}\sum_{j=1}^{2}[g_{ij}z_{ij} + (1 - g_{ij})y_j].$$

Thus, setting $z_i = z_{i1} - z_{i2}$, $y = y_1 - y_2$, we find

$$
\sum_{i,j=1}^{2} J_{ij}\cdot(z_{ij} - z_{im}) = \frac{1}{2}[\sum_{i,j=1}^{2} G_i J_{ij}\cdot z_{ij}
$$
$$
+ \sum_{i=1}^{2}(g_{i2}J_{i1} - g_{i1}J_{i2})\cdot z_i - \sum_{i=1}^{2}(J_{i1} + J_{i2})\cdot\sum_{j=1}^{2}(1 - g_{ij})y_j]. \tag{4.5}
$$

We are now going to estimate the right-hand side of (4.5) term by term. To this end we drop the subscript i for convenience. From (2.1) (4.1) and (A1)-(A7) we see that

$$GJ_j\cdot z_j \leq -G(a_j + c_0 m_\gamma)|z_j|^2 = -G(a_j + \delta)|z_j|^2,   \delta > 0. \tag{4.6}$$

Next, setting $g_m = g((u_1 + u_2)/2)$ and using (A6), (A7), we get

$$
\begin{aligned}
(g_2 J_1 - g_1 J_2)\cdot z &= -[g_2(a_1 z_1 + b_1\beta_1 + g_m g_1\gamma_1) - g_1(a_2 z_2 + b_2\beta_2 + g_m g_2\gamma_2)]\cdot z \\
&= -[g_2 a_1 z_1 - g_1 a_2 z_2 + \frac{1}{2}(g_2 b_1 + g_1 b_2)(\beta_1 - \beta_2) \\
&\quad + \frac{1}{2}(g_2 b_1 - g_1 b_2)(\beta_1 + \beta_2) + g_m g_1 g_2(\gamma_1 - \gamma_2)]\cdot z \\
&\leq -(g_2 a_1 z_1 - g_1 a_2 z_2)\cdot z + c|v||z| - 2\delta|z|^2 \\
&\leq -(g_2 a_1 z_1 - g_1 a_2 z_2)\cdot z + c|v|^2 - \delta|z|^2.
\end{aligned}
\tag{4.7}
$$

Now, setting

$$A = (G + g_2)a_1,   B = (G + g_1)a_2,   C = g_2 a_1 + g_1 a_2,$$

and noting that by (A4)

$$4AB - C^2 = 8Ga_1 a_2 - (g_1 a_2 - g_2 a_1)^2 \geq \delta G,$$

we find by (A4)

$$G\sum_{j=1}^{2} a_j|z_j|^2 \; + \; (g_2a_1z_1 - g_1a_2z_2)\cdot z = A|z_1|^2 + B|z_2|^2 - Cz_1\cdot z_2$$

$$\geq \; A|z_1|^2 + B|z_2|^2 - \frac{C}{2}((A/B)^{1/2}|z_1|^2 + (B/A)^{1/2}|z_2|^2)$$

$$= \; \left(1 - (1 - \frac{4AB - C^2}{4AB})^{1/2}\right)(A|z_1|^2 + B|z_2|^2)$$

$$\geq \; \frac{4AB - C^2}{8}(|z_1|^2/B + |z_2|^2/A)$$

$$\geq \; \delta G(|z_1|^2 + |z_2|^2).$$

Thus it follows from (4.6) and (4.7)

$$G\sum_{j=1}^{2} J_j\cdot z_j + (g_2J_1 - g_1J_2)\cdot z \leq c|v|^2 - \delta(G\sum_{j=1}^{2}|z_j|^2 + |z|^2). \tag{4.8}$$

We turn now to the remaining term from (4.5). Using (2.2), (A6), (A7) and setting $g = g_1 - g_2$, we get

$$|(J_1 + J_2)\cdot\sum_{j=1}^{2}(1 - g_j)y_j| \; = \; |\frac{1}{2}(J_1 + J_2)\cdot(G\sum_{j=1}^{2}y_j - gy)|$$

$$\leq \; \frac{1}{2}|J_1 + J_2|(G\sum_{j=1}^{2}|y_j| + |g||y|)$$

$$\leq \; c(1 + |z_1| + |z_2|)(G\sum_{j=1}^{2}|y_j| + G^{1/2}|y|) \tag{4.9}$$

$$\leq \; \delta G(|z_1|^2 + |z_2|^2) + c(\delta)(G(|y_1|^2 + |y_2|^2) + |y|^2 + |v|^2).$$

Setting $\sigma = \sum_{j=1}^{2}\|y_j\|_q^{2q/(q-N)} \in L_1(S)$, we find by means of the inequalities of Hölder, Gagliardo-Nirenberg [14] and Young [10]

$$\int G|y_j|^2\, d\Omega \; \leq \; \|G\|_{q/(q-2)}\|y_j\|_q^2 \leq c\|v\|_{2q/(q-2)}^2\|y_j\|_q^2$$

$$\leq \; c\|\nabla v\|^{2N/q}\|v\|^{2(q-N)/q}\|y_j\|_q^2 \leq \frac{\delta}{2}\|\nabla v\|^2 + c(\delta)\sigma\|v\|^2 \tag{4.10}$$

$$\leq \; \delta(\|z\|^2 + \|y\|^2) + c\sigma\|v\|^2.$$

Now the lemma follows from (4.4), (4.5) and (4.8)–(4.10) ∎

We are now able to state our main result.

Theorem 4.1 *Let* v_j, $j = 1, 2$, *be (weak) solutions of* (1.2)–(1.5) *such that the regularity assumption* (4.2) *is satisfied. Let in addition*

$$\|v_{ij}\|_{L_\infty(Q)} \leq K, \quad \|v_{i1} - v_{i2}\|_{L_\infty(Q)} \leq \kappa \qquad (4.11)$$

and (A1)–(A9) *be satisfied. Then* $v_1 = v_2$.

Proof.

The theorem is a consequence of Lemma 4.1, Lemma 4.2 and Gronwall's Lemma (comp. [10]) ■

Remark 4.1 *In view of the initial condition* (1.4) *the second part of* (4.11) *can be replaced by supposing the solutions* v_j *to be continuous in time and space.*

5   Applications

As to the existence of solutions v in the sense of Definition 2.1, we refer to [4]. Further, the electrostatic potential $v_0(t), t > 0$, turns out to be weak solution of Poisson's equation (1.1), provided the initial value $v(0) = (v_{l0})$ satisfies (1.1). Thus standard regularity results for weak solutions to linear elliptic equations can be applied to verify the condition (4.2). In particular, under quite general conditions it can be proved [7] that $|\nabla v_0(t)| \in L_q$ for some $q > 2$. Concerning conditions guaranteeing $q > N = 3$, we refer to [16]. Finally, results implying continuity of solutions to drift–diffusion–equations in the spatially two-dimensionally case can be found in [8].

We turn to the verification of the basic assumptions and restrict us to the most involved conditions (A2) and (A4).

Condition (A2)

A sufficient condition for (A2) is:

Lemma 5.1 *Let* $f \in C^3(\mathbb{R})$ *satisfy*

$$f' > 0, \quad f''^2 > f' f'''.$$

Then $g = f' \circ f^{-1}$ *is uniformly concave.*

Proof. For $s_1, s_2 \in \mathbb{R}$, $s = \frac{s_2 - s_1}{2}$, we have

$$2g\left(\frac{s_1 + s_2}{2}\right) - g(s_1) - g(s_2) = -s^2 \int_0^1 \int_0^1 g''(s_1 + (\tau + \eta)s) \, d\tau \, d\eta.$$

Since

$$g''(s) = \frac{f' f''' - f''^2}{f'^3} \circ f^{-1},$$

the lemma follows ∎

Using this lemma, it is easy to show that the function

$$f_a(s) = \eta^{3/2} \;, \; \eta = c \log(1 + \frac{\tau}{c}) \;, \; \tau = \exp(2s/3) \;, \; c = \left(\frac{6}{\pi}\right)^{1/3} \;, \tag{5.1}$$

satisfies (A2) for arbitrary $0 < \kappa, K < \infty$. f_a serves in some situations as simple but sufficient approximation of the Fermi function

$$f_i(s) = \mathcal{F}(s) = \frac{2}{\sqrt{\pi}} \int_0^\infty \frac{\sqrt{t}\, dt}{1 + \exp(t - s)} \;, \tag{5.2}$$

which is the favoured example from the physical point of view, since Fermi–Dirac statistics is based on it. Unfortunately, it seems to be difficult to prove rigorously that the Fermi function satisfies (A2). However, computer calculations make this evident (comp. [2]).

In many physical situations Boltzmann statistics is considered as sufficient approximation for Fermi–Dirac statistics. Boltzman statistics is based on

$$f_i(s) = f(s) = exp(s), \tag{5.3}$$

which implies

$$g_i(s) = s, \quad G_i = 0. \tag{5.4}$$

Thus (5.3) violates (A2) (and also (A4)) and can be looked at merely as (nonallowed) limit case. However, uniqueness results under Boltzmann statistics have been proved in [4] for the special case $\Gamma = \emptyset$, using only the chemical part of the energy functional, i. e.

$$F_c(u) = \int \sum_{i=n}^p \int_0^{u_i} f_i^{-1}(s) \, ds \, d\Omega$$

for defining a distance. Moreover, as a pleasant consequence of (5.4) even the regularity condition (4.2) could be removed in [4] .

Condition (A4)

There are different options for a_i. A first physically interesting choice is [15]

$$a_i = \mu_i f_i', \quad \mu_i = const.,$$

which verifies (A4) trivially with $\rho = 0$, provided g_i is strictly concave. A frequently used choice is [4]

$$a_i = \mu_i f_i, \quad \mu_i = const.. \tag{5.5}$$

Of course, both choices coincide under Boltzmann statistics.

By computing niveau levels of the function $\rho(s_1, s_2)$ corresponding to (5.2) and (5.5) it can be made evident (comp. [2]) that the condition (A4) holds for finite K and suitable κ. In particular, the trace of this function given by

$$\rho(s) = \lim_{s_2 \to s_1 = s} = \frac{(ff'' - f'^2)^2}{2f^2(f''^2 - f'f''')}(s) = \frac{-[(\log(\log f)')')]^2}{2(\log f')''}(s),$$

satisfies

$$\rho(s) < 1, \quad s \in \mathbb{R}. \tag{5.6}$$

For the approximation (5.1), one gets rigorously

$$\lim_{s \to -\infty} = 0 < \rho_{f_a}(s) = \frac{(\tau - \eta)^2}{\tau(\tau - \eta + 2\eta^2/c)} < \lim_{s \to \infty} = 1.$$

Remark 5.1 *The 'local' condition (5.6) along with the continuity of the function ρ imply the 'global' condition (A4) for sufficiently small κ.*

References

[1] Gajewski, H., On a variant of monotonicity and its application to differential equations, Nonlinear Analysis: Theory, Methods & Applications (to appear).

[2] Gajewski, H., On uniqueness of solutions to the drift–diffusion–model of semiconductor devices, Preprint (1992).

[3] Gajewski, H., K. Gröger, On the basic equations for carrier transport in semiconductors, Journal of Mathematical Analysis and Applications 113 (1986), 12–35.

[4] Gajewski, H., K. Gröger, Semiconductor equations for variable mobilities based on Boltzmann statistics or Fermi–Dirac statistics, Math. Nachr. 140 (1989), 7 – 36.

[5] Gajewski, H., K. Gröger, K. Zacharias, Nichtlineare Operatorgleichungen und Operatordifferentialgleichungen, Akademie–Verlag Berlin (1974).

[6] Gilbarg, D., N.S. Trudinger, Elliptic partial differential equations of second order, Springer–Verlag, Berlin, Heidelberg (1983).

[7] Gröger, K., A $W^{1,p}$–estimate for solutions to mixed boundary value problems for second order elliptic differential equations, Math. Anal. 283 (1989), 679–687.

[8] Gröger, K., J. Rehberg, On the regularity of solutions to parabolic equations in case of mixed boundary conditions, (submitted to J. Reine und Angewandte Mathematik).

[9] Gröger, K., J. Rehberg, Uniqueness for the two-dimensional semiconductor equations based on Fermi–Dirac statistics, (to appear).

[10] Ladyshenskaja, O. A., Mathematical problems of the dynamic of viscose incompressible fluids, Moskau (1970) (russ).

[11] Markowich, P. A., The stationary semiconductor device equations, Springer Series on Computational Microelectronics, Springer, Wien–New York 1986.

[12] Mock, M. S., An initial value problem from semiconductor device theory, SIAM J. Math. Anal. 5 (1974), 597–612.

[13] Mock, M. S., Analysis of mathematicals models of semiconductor devices, Boole press, Dublin (1983).

[14] Nirenberg, L., On elliptic partial differential equations, Ann Scuola Norm. Sup. Pisa, Ser. III 13 (1959), 115-162.

[15] Platen, E., A stochastic approach to hopping transport in semiconductors, Journal of Statistical Physics, 59 (1990), 1329–1353.

[16] Shamir, E., Regularization of mixed second order elliptic problems, Israel Journal of Mathematics 6 (1968), 150–168.

[17] Van Roosbroeck, W., Theory of the flow of electrons and holes in germanium and other semiconductors, Bell Syst. Tech. J. 29 (1950), 560–607.

[18] Zeidler, E., Nonlinear functional analysis and its applications II/3 (Nonlinear monotone operators), Springer–Verlag, Berlin, Heidelberg (1990).

H. Gajewski
Institut für Angewandte Analysis und Stochastik
Mohrenstrasse 39, O–1086 Berlin, Germany

International Series of Numerical Mathematics, Vol. 117, © 1994 Birkhäuser Verlag Basel

ON RESTRICTIONS FOR DISCRETIZATIONS OF THE SIMPLIFIED LINEARIZED VAN ROOSBROECK'S EQUATIONS

K. GÄRTNER*

Abstract. The system of the van Roosbroeck's equations, often used to describe the electrical behaviour of semiconductors, is used in a simplified version and without recombination to demonstrate restrictions on discretizations arising from the coupling of the equations: the Jacobi matrix of the discrete current continuity equation with respect to the electrostatic potential should be a definite matrix. It is shown that the Scharfetter-Gummel discretization with common restrictions on the geometry of the grid fulfills this requirement in any space dimension. An example with multiple solutions from the literature has a unique solution if the Scharfetter-Gummel discretization is used.

1. Van Roosbroeck's equations and its linearization. With a convenient scaling and the notation in [1] one has the following system of Poisson's equation and continuity equations for electrons and holes:

$$
\begin{aligned}
-\Delta u &= f - n + p, \\
-\nabla \cdot J_n &= -R, \; J_n = \mu_n(\nabla n - n\nabla u), \\
\nabla \cdot J_p &= -R, \; J_p = -\mu_p(\nabla p - p\nabla u) \;\; in \; G,
\end{aligned}
$$

(1)

$$
u = u^0, \; n = n^0, \; p = p^0 \; on \; \Gamma_0, \;\; \nu \cdot u + \alpha(u - u^1) = \nu \cdot J_n = \nu \cdot J_p = 0 \;\; on \; \Gamma_1.
$$

Here:

$G \subset \mathbb{R}^d$, $1 \le d \le 3$, is a Lipschitzian domain with boundary Γ;

$\Gamma = \Gamma_0 \cup \Gamma_1$, where Γ_0, Γ_1 are disjoint, Γ_0 is closed and has a positive surface measure;

ν is the outer unit normal at any point of Γ;

u, n, p are the unknown functions: electrostatic potential, densities of electrons and holes;

J_n and J_p denote the electron and hole current density;

f is a given density of impurities;

u^0, n^0, p^0 represent boundary values at ohmic (Dirichlet) contacts;

u^1 and α are given functions modelling gate contacts;

---

* IPS, ETH–Zürich

μ_n, μ_p are the mobilities, they are assumed to be given space dependent functions; $R(n,p)$ is the recombination/generation term.

Neglected recombination is assumed throughout the paper although $R(n,p)$ is introduced in a numerical example to demonstrate nonuniqueness.

First, from the literature for system (1) and its linearization, we cite some results and properties which carry over to the discrete case or can be used as a guideline for its discussion. Let $L_q(G)$, $1 < q \le \infty$, be the Lebesgue spaces with norms

$$\|h\|_q = \left(\int h^q \, dG \right)^{1/q}, \quad \|h\|_2 = \|h\|,$$

and let $W_2^1(G)$ be the Sobolev space of all functions $h \in L_2(G)$ such that $|\nabla h| \in L_2(G)$. Define further

$$X = L_\infty(G) \cap W_2^1(G), \quad H = \{h \in W_2^1(G) : h = 0 \text{ on } \Gamma_0\}.$$

Denote by H^* the dual of H and by $(\cdot, \cdot)$ the scalar product in $L_2(G)$ as well as the pairing between H^* and H. Assume that

$$u^0, \ u^1, \ n^0, \ p^0 \ \in \ X, \quad 0 \le \check{m}^0 \le n^0, p^0 \le \hat{m}^0, \quad f, \ \mu_n, \ \mu_p, \ R \ \in \ L_\infty,$$

define the affine space

$$M = \{u, \ n, \ p \ \in \ X, \ u = u^0 + x, \ n = n^0 + y, \ p = p^0 + z, \ x, \ y, \ z \ \in \ H\},$$

and introduce operators A, $B(u)$, $C(u)$ with values in H^* by

$$(Au, x) = \int \nabla u \cdot \nabla x \, dG + \int \alpha u x \, d\Gamma_1 \quad \forall x \in H,$$

$$(B(u)n, y) = \int J_n(u)n \cdot \nabla y \, dG \quad \forall y \in H,$$

$$(C(u)p, z) = -\int J_p(u)p \cdot \nabla z \, dG \quad \forall z \in H, \quad (u, n, p) \in M.$$

Then the functional analytic version of (1) is

(2)
$$\begin{aligned} Au &= f_1 + p - n, \ (f_1, h) := (f, h) + \int \alpha u^1 h \, d\Gamma_1, \\ B(u)n &= C(u)p = 0. \end{aligned}$$

Using the (weak) maximum principle [3] and the Leray-Schauder theorem it can be shown [4] that (2) has at least one solution $(u, n, p) \in M$ such that

$$\check{u} \le u \le \hat{u}, \ 0 < \check{m} \le n, \ p \le \hat{m} \ \textit{in } G,$$

where the constants $\check{u}$, $\hat{u}$, $\check{m}$, $\hat{m}$ can be estimated by $\check{m}^0$, $\hat{m}^0$, u^0, u^1, and f_1.

According to the Boltzmann statistics, so called quasi-Fermi levels v and w can be introduced by setting

$$n = e^{u-v}, \ p = e^{w-u}.$$

In terms of the Slotboom variables e^{-v} and e^w the continuity equations can be expressed by linear elliptic operators

(3)
$$- \ \nabla \cdot \mu_n(x)e^u \nabla e^{-v} = 0,$$
(4)
$$- \ \nabla \cdot \mu_p(x)e^{-u} \nabla e^w = 0,$$

and the appropriately transformed boundary conditions.

Assume that $(U, \ N, \ P) \in M$ is a given approximate solution of the system (2). Choosing $(U, \ N, \ P)$ as linearization point one looks for a solution $(u, \ n, \ p) \in M$ of the following linearized system

(5)
$$
\begin{aligned}
Au &= f_1 + p - n, \\
B(U)n + B'_u(N)(u - U) &= 0, \\
C(U)p + C'_u(P)(u - U) &= 0,
\end{aligned}
$$

where the derivatives B'_u and C'_u of B and C with respect to u, mapping H into H^* are defined by

$$(B'_u(N)x, y) = - \int N \nabla x \cdot \nabla y \ dG, \quad x, y \in \ H,$$

$$(C'_u(P)x, z) = \int P \nabla x \cdot \nabla z \ dG, \quad x, z \in \ H.$$

Remarks:

- The quadratic forms $(-B'_u(N)x, x)$, $(C'_u(P)x, x)$ are positive definite.
- In the case of a one–carrier density model $d = 1$ and $R = 0$, (1) has a unique solution (see [2]).
- The linearized system (5) can be rewritten as one relation for x (a sum of products of positive definite operators), in full analogy to the discrete system (19).

2. The discretized system. Let us start with an example – the simple and often used discretization of the continuity equations (2) – (4) is the so-called Scharfetter-Gummel-scheme [6]. In the following, for simplicity, in the example discussed in the next section, a box integration variant for tensor product grids in $\mathbb{R}^d$ is used. The results can be carried over directly to box integration on the dual Voronoi grid with the usual restrictions on the interior angles of a triangulation of Γ. We assume that the grid points coincide with Γ and the material interfaces.

Integration of (3) with a test function $\psi_j = \{^{1 \ in \ G_j,}_{0 \ elsewhere}$ (where G_j is the control volume at the discretization point j) and assuming a constant current J_{edge} along an edge between the next neighbours $j, j+1$ yields (with R sufficiently small, U given and piecewise linear, $U(x) = U(x_j) + \xi(x_{j+1} - x_j), \mu = \bar{\mu} = const$)

$$J_{edge} = -\bar{\mu}\xi \frac{e^{-v}|_{x_j}^{x_{j+1}}}{e^{-U}|_{x_j}^{x_{j+1}}}.$$

So in the electron case (3) is approximated by

$$-\int \psi_j \nabla \cdot \mu_n(x) e^u \nabla e^{-v} dV \approx \sum_{all \ edges} a_{ij} J_{edge}.$$

This yields in the discrete Slotboom variables e^{-v} for the direction k of the tensor product grid

$$\tilde{B}_k e^{-v} + R_k = 0, \quad \tilde{B} = \sum_k^d \tilde{B}_k,$$

with one row of the matrix $\tilde{B}_k$ at point j with neighbours $j-1, j+1$

$$(6) \qquad \tilde{B}_k = (-\gamma^- s_j^-, \ \gamma^- s_j^- + \gamma^+ s_j^+, \ -\gamma^+ s_j^+),$$

$$(7) \qquad s_j^+ = \frac{e^{(U_j + U_{j+1})/2}}{sh((U_j - U_{j+1})/2)}, \quad s_j^- = s_{j-1}^+,$$

$$(8) \qquad \gamma_j^+ = \bar{\mu}_{j,j+1} \frac{a_{j,j+1}}{\Delta x_{j,j+1}}, \quad \gamma_j^- = \gamma_{j-1}^+, \gamma^\pm = \gamma_j^\pm,$$

$a_{j,j+1}$ is the signed surface measure, $\Delta x_{j,j+1} = x_{j+1} - x_j$. The function sh and the related Bernoulli function are defined as

$$sh(x) = \frac{\sinh(x)}{x}, \quad b(x) = \frac{x}{e^x - 1}$$

and fulfil the relations

$$(9) \qquad b(x) - b(-x) = -x, \quad b(x) > 0,$$

$$(10) \qquad b'(x) + b'(-x) = -1, \quad b'(x) < 0,$$

$$(11) \qquad \frac{e^{-x}}{sh(x)} = b(2x), \quad sh(x) = sh(-x).$$

Remark: $\tilde{B} = \tilde{B}^T$ is weakly diagonally dominant and positive definite ($\Gamma_0 \neq 0$).

Transformation by the diagonal matrices e^U yields

$$\tilde{B}(U)e^{-U}e^{U}e^{-v} + R = 0,$$

$$\tilde{B}e^{-U}n + R = 0, \quad \tilde{B}e^{-U} = B = \sum_{k=1}^{d} B_k$$

for the discrete electron density vector n. In analogy to (6), one row of B is given by

$$B_j = \left(-\gamma^- b(-\Delta u_{j-1}), \quad \gamma^- b(\Delta u_{j-1}) + \gamma^+ b(-\Delta u_j), \quad -\gamma^+ b(\Delta u_j) \right),$$

which is the usual expression for the discrete continuity equations. The discrete variables n_j, p_j at point j are approximated by

$$n_i = \int n \, dG_i, \quad p_i = \int p \, dG_i, \quad R(n,p) \approx R_j(n_j, p_j).$$

For the Jacobi matrix with respect to the electrostatic potential and $i = j - 1, \ j, \ j + 1$ holds

$$(12) \qquad B'_{u_j} = \frac{\partial B(u) N|_j}{\partial u_i} = (-\beta_j^-, \ \beta_j^- + \beta_j^+, \ -\beta_j^+),$$

$$(13) \qquad \beta_j^- = \gamma^- b'(-\Delta u_{j-1}) N_{j-1} + \gamma^- b'(\Delta u_{j-1}) N_j,$$

$$(14) \qquad \beta_j^+ = \gamma^+ b'(-\Delta u_j) N_j + \gamma^+ b'(\Delta u_j) N_{j+1}.$$

LEMMA 2.1. $-B'$ is a weakly diagonally dominant symmetric M-matrix for $N > 0$.

Proof. For any given N the row sum vanishes for every β^- or β^+ separately on the inner points. Symmetry follows from $\beta_j^- = \beta_{j-1}^+$. For $N > 0$ the inequalities $\beta_j^- < 0$, $\beta_j^+ < 0$ hold, because $b'(x) < 0$. On Γ_0, homogeneous Dirichlet conditions are fulfilled resulting in a positive row sum, and on Γ_1, $\beta^+ = 0$ holds because of $\gamma^+ J_n(N)|_{\Gamma_1} = 0$. $\square$

Returning to the system (5) again, one observes that this scheme for discretizing one single continuity equation fulfills all conditions imposed by uniqueness requirements for a one–carrier density model:
expressing the linearized system in the variables

$$u = U + x, \quad n = N(1 + x - y), \quad p = P(1 + z - x)$$

and setting

$$\bar{B}'_u = B'_u - R_p P, \quad \bar{B} = B + R_n, \quad \bar{C}'_u = C'_u + R_n N, \quad \bar{C} = C + R_p$$

it is easy to check that Newton's system can be written as

$$(A + P + N)x - Ny - Pz = rhs_1, \tag{15}$$
$$-(-\bar{B}'_u)x + \bar{B}(Nx - Ny) + R_pPz = rhs_2, \tag{16}$$
$$\bar{C}'_ux - \bar{C}(Px - Pz) - R_nNy = rhs_3, \tag{17}$$

where rhs_i stands for transformed right hand sides. They are of no interest here. Multiplication of (16) and (17) with $\bar{B}^{-1}$ and $\bar{C}^{-1}$, respectively, and substitution of $(Nx - Ny)$, $(Px - Pz)$ into (15) leads to

$$Ax + \bar{B}^{-1}((-\bar{B}'_u)x - R_pPz) + \bar{C}^{-1}(\bar{C}'_ux - R_nNy) = rhs_4. \tag{18}$$

If we suppose $R = g(n,p)(np-1)$, $R_n \geq 0$, $R_p \geq 0$ (this is true for Shockley-Read-Hall recombination $(g = 1/(\tau_p(n+1) + \tau_n(p+1)))$ the inverse matrices $\bar{B}^{-1}$ and $\bar{C}^{-1}$ exist, they are nonnegative and similar to symmetric positive definite ones.
For $R = 0$, the coupling terms in (18) vanish – so the local uniqueness problem for the van Roosbroeck system (1) is reduced to that of the equation for x with the typical structure, in full analogy to the analytic case:

$$Ax + B^{-1}(-B'_u)x + C^{-1}C'_ux = rhs_5. \tag{19}$$

For some triple $(U, N > 0, P > 0)$ the matrix $A + B^{-1}(-B'_u) + C^{-1}C'_u$ may become singular – but a natural requirement is to have uniqueness in the case of a one–carrier density model again (here $C = 0, p = 0$).

THEOREM 2.2. *Suppose Poisson's equation is discretized in the standard manner, so A is symmetric and positive definite. Then, if $(-B'_u)$ and $\bar{B}$ are positive definite, uniqueness is guaranteed for the one-carrier density model; hence for the discretization discussed above.*

Proof. It suffices to show that G, defined by

$$G = A^{-1/2}B^{-1}(-B'_u)A^{-1/2},$$

has nonnegative eigenvalues γ_i. G is congruent or similar, respectively, to the following matrices:

$$e^U \bar{B}^{-1} \ (-B'_u),$$
$$[e^{U/2}\bar{B}^{-1}e^{U/2}] \ [e^{-U/2} \ (-B'_u) \ e^{-U/2}],$$
$$[e^{U/2}\bar{B}^{-1}e^{U/2}]^{1/2} \ [e^{-U/2} \ (-B'_u) \ e^{-U/2}][e^{U/2}\bar{B}^{-1}e^{U/2}]^{-1/2},$$
$$[e^{-U/2} \ (-B'_u) \ e^{-U/2}],$$
$$(-B'_u).$$

But $(-B'_u)$ is positive definite by the lemma stated above; hence $\gamma_i > 0$ (of course it would be sufficient to have $-B'_u$ semi definite). □

3. Some remarks on Mock's example of nonuniqueness. Mock [5] has stated the example of multiple solutions with $R = 0$. The discretization differs from that given above by taking into account next neighbours for the zero order terms in the potential equation. The problem of this scheme is to guarantee the qualitative behaviour for a finite step size h and typically nonsmooth functions U. Poisson's equation is discretized in the following manner

$$Au = f - n + \tilde{A}^n(u)n + p - \tilde{A}^p(u)p,$$

with one row of the matrix $\tilde{A}^n(u)$ at point j

$$\tilde{A}^n(u) = (-f_m(\Delta u_{j-1}), \quad f_m(-\Delta u_{j-1}) + f_m(\Delta u_j), \quad -f_m(\Delta u_j)),$$

where f_m is defined by

$$f_m(x) = \frac{e^x - 1 - x - x^2/2}{x^2(e^x - 1)}, \quad f_m(0) = 1/6, \ 0 \le f_m(x) \le 1/2, \ f'_m(x) < 0.$$

Linearization yields (with E the identity matrix, (U, N, P) diagonal matrices or vectors of (u, n, p) at the linearization point, (x, y, z) as before)

(20) $$Ax + (E - \tilde{A}^n(U))(Nx - Ny) + (E - \tilde{A}^p(U))(Nx - Nz) \\ - \tilde{A}^{n\prime}_u(U,N)x + \tilde{A}^{p\prime}_u(U,P)x = rhs$$

where

$$\tilde{A}^{n\prime}_u(U,N) = (\tilde{a}_-, \quad -\tilde{a}_- - \tilde{a}_+, \quad \tilde{a}_+),$$

$$\tilde{a}_- = -f'_m(-\Delta U_{j-1})N_{j-1} + f'_m(-\Delta U_{j-1})N_j,$$

$$\tilde{a}_+ = +f'_m(\Delta U_j)N_j - f'_m(\Delta U_j)N_{j+1},$$

and

$$\text{sign}(\tilde{A}^{n\prime}_u(U,N)_{i,i-1}) = -\text{sign}(\tilde{A}^{n\prime}_u(U,N)_{i-1,i}),$$

$$\sum_j \tilde{A}^{n\prime}_u(U,N)_{i,j} = 0$$

at inner points. Elimination of $(Nx - Ny)$, $(Nx - Nz)$ can be done again with additional factors $(E - \tilde{A}^n(u))$, $(E - \tilde{A}^p(u))$:

(21) $$Ax + (E - \tilde{A}^n(U))B^{-1}(-B'_u)x + (E - \tilde{A}^p(U))C^{-1}C'_u x \\ - \tilde{A}^{n\prime}_u(U,N)x + \tilde{A}^{p\prime}_u(U,P)x = rhs_6.$$

The properties of $B^{-1}(-B'_u)$ etc. remain unchanged. The additional term $-\tilde{A}^{n\prime}_u(U,N)+\tilde{A}^{p\prime}_u(U,P)$ may be indefinite. The properties of the factor $(E-\tilde{A}^n(U))$ can be roughly characterized as follows: if we have Dirichlet boundary conditions only, it is easy to see that $\tilde{A}^n(U)$ is a nonsymmetric M-matrix with $\text{diag}\,\tilde{A}^n(U)<1$ for finite jumps in U. For a row j with a sufficiently large jump in U we have

$$\lim_{\Delta U_{j-1}\to-\infty}(E-\tilde{A}^n(U))=(1/2,\ 1/3,\ 1/6)$$

and in the limit of resolved boundary layers

$$\lim_{\Delta U_j\to0}(E-\tilde{A}^n(U))=(1/6,\ 2/3,\ 1/6).$$

For a one–carrier density model neglecting $\tilde{A}^{n\prime}_u(U,N)$, nonuniqueness may appear if negative real parts for the eigenvalues of $(E-\tilde{A}^n(U))$ occur. The corresponding jump U_{crit} in a step function U can be estimated from Gershgorin circles

$$-f_m(U_{crit})+1-f_m(U_{crit})-1/6-1/6>0.$$

For $U_{crit}\ll-1$ one has $f(-U_{crit})\approx1/2-1/U_{crit}+1/U^2_{crit}$, and it follows that $U_{crit}\approx-3-\sqrt{3}$, in good agreement with the numerically evaluated value $U_{crit}\approx-4.6246$. However, this value is much too small for cases of practical interest. If the doping f (and hence U) is sufficiently nonsmooth, one should expect multiple solutions not far from thermodynamic equilibrium. For higher bias voltages and large currents the electrostatic potential U will be much smoother and the solution should approximate the true one — especially if one is starting with a smooth initial guess related to a high current state.

Mock's thyristor example, which features very high doping concentrations with jumps, is exactly of this type. Unfortunately the applied smoothing procedure can not be verified, and the device length can be evaluated only approximately since the value of one constant is missing. U_{crit} may be exceeded by a factor of 5 or 10. An example with qualitatively the same data as in [5] was solved with two different arithmetic precisions (VAX DFloating REAL*8, REAL*16), a Newton and a Gummel type algorithm for the standard discretization, and a Gummel procedure for the discretization used by Mock with identical data. The results are shown in the following two figures (the upper branch given by Mock was not computed):

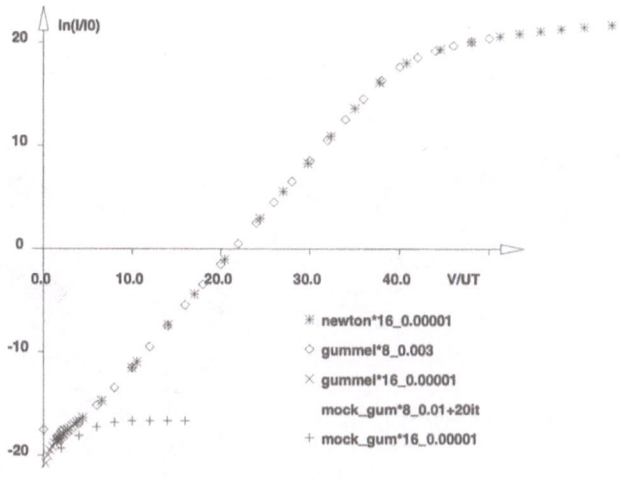

"Mock's-Thyristor", standard discretization with local and

Mock's one with non local approximation of n, p (real*8, *16)

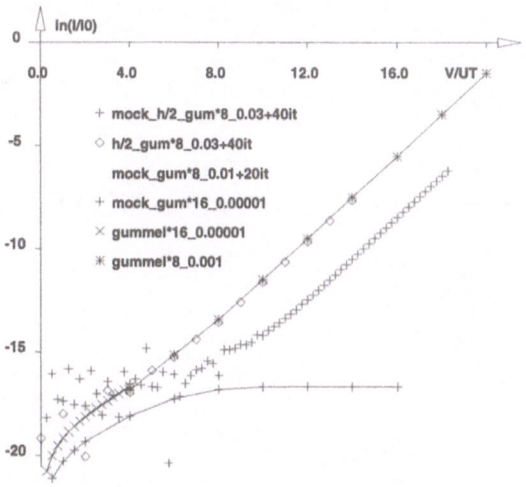

"Mock's-Thyristor", standard discretization with local and

Mock's one with non local approximation of n, p; h and h/2 grid

To illustrate the problem once more, results for another thyristor with recombination are shown. A nice dependence on the parameter m multiplying the recombination term occurs:

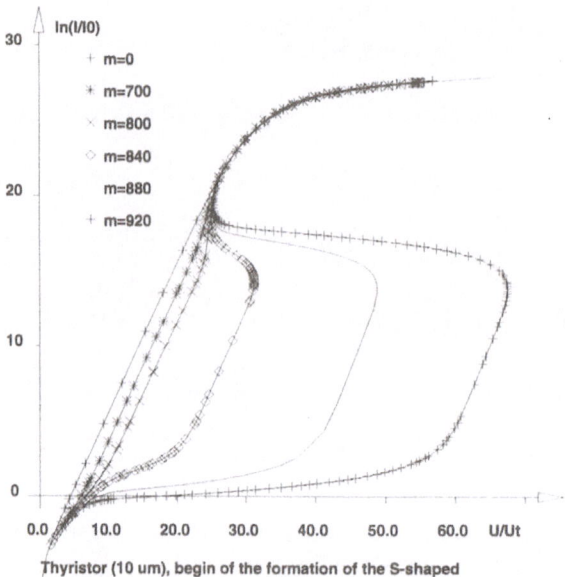

Thyristor (10 um), begin of the formation of the S-shaped
characteristics, recombination: m * (n*p-1)/(tp*n+tn*p+c)

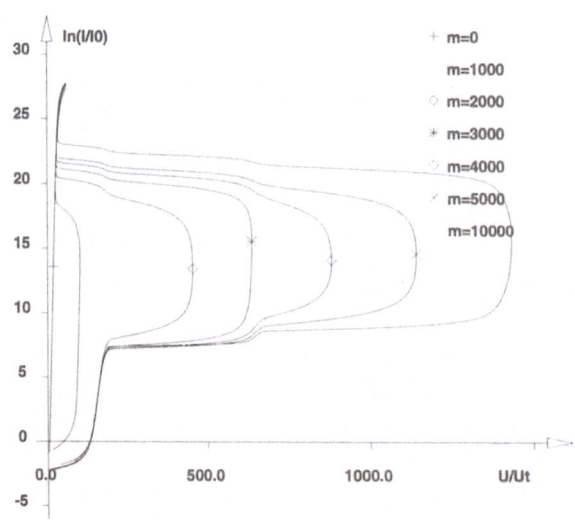

Thyristor (10 um), the formation of the S-shaped
characteristics, recombination: m * (n*p-1)/(tp*n+tn*p+c)

Conclusions and Acknowledgments. The one–carrier model and it's uniqueness can be used to investigate some aspects of discretizations of the semiconductor device equations. The definiteness requirement for the Jacobi matrix of the discrete current continuity equation with respect to the electrostatic potential may be restrictive for higher order approximation schemes.

The author is indebted to his former and present colleagues at Berlin and Zürich for many discussions and support. Part of this work was supported by the Swiss National Science Foundation.

References.

[1] Gajewski, H., Gärtner, K.: On the iterative solution of van Roosbroeck's equations. ZAMM 72 (1992), 19-28.

[2] Gajewski, H., Sommrey, J.: On the Uniqueness of Solutions of van Roosbroeck Equations. ZAMM 72 (1992), 151-153.

[3] Gilbarg, D., Trudinger, N.S.: Elliptic partial differential equations of second order. Springer, Berlin-Heidelberg-New York 1977.

[4] Markowich, P.A.: The stationary semiconductor device equations. Springer Series on Computational Microelectronics, Springer, Wien-New-York 1986.

[5] Mock, M.S.: An example of nonuniqueness of stationary solution in semiconductor device models. Compel 1 (1982), 165-175

[6] Scharfetter, D.L., Gummel, H.K.: Large-signal analysis of a silicon read diode oscillator. IEEE Trans. Electr. Dev. ED-16 (1969), 64 - 77

International Series of Numerical Mathematics, Vol. 117, © 1994 Birkhäuser Verlag Basel

MIXED FINITE ELEMENT DISCRETIZATION OF CONTINUITY EQUATIONS ARISING IN SEMICONDUCTOR DEVICE SIMULATION

R. HIPTMAIR AND R.H.W. HOPPE *

Summary. In the wake of decoupling and linearization semiconductor device simulation based on van Roosbroecks's equations requires the solution of convection–diffusion equations. It is well known that due to the occurrence of local regions of strong convection standard discretizations do not behave properly. As an alternative among others, mixed methods have been suggested having their roots in the dual variational formulation of the convection–diffusion problem. Their efficient implementation has to make use of Lagrangian multipliers. In a novel approach we already introduce the multiplier prior to discretizing, through a process called hybridization. In the sequel we use the resulting variational problem to develop a new discretization scheme. Next, we outline how to implement a standard mixed scheme and investigate some of its aspects. Finally, the behaviour of the mixed method is illustrated by a series of numerical experiments.

Key words. convection–diffusion problem, flux oriented schemes, hybridization, Lagrangian multipliers, mixed finite elements, Raviart–Thomas elements

MSC subject classifications. Primary 65N30; secondary 35J20

1. Introduction. We consider the linearized current continuity equation in 2–D semiconductor device simulation stated as a drift–diffusion equation for the carrier concentration n

$$
\begin{aligned}
\operatorname{div}(\nabla n - n\nabla\Psi) - Rn &= f && \text{in } \Omega \\
n &= g && \text{on } \Gamma_D \subset \partial\Omega \\
\tfrac{\partial n}{\partial \nu} - n\tfrac{\partial \Psi}{\partial \nu} &= 0 && \text{on } \Gamma_N \subset \partial\Omega
\end{aligned}
\tag{1}
$$

Here $\Omega \subset \mathbf{R}^2$ stands for the cross section of the device, Ψ is the electric potential, R refers to the differential net recombination rate and f denotes a source term. Further, we assume inhomogeneous Dirichlet boundary conditions at the Ohmic contacts $\Gamma_D \subset \Gamma := \partial\Omega$ and homogeneous Neumann boundary conditions at the insulation part Γ_N of the boundary ($\Gamma_N \cup \Gamma_D = \Gamma, \Gamma_N \cap \Gamma_D = \emptyset$). The familiar standard variational (primal) formulation of (1) is:

$$
\text{Find } n \in n_0 + H^1_{\Gamma_D}(\Omega) \text{ such that } \quad a_{std}(n,v) = f_{std}(v), \quad v \in H^1_{\Gamma_D}(\Omega)
\tag{2}
$$

---

* Mathematisches Institut, Technische Universität München, Arcisstraße 21, D-8000 München 2,e–mail: rohop@mathematik.tu-muenchen.de, hiptmair@mathematik.tu-muenchen.de. The first author was supported by FORTWIHR, Bavarian Consortium for High Performance Scientific Computing

where $H^1_{\Gamma_D}(\Omega)$ contains the functions of $H^1(\Omega)$ with zero trace on Γ_D and $n_0 \in H^1(\Omega)$, $n_0|_{\Gamma_D} = g$ (in the sense of a trace). Additionally we used the bilinear form

$$
a_{std} : \left\{ \begin{array}{ccc} H^1(\Omega) \times H^1(\Omega) & \mapsto & \mathbf{R} \\ (n, v) & \mapsto & \int_\Omega e^\Psi \left\langle \nabla(e^{-\Psi} n), \nabla v \right\rangle d\mathbf{x} + \int_\Omega Rnv \, d\mathbf{x} \end{array} \right.
\tag{3}
$$

and right–hand side

$$
f_{std} : H^1(\Omega) \mapsto \mathbf{R} \quad ; \quad v \mapsto - \int_\Omega fv \, d\mathbf{x}
\tag{4}
$$

In view of the identity $e^\Psi \nabla(e^{-\Psi} n) = \nabla n - n \nabla \Psi$ we may introduce the Slotboom variable $u := e^{-\Psi} n$ and get the problem:

$$
\begin{array}{rclcl}
\operatorname{div}(e^\Psi \nabla u) - e^\Psi Ru & = & f & & \text{in } \Omega \\
u & = & e^{-\Psi} g & & \text{on } \Gamma_D \subset \partial\Omega \\
\frac{\partial u}{\partial \boldsymbol{\nu}} & = & 0 & & \text{on } \Gamma_N \subset \partial\Omega
\end{array}
\tag{5}
$$

In this paper we are going to switch frequently between both unknowns. To guide the reader they are invariably denoted by u and n.

By virtue of its symmetric structure (5) can be formulated as an unconstrained minimization problem for the total energy over the space $H^1(\Gamma_D, e^{-\Psi} g) := \{v \in H^1(\Omega); v_{|\Gamma_D} = e^{-\Psi} g\}$:

Find u in $H^1(\Gamma_D, e^{-\Psi} g)$ such that

$$
J(u) := \inf_{v \in H^1(\Gamma_D, e^{-\Psi} g)} J(v)
\tag{6}
$$

where

$$
J(v) := \frac{1}{2} \int_\Omega e^\Psi \left(|\nabla v|^2 + R|v|^2 \right) dx + \int_\Omega fv \, d\mathbf{x}.
$$

Usually (2) is the starting point for a Galerkin approximation or, more preferably, a Petrov–Galerkin approach which is more suited in case of strong convection because of its upwind features (see for example the streamline diffusion method developed by Hughes et al. in [10]).

On the other hand, for problems where the flux is of primary interest the above formulation is less appropriate, since the current $\mathbf{j} := e^\Psi \nabla u$ must be calculated by differentiation thus resulting in a loss of accuracy. Moreover, for problems with regions of strong convection it may be more advisable to use flux–oriented variational principles that are based on the conservation of the current. Such variational principles can be obtained by duality arguments. In particular, it is well known from convex analysis (see for instance [9]) that the unconstrained minimization of a convex objective functional is equivalent to a saddle point problem in terms of the convex conjugate which constitutes the so–called dual problem (as opposed to the primal problem (2)). For the energy functional J in (6) we can use the obvious identity

$$\frac{1}{2}\int_\Omega |\nabla v|^2\, dx = \sup_{\mathbf{q}\in(L^2(\Omega))^2} \left\{\int_\Omega \langle \mathbf{q}, \nabla v\rangle\, dx - \frac{1}{2}\int_\Omega |\mathbf{q}|^2\, dx\right\}$$

together with Green's formula

$$\int_\Omega v\,\mathrm{div}\,\mathbf{q}\, dx + \int_\Omega \langle \nabla v, \mathbf{q}\rangle\, dx = \int_\Gamma v\,\langle \mathbf{q}, \boldsymbol{\nu}\rangle\, d\Gamma, \quad v\in H^1(\Omega) \quad \mathbf{q}\in \mathbf{H}(\mathrm{div},\Omega)$$

where $\mathbf{H}(\mathrm{div},\Omega) := \{\mathbf{q}\in (L^2(\Omega))^2\,;\mathrm{div}\,\mathbf{q}\in L^2(\Omega)\}$ (We use bold type for vector–valued variables). Setting $\mathbf{H}(\mathrm{div},\Omega,\Gamma_N) := \{\mathbf{q}\in \mathbf{H}(\mathrm{div},\Omega); \langle \mathbf{q}, \boldsymbol{\nu}\rangle_{|\Gamma_N} = 0\}$, we thus get the dual problem as the following *saddle point problem*:

Find $(\mathbf{j}, u)\in \mathbf{H}(\mathrm{div},\Omega,\Gamma_N)\times L^2(\Omega)$ such that

$$L(\mathbf{j}, u) = \inf_{\mathbf{v}\in\mathbf{H}(\mathrm{div},\Omega,\Gamma_N)} \sup_{w\in L^2(\Omega)} L(\mathbf{v}, w) \tag{7}$$

where

$$L(\mathbf{v}, w) := \frac{1}{2}\int_\Omega e^{-\Psi}|\mathbf{v}|^2\, dx + \int_\Omega \left(\mathrm{div}\,\mathbf{v} - \frac{1}{2}e^\Psi R w - f\right) w\, dx - \int_{\Gamma_D} e^{-\Psi} g\,\langle \mathbf{v}, \boldsymbol{\nu}\rangle\, d\Gamma$$

The dual problem (7) is also called the mixed formulation of (5).

A widely used approach to the numerical solution of that mixed formulation is to use mixed finite elements (cf. [7], [8]), a specimen of which are the lowest order Raviart–Thomas elements presented in [11]. Following De Veubeke's smart idea in [13], one subsequently eliminates the continuity constraints for the normal component of the discrete fluxes at the interelement boundaries from the discrete flux space by means of appropriate *Lagrangian multipliers*. Static condensation and rescaling then leads to a Schur complement system which is related to a specific nonconforming finite element method. This hybridization of the mixed discretization has been theoretically analyzed by Brezzi at al. (cf. [6], [1]) and has recently been implemented by Reusken (cf. [12]) within a multigrid framework.

So far De Veubeke's trick has only been employed after discretization had already been completed. To our mind a quite appealing alternative is to apply hybridization directly to the dual formulation which results in some sort of continuous analogue of the Schur complement. As the side–effects of a particular discretization do not have to be taken into account this approach may reward us with additional insights into the genuine properties of the problems. As the continuous problem is subject to spectral analysis, prospects arise to gain important information about appropriate preconditioning. Note that in classical domain decomposition a related approach, namely hybridization of the primal problem has been investigated by Bjørstad and Widlund in [4].

The remainder of the paper is organized as follows: In section 2 we will give the mixed formulation of (1) and present the details of dual hybridization. A spectral analysis of the dual hybrid operator is then carried out for a simple model problem. The next section is devoted to a special flux–oriented upwind scheme that is related to the discretization technique of Baliga–Patankar (cf. [2]). In section 4 we are reviewing some aspects of the standard mixed discretization and its relationship to inverse average type nonconforming Petrov–Galerkin methods. Finally, in section 5 we are presenting the results of a couple of numerical experiments which center on the performance of the mixed method in special situations.

2. Dual Hybridization. The variational equations arising from the mixed saddle point problem (7) by means of Gâteaux differentiation of the functional L with respect to both variables read:

Find $(\mathbf{j}, u) \in \mathbf{H}(\mathrm{div}, \Omega, \Gamma_N) \times L^2(\Omega)$ such that

$$
\begin{aligned}
\int_\Omega e^{-\Psi} \langle \mathbf{j}, \mathbf{v} \rangle \, d\mathbf{x} \; &+ \; \int_\Omega u \, \mathrm{div}\, \mathbf{v} \, d\mathbf{x} \; = \; \int_{\Gamma_D} e^{-\Psi} g \, \langle \mathbf{v}, \boldsymbol{\nu} \rangle \, d\Gamma, \quad \mathbf{v} \in \mathbf{H}(\mathrm{div}, \Omega, \Gamma_N) \\
\int_\Omega w \, \mathrm{div}\, \mathbf{j} \, d\mathbf{x} \; &- \; \int_\Omega e^{\Psi} Ruw \, d\mathbf{x} \; = \; \int_\Omega fw \, d\mathbf{x}, \qquad\qquad w \in L^2(\Omega)
\end{aligned}
\tag{8}
$$

Another derivation of (8) immediately replaces $e^\Psi \nabla u$ in (5) by the new variable $\mathbf{j}$ (the flux) and thus converts (5) into a first order system of partial differential equations:

$$
\left.\begin{aligned}
\mathbf{j} &= e^\Psi \nabla u \\
\mathrm{div}\, \mathbf{j} - e^\Psi Ru &= f
\end{aligned}\right\} \text{ in } \Omega
\qquad
\begin{aligned}
u &= e^{-\Psi} g && \text{on } \Gamma_D \\
\langle \mathbf{j}, \boldsymbol{\nu} \rangle &= 0 && \text{on } \Gamma_N \ .
\end{aligned}
$$

Both equations are now written in variational form and applying Green's formula we arrive at (8). We can take our cue from these considerations to prove that both the mixed problem (8) and its standard counterpart (2) do yield the same solutions apart from scaling.

The main objective of hybridization is to get rid of the global space $\mathbf{H}(\mathrm{div}, \Omega, \Gamma_N)$. It is a common experience — see for instance [11] —, that its discrete analogues are not easily handled. The tool of a Lagrangian multiplier enables us to break up the space into less bulky pieces. The splitting of the space $\mathbf{H}(\mathrm{div}, \Omega, \Gamma_N)$ is based on a decomposition of Ω, a "generalized triangulation" $\mathcal{T}_h := \{T_i\}_{i=1}^N$ ($N \in \mathbb{N}$): The patches T_i are open, nonempty convex subsets, mutually disjoint and the union of their closures entirely covers Ω. Furthermore we demand that they align with the boundary parts. A general triangulation of this kind is feasible as long as Ω has polygonal boundary and all internal boundaries have to be straight lines. In the sequel this is always taken for granted. For convenience we write $\mathcal{E}$ to refer to the union of edges of the T_i.

A fundamental result (cf. [6]) now states that a "fragmented" vector field in $\mathcal{X} := \bigotimes_i \mathbf{H}(\mathrm{div}, T_i)$ belongs to $\mathbf{H}(\mathrm{div}, \Omega, \Gamma_N)$ if and only if its normal components at the internal boundaries of $\mathcal{T}$ are continuous and do vanish on Γ_N. Hence, any Lagrangian multiplier meant to single out members of $\mathbf{H}(\mathrm{div}, \Omega, \Gamma_N)$ has to enforce the continuity of the normal components across the joint edge of two patches. We do not want to dip into details here: A careful analysis reveals that $\mathcal{M} := H^1_{\Gamma_D}(\Omega)$ is a suitable space of multipliers. Of course, for an $m \in \mathcal{M}$ only the traces on $\mathcal{E}$ are relevant. Now, by means of the multiplier space $\mathcal{M}$ an element $\mathbf{v} \in \mathcal{X}$ lives in $\mathbf{H}(\mathrm{div}, \Omega, \Gamma_N)$ if and only if

$$l(\mathbf{v}, m) = 0, \quad m \in \mathcal{M} \tag{9}$$

with $l : \mathcal{X} \times \mathcal{M} \mapsto \mathbb{R}$ denoting the bilinear mapping

$$l(\mathbf{v}, m) := -\sum_{i=1}^{N} \int_{\partial T_i} \left\langle \mathbf{v}_{|T_i}, \boldsymbol{\nu} \right\rangle m \, d\Gamma .$$

(7) and (9) are now merged into an augmented saddle point problem: Find $(\mathbf{j}, u, p) \in \mathcal{X} \times \mathcal{W} \times \mathcal{M}$ such that

$$\tilde{L}(\mathbf{j}, u, p) = \sup_{w \in \mathcal{W}} \inf_{\mathbf{v} \in \mathcal{X}} \sup_{m \in \mathcal{M}} \tilde{L}(\mathbf{v}, w, m) \tag{10}$$

where $\tilde{L}(\mathbf{v}, w, m,) := L(\mathbf{v}, w) + l(\mathbf{v}, m)$ and $\mathcal{W}$ served as an abbreviation for $L^2(\Omega)$. For the sake of accuracy we should point out that L (cf. (7)) is assumed to be extended to functions in $\mathcal{X}$ in a canonical fashion. Once more, Gâteaux differentiation of (10) is quickly done and yields the *mixed hybrid problem*

Find $(\mathbf{j}, u, p) \in \mathcal{X} \times \mathcal{W} \times \mathcal{M}$ such that

$$
\begin{aligned}
a(\mathbf{j}, \mathbf{v}) \;+\; b(\mathbf{v}, u) \;+\; l(\mathbf{v}, p) &= g(\mathbf{v}), & \mathbf{v} \in \mathcal{X} \\
b(\mathbf{j}, w) \;-\; d(u, w) & = f(w), & w \in \mathcal{W} \\
l(\mathbf{j}, m) & = 0, & m \in \mathcal{M}
\end{aligned}
\tag{11}
$$

where

$$a(\mathbf{j}, \mathbf{v}) := \int_{\Omega} e^{-\Psi} \langle \mathbf{j}, \mathbf{v} \rangle \, d\mathbf{x}, \quad b(\mathbf{v}, u) := \sum_i \int_{T_i} u \, \mathrm{div}\, \mathbf{v} \, d\mathbf{x}, \quad d(u, w) := \int_{\Omega} e^{\Psi} Ruw \, d\mathbf{x}$$

and

$$g(\mathbf{v}) := \int_{\Gamma_D} e^{-\Psi} g \langle \mathbf{v}, \boldsymbol{\nu} \rangle \, d\Gamma , \quad f(w) := \int_{\Omega} fw \, d\mathbf{x} \quad .$$

Standard techniques developed for general saddle point problems (cf. [6]) establish existence and uniqueness of a solution of (11); uniqueness of p is to be understood as a statement about its traces on $\mathcal{E}$.

As an elementary but interesting feature of (11) we wish to mention that the components u and p of the solution have the same trace on $\mathcal{E}$. First of all, this fact teaches us how to write (11) for the original unknown n: Just replace u and p by $e^{\Psi}u$ and $e^{\Psi}p$, respectively. This observation is also numerically significant: As soon as an approximation of p is available, one of u is also known on the edges.

Each bilinear form of (11) can be associated with a Riesz operator denoted by the respective capital letter. Their adjoints are marked by an asterisk. Writing (11) as an equation for operators in function spaces we then obtain the following system:

$$\begin{pmatrix} A & B^* & L^* \\ B & -D & 0 \\ L & 0 & 0 \end{pmatrix} \begin{pmatrix} \mathbf{j} \\ u \\ p \end{pmatrix} = \begin{pmatrix} g \\ f \\ 0 \end{pmatrix} \tag{12}$$

2.1. The Dual–Hybrid Variational Problem.

Due to its special structure the system (12) lends itself to static condensation in a wider sense: Through a process similar to block elimination of linear systems we seek to dump all unknowns except for p. Doing so produces the following equation:

$$\begin{pmatrix} L \\ 0 \end{pmatrix} \begin{pmatrix} A & B^* \\ B & -D \end{pmatrix}^{-1} \begin{pmatrix} L \\ 0 \end{pmatrix}^* p = \begin{pmatrix} L \\ 0 \end{pmatrix} \begin{pmatrix} A & B^* \\ B & -D \end{pmatrix}^{-1} \begin{pmatrix} g \\ f \end{pmatrix}$$

which we are going to write as $Sp = q$.

To go beyond playing with symbols we are now scrutinizing the related variational problem: Find $p \in \mathcal{M}$ such that

$$\langle Sp, m \rangle_{\mathcal{M}' \times \mathcal{M}} = q(m), \quad m \in \mathcal{M} \tag{13}$$

by untangling the left hand side: (As usual, dual spaces are marked by a prime ')

- First $\begin{pmatrix} L \\ 0 \end{pmatrix}^* p \in (\mathcal{X} \times \mathcal{W})'$ for $p \in \mathcal{M}$ is given by

$$\begin{pmatrix} L \\ 0 \end{pmatrix}^* p \begin{pmatrix} \mathbf{v} \\ w \end{pmatrix} = -\sum_{i=1}^{N} \int_{\partial T_i} p \left\langle \mathbf{v}_{|T_i}, \boldsymbol{\nu} \right\rangle d\Gamma = l(\mathbf{v}, p), \qquad \begin{pmatrix} \mathbf{v} \\ w \end{pmatrix} \in \mathcal{X} \times \mathcal{W}$$

- Evaluating $\begin{pmatrix} A & B^* \\ B & -D \end{pmatrix}^{-1} \begin{pmatrix} h_1 \\ h_2 \end{pmatrix} \in \mathcal{X} \times \mathcal{W}$ for $h_1 \in \mathcal{X}', h_2 \in \mathcal{W}'$ is equivalent to the variational problem:

    Find $(\mathbf{j}, u) \in \mathcal{X} \times \mathcal{W}$ such that

$$\begin{aligned} a(\mathbf{j}, \mathbf{v}) + b(\mathbf{v}, u) &= h_1(\mathbf{v}), & \mathbf{v} \in \mathcal{X} \\ b(\mathbf{j}, w) - d(u, w) &= h_2(w), & w \in \mathcal{W} \end{aligned} \tag{14}$$

Now, we are witnessing the crucial benefit of hybridization: The main purpose of introducing a multiplier was *decoupling*: Writing $\mathbf{v} = \sum_i \mathbf{v}_i, \mathbf{v}_i \in \mathbf{H}(\mathrm{div}, T_i, \Gamma_N)$ and $w = \sum_i w_i, w_i \in L^2(T_i)$ (14) spawns a host of variational problems:

Find $(\mathbf{j}_i, u_i) \in \mathbf{H}(\mathrm{div}, T_i, \Gamma_N) \times L^2(T_i)$ such that

$$
\begin{aligned}
a(\mathbf{j}_i, \mathbf{v}_i) \; + \; b(\mathbf{v}_i, u_i) \; &= \; \tilde{h}_1(\mathbf{v}_i), && \mathbf{v}_i \in \mathbf{H}(\mathrm{div}, T_i, \Gamma_N) \\
b(\mathbf{j}_i, w_i) \; - \; d(u_i, w_i) \; &= \; \tilde{h}_2(w_i), && w_i \in L^2(T_i)
\end{aligned}
\qquad 1 \leq i \leq N \;\; (15)
$$

Since we have disposed of all the continuity ties between the $\mathbf{v}_i$, these problems can be solved independently.

- Finally, $\begin{pmatrix} L \\ 0 \end{pmatrix}\begin{pmatrix} v \\ w \end{pmatrix}$ with $\begin{pmatrix} v \\ w \end{pmatrix} \in \mathcal{X} \times \mathcal{W}$ is a linear form on $\mathcal{M}$:

$$
\begin{pmatrix} L \\ 0 \end{pmatrix} \begin{pmatrix} \mathbf{v} \\ w \end{pmatrix} p = l(\mathbf{v}, p) = -\sum_{i=1}^{N} \int_{\partial T_i} p \, \langle \mathbf{v}, \boldsymbol{\nu} \rangle \, d\Gamma, \quad p \in \mathcal{M} \quad .
$$

In sum, the evaluation of $\langle Sp, m \rangle_{\mathcal{M}' \times \mathcal{M}}$ requires basically two steps:

1. As a first step for each $T_i \in \mathcal{T}$ we have to solve

    Find $(\mathbf{j}_i, u_i) \in \mathbf{H}(\mathrm{div}, T_i, \Gamma_N) \times L^2(T_i)$ such that

$$
\begin{aligned}
a(\mathbf{j}_i, \mathbf{v}_i) \; + \; b(\mathbf{v}_i, u_i) \; &= \; -l(\mathbf{v}_i, p) \,, && \mathbf{v}_i \in \mathbf{H}(\mathrm{div}, T_i, \Gamma_N) \\
b(\mathbf{j}_i, w_i) \; - \; d(u_i, w_i) \; &= \; 0 \,, && w_i \in L^2(T_i)
\end{aligned}
\qquad 1 \leq i \leq N(16)
$$

    Recalling (7) these problems turn out to be local Dirichlet problems for the equation (5) with vanishing source term f and boundary values provided by the multiplier.

2. Using the result of the previous step this one reduces to the calculation of

$$
\sum_{i=1}^{N} \int_{\partial T_i} m \, \langle \mathbf{j}_i, \boldsymbol{\nu} \rangle \, d\Gamma \qquad \text{for} \quad m \in \mathcal{M} \quad .
$$

In order to condense the remaining steps into a single formula we resort to *Poincaré–Steklov operators*, defined to be the Dirichlet–Neumann operators

$$
T_{\mathrm{DN}}^i : H^1(T_i) \mapsto H^{-\frac{1}{2}}(\partial T_i) \quad ; \quad p \mapsto e^{\Psi} \frac{\partial u_i}{\partial \nu}
$$

They map Dirichlet data on the boundary of a patch T_i via the local problem (16) to the resulting flux through the edges of T_i. The desired formula now reads:

$$
\langle Sp, m \rangle_{\mathcal{M}' \times \mathcal{M}} = \sum_{i=1}^{N} \int_{\partial T_i} T_{\mathrm{DN}}^i p \cdot m \, d\Gamma \quad , \quad p, m \in H^1_{\Gamma_D}(\Omega) =: \mathcal{M}
$$

The right hand side q can be treated in a similar fashion.

2.2. Spectral Analysis of a Model Problem. The mixed hybrid problem for the Lagrangian multiplier is now more closely examined in the case of a simple model problem which permits an explicit calculation of eigenfunctions. The setting is as follows (see Figure 1):

- Ω is the square $]0, \pi[^2$
- The triangulation $\mathcal{T}$ of Ω consists of two patches $T_1 =]0, \pi[\times]0, h[$ and $T_2 =]0, \pi[\times]h, \pi[$ for $h \in]0, \pi[$ with joint edge $\Gamma_3 := \partial T_1 \cap \partial T_2$
- The analysis is confined to piecewise linear potential, meaning that $\Psi(\mathbf{x}) = \langle \mathbf{c}, \mathbf{x} \rangle\ \forall \mathbf{x} \in \Omega$ with a constant vector $\mathbf{c} := \binom{c_1}{c_2} \in \mathbf{R}^2$.
- We assume $R(\mathbf{x}) = 0$ for all $\mathbf{x} \in \Omega$
- We have homogeneous Dirichlet conditions all over $\partial\Omega$.

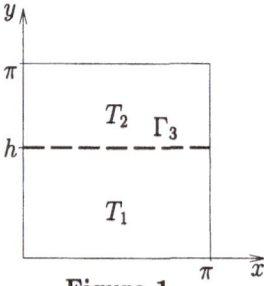

Figure 1

We regard S of (13) as an operator in a space of functions defined on Γ_3 and we are now looking for its eigenfunctions. Since convection–diffusion problems on each subdomain T_1, T_2 are involved at first we have to focus on the operator

$$L : \begin{cases} H^1(D) & \mapsto & H^{-1}(D) \\ n & \mapsto & \Delta n - \langle \mathbf{c}, \nabla n \rangle \end{cases}$$

for an arbitrary domain D. Straightforward calculation shows that for any $\kappa \in \mathbf{R}$ the functions

$$\Psi_\kappa(\mathbf{x}) = e^{\frac{1}{2}\langle \mathbf{c}, \mathbf{x} \rangle} \sin(\kappa x) \sinh\left(\sqrt{\kappa^2 + \frac{1}{4}|\mathbf{c}|^2}\, y\right), \quad \mathbf{x} = \binom{x}{y} \in \Omega, \quad \kappa \in \mathbf{R}$$

belong to the kernel of L. Thanks to this result the boundary problems (16) we face as a first step in the evaluation of Sp yield an analytic solution, if the multiplier p on Γ_3 is of the form

$$p_\kappa(x) = e^{\frac{1}{2}c_1 x} \sin(\kappa x), \quad x \in [0, \pi], \quad \kappa \in \mathbf{N}$$

These functions also satisfy the boundary conditions imposed by our current settings. The main difficulty is now overcome; it just takes mere calculation to show that those $p_\kappa(x)$ are eigenfunctions of S. Moreover, we also get the eigenvalues

$$\lambda_\kappa = \sqrt{\kappa^2 + \frac{1}{4}|\mathbf{c}|^2} \left(\frac{\cosh\left(\sqrt{\kappa^2 + \frac{1}{4}|\mathbf{c}|^2}\, h\right)}{\sinh\left(\sqrt{\kappa^2 + \frac{1}{4}|\mathbf{c}|^2}\, h\right)} + \frac{\cosh\left(\sqrt{\kappa^2 + \frac{1}{4}|\mathbf{c}|^2}\, (\pi - h)\right)}{\sinh\left(\sqrt{\kappa^2 + \frac{1}{4}|\mathbf{c}|^2}\, (\pi - h)\right)} \right), \quad \kappa \in \mathbf{N}$$

Oddly enough, this lengthy expression deserves further attention. First the p_κ represent a complete set of eigenfunctions for the one–dimensional convection–diffusion operator acting on $H^1(\Gamma_3)$, homogeneous boundary conditions provided. Its corresponding eigenvalues are given by $\mu_\kappa := \kappa^2 + \frac{1}{4}|c|^2$: for large κ λ_κ is about twice the square root of μ_κ. Loosely speaking, S behaves like the square root of a 1D convection–diffusion operator. As it was successfully done in the case of purely elliptic problems (see e.g. [4]), these findings can be exploited to construct a preconditioner for the Schur complement system in a domain decomposition framework.

3. Flux–oriented discretization scheme. We now pursue the intriguing idea to find a discretization scheme for problem (1) based on its mixed hybrid formulation (13). In principle we stick to the customary finite element approach of replacing the function spaces by finite dimensional analogues. We have to supply both a finite dimensional space $\mathcal{M}_h$ of ansatz functions and a space $\tilde{\mathcal{M}}_h$ of test functions. They may but do not need to coincide as we do not want to rule out Petrov–Galerkin techniques. In the usual manner approximating finite element spaces $\mathcal{M}_h, \tilde{\mathcal{M}}_h$ are built upon a triangulation $\mathcal{T}$ of Ω. So far any kind of polygonal patches was admitted but from now on we restrict their shapes to triangles and are calling them elements. All other requirements upon $\mathcal{T}$ remain in effect.

We continue with the local definition of approximating finite element functions, developing a representation for a single element. The rash approach to fix the ansatz for the discrete multiplier $p_h \in \mathcal{M}_h$ in the first place is certainly doomed in practice: it is all but impossible to obtain solutions of the local problems (16) that bulked large in the definition of the mixed hybrid problem. A promising remedy lies in role reversal: we recommend to determine a simple solution of (16) first and then to take its values on ∂T_i to build $\mathcal{M}_h$.

We now assume that R vanishes in all of Ω and that the potential Ψ varies only linearly over each element, i.e. $\Psi(\mathbf{x}) = \langle \mathbf{c}_i, \mathbf{x} \rangle$, $\mathbf{x} \in T_i$ with constant $\mathbf{c}_i \in \mathbf{R}^2$. Then, provided that $\mathbf{c}_i \neq 0$, it makes sense to choose

$$\psi_{i,h}(\mathbf{x}) = \alpha \left(\frac{e^{\langle \mathbf{c}_i, \mathbf{x} \rangle} - 1}{|\mathbf{c}_i|} \right) + \beta \langle \mathbf{r}_i, \mathbf{x} \rangle + \gamma, \quad \mathbf{x} \in T_i \qquad (17)$$

(in the trivial case $\mathbf{c}_i = 0$, $\psi_{i,h}$ is taken to be an ordinary linear function), with $\mathbf{r}_i$ being any nonzero vector perpendicular to $\mathbf{c}_i$, as a local ansatz for a function in $\mathcal{M}_h$. The very same approximation of the local flux has already been used by Baliga and Patankar in a finite volume scheme (cf. [2]). Our decision in favour of (17) is backed by several observations: Firstly, the discrete flux

$$\mathbf{j}_i(\mathbf{x}) = e^\Psi \nabla(e^{-\Psi} \psi_{i,h})(\mathbf{x}) = \beta \mathbf{r}_i - (\beta \langle \mathbf{r}_i, \mathbf{x} \rangle + \gamma - \alpha/|\mathbf{c}_i|) \, \mathbf{c}_i$$

is a "smooth" function and does not change in the direction of $\mathbf{c}_i$. Remember that $\mathbf{c}_i$ is parallel to the electric field to see that $\mathbf{j}_i$ neatly fits the current which does not vary much along electric field lines either. Secondly, the ansatz also meets our original goal of satisfying the local problem what $\operatorname{div} \mathbf{j}_i = 0$ gives evidence for.

Given $\operatorname{div}\mathbf{j}_\imath = 0$, the current ansatz is reminiscent of one proposed by Bank et al. in [3], for it is obviously *divergence-free* and this property carries important consequences: By means of Green's formula we see for $\psi_{\imath,h}$ of type (17) and arbitrary $\phi \in H^1(\Omega)$

$$
\begin{aligned}
\langle S\psi_{\imath,h}, \phi\rangle_{\mathcal{M}'(T_\imath)\times\mathcal{M}(T_\imath)} &= \int_{T_\imath} \underbrace{\operatorname{div}(e^\Psi \nabla(e^{-\Psi}\psi_{\imath,h}))}_{=0}\, d\mathbf{x} + \int_{T_\imath} \left\langle e^\Psi \nabla(e^{-\Psi}\psi_{\imath,h}), \nabla\phi\right\rangle d\mathbf{x} \\
&= \int_{T_\imath} e^\Psi \left\langle \nabla(e^{-\Psi}\psi_{\imath,h}), \nabla\phi\right\rangle d\mathbf{x} \\
&= a_{std}(\psi_{\imath,h}, \phi)_{|T_\imath} \quad \text{in} \quad T_\imath
\end{aligned}
$$

with $a_{std}: H^1(\Omega) \times H^1(\Omega) \mapsto \mathbf{R}$ as defined in (3). A related equation can be established for the right hand side of (13): It relies on $\rho_\imath \in H_0^1(T_\imath)$ satisfying

$$
\begin{aligned}
\operatorname{div}(e^\Psi \nabla(e^{-\Psi}\rho_\imath)) &= f \quad \text{in} \quad T_\imath \\
\rho_\imath &= 0 \quad \text{on} \quad \partial T_\imath \; .
\end{aligned}
$$

Due to its shape we dub $\rho_\imath$ "bubble". Again an application of Green's formula shows:

$$
q(\psi_{\imath,h})_{|T_\imath} = f_{std}(\psi_{\imath,h})_{|T_\imath} - a_{std}(\rho_\imath, \psi_{\imath,h})_{|T_\imath} \quad \text{in} \quad T_\imath \; .
$$

If we set $\mathcal{M}_h$ to be the space spanned by the functions (17) over all elements, a preliminary result is that the solutions p_h and n_h of the variational problems

$$
\begin{aligned}
&\text{Find } p_h \in \mathcal{M}_h \text{ such that } \langle Sp_h, m_h\rangle_{\mathcal{M}'\times\mathcal{M}} = q(m_h)\,, \quad m_h \in \mathcal{M}_h \\
&\text{Find } n_h \in \mathcal{M}_h \text{ such that } a_{std}(n_h + \rho, v_h) = f_{std}(v_h)\,, \quad v_h \in \mathcal{M}_h
\end{aligned}
$$

have common values on the edges.

Up to now we have ignored that $\mathcal{M}_h \subset \mathcal{M}$ is violated: Through opting for a simple flux we have sacrificed continuity at interelement boundaries, for tightly welding the $\psi_{\imath,h}$ together would cost us almost all degrees of freedom. Thus we inevitably have to put up with a *nonconforming* method in a narrow sense. Apart from pure continuity there is also a less phenomenological criterion to categorize methods. A discretization is regarded as by nature conforming if source terms are taken into account by jumps of the normal component of the flux across interelement boundaries. Conversely, nonconforming methods usually trade continuity of the solution for n for being more or less flux–conserving. For a reasonably manageable scheme it seems impossible to accommodate both features. In this perspective our divergence–free ansatz can be assigned to the conforming class. Since the flux is the crucial unknown of the continuity equation, we are eager to add some nonconforming flavour. Incorporating the nonconforming principle can be accomplished by enlarging the space. The above calculations provide a clue: They suggest that we add the bubbles $\rho_\imath$ as an additional local basis function. In other words, we try $\mathcal{MB}_h := \mathcal{M}_h + \mathcal{B}_h$ with $\mathcal{B}_h := \{\beta_h : \Omega \mapsto \mathbf{R}\,;\, \beta_{h|T_\imath} \in \operatorname{Span}\{\rho_\imath\}\,, T_\imath \in \mathcal{T}_h\}$ as new ansatz space. But the particular appeal of this choice emerges no sooner than one discovers the relation $a_{std}(m_h, \rho_i) = 0$, $m_h \in \mathcal{M}_h$ meaning that the arising stiffness matrix possesses upper triangular structure if the basis functions are properly arranged. This paves the way for processing source terms independently on each element in advance to calculating the solution on the edges.

We must not gloss over the glaring flaw of the approach described so far: the ρ_i remain elusive. As we are dealing with approximate solutions anyway, we hope to find a simple replacement sufficiently close to ρ_i. For example, the classical cubic bubble scaled by the exponential $e^{1/2\langle \mathbf{c}_i, \mathbf{x}\rangle}$ might be suitable. Also the test space still needs to be specified; pondering experiences obtained for other discretizations we guess that linear test functions can be recommended leading to a Petrov–Galerkin method. All this badly needs empirical underpinning and the results of numerical experiments will be reported in a forthcoming paper.

4. Standard Mixed Finite Element Discretization. Again for simplicity we assume Ω to be a bounded polygonal domain and seek to discretize (11) immediately by means of finite elements. Then with respect to a simplicial triangulation $\mathcal{T} := \{T_i\}_i$ of Ω an appropriate local approximation of the flux space $\mathbf{H}(\mathrm{div}, \Omega, \Gamma_N)$ can be obtained by means of the lowest order Raviart–Thomas element given by

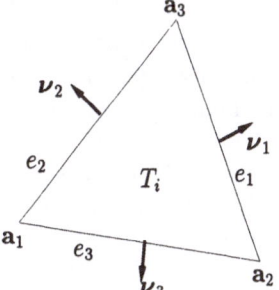

$$\mathcal{RT}_h(T_i) := \{\mathbf{x} \mapsto \mathbf{a} + \beta\mathbf{x} \, ; \, \mathbf{x} \in T_i \, , \, \mathbf{a} \in \mathbf{R}^2 \, , \, \beta \in \mathbf{R}\}$$

Then we set $\mathcal{X}_h := \bigotimes_i \mathcal{RT}_h(T_i)$ to be the "chopped up" global approximation space for the flux. For each $\mathbf{v}_h \in \mathcal{RT}_h(T_i)$ the divergence $\mathrm{div}\,\mathbf{v}_h$ is constant in T_i and so are its normal traces $\langle \mathbf{v}_h, \boldsymbol{\nu}\rangle_{|e_{i,k}}$ along the edges $e_{i,k}$, $k = 1, 2, 3$ of T_i. For the meaning of the symbols we refer to the figure beside.

According to famous theoretical results on saddle point problems (Babuška–Brezzi condition, see [5]), we have to choose a locally constant ansatz for $\mathcal{W}_h$. One can easily check that $\mathbf{v}_h \in \mathcal{RT}_h(T_i)$ is uniquely determined by its normal traces on ∂T_i. This hints that the discrete multipliers need only be constant along edges. It takes little more than counting dimensions of the spaces involved to verify that such an $\mathcal{M}_h$ complies with (9). So we have made sure existence and uniqueness of a solution of

Find $(\mathbf{j}_h, n_h, p_h) \in \mathcal{X}_h \times \mathcal{W}_h \times \mathcal{M}_h$ such that

$$
\begin{aligned}
a(\mathbf{j}_h, \mathbf{v}_h) \;+\; b_1(\mathbf{v}_h, n_h) \;+\; l(\mathbf{v}_h, p_h) &= g(\mathbf{v}_h) \,, & \mathbf{v}_h \in \mathcal{X}_h \\
b_2(\mathbf{j}_h, w_h) \;-\; d(n_h, w_h) \phantom{+ l(\mathbf{v}_h, p_h)} &= f(w_h) \,, & w_h \in \mathcal{W}_h \qquad (18)\\
l(\mathbf{j}_h, m_h) \phantom{+ b_1(\mathbf{v}_h, n_h) + l(\mathbf{v}_h, p_h)} &= 0 \,, & m_h \in \mathcal{M}_h
\end{aligned}
$$

If we partition a comprehensive basis of all approximating spaces according to $(\mathcal{X}_h \times \mathcal{W}_h) \times \mathcal{M}_h$ the associated linear system takes the form

$$\underbrace{\begin{pmatrix} A & L_1^T \\ L_2 & 0 \end{pmatrix}}_{\text{stiffness matrix}} \begin{pmatrix} \vec{z}_h \\ \vec{p}_h \end{pmatrix} = \underbrace{\begin{pmatrix} \vec{r} \\ 0 \end{pmatrix}}_{\text{load vector}} \tag{19}$$

$\vec{z}_h, \vec{p}_h$ are the vectors of unknowns belonging to basis functions in $\mathcal{X}_h \times \mathcal{W}_h$ and $\mathcal{M}_h$, respectively. Please note that now *Lemat* $\neq$ *Lzmat* since the problem in the original unknown n is not symmetric.

4.1. Discrete Hybridization: Implementational Point of View.
As pointed out at length the natural treatment of (19) is *static condensation*. What remains is the smaller linear problem

$$S\vec{p}_h = L_2 A^{-1} \vec{f} =: \vec{t} \quad \text{with} \quad S := L_2 A^{-1} L_1^T \tag{20}$$

Thanks to decoupling through the multiplier, A is blockdiagonal and hence cheaply invertible. With the unknowns in $\vec{z}_h$ being eliminated the choice of a particular basis for $\mathcal{X}_h$ and $\mathcal{W}_h$ does not affect S. A basis of $\mathcal{M}_h$ is constructed in a canonical way: Each of its members has support equal to a single edge in $\mathcal{E}_0$.

We do not need to deal with S as a whole: As the Lagrangian multiplier is the only unknown tying together adjoining elements with respect to (19), static condensation can be carried out on the elements' level, provided the bases of $\mathcal{W}_h$ and $\mathcal{X}_h$ are purely local. Our task is thus reduced to calculating 7×7–stiffness matrices and, on these, doing a block elimination of the unknowns belonging to j_h, n_h. After this cumbersome procedure has been finished we have the 3×3–element stiffness matrix for the multipliers, in symbolic notation given by

$$S_{loc} = -\epsilon^{-1} \begin{pmatrix} |e_1| & & \\ & |e_2| & \\ & & |e_3| \end{pmatrix} U \begin{pmatrix} \theta_1 & & \\ & \theta_2 & \\ & & \theta_3 \end{pmatrix}$$

where we used the abbreviations

$$U := (\langle \nu_l, \nu_j \rangle)_{1 \le l, j \le 3} \quad , \quad \epsilon := \int_{T_i} e^{-\Psi} d\mathbf{x} \quad , \quad \theta_j := \int_{e_j} e^{-\Psi} d\Gamma$$

We remark that the formulas above are only valid in the case $R = 0$. For ease of presentation we also forgo the separate treatment of elements attached to the Dirichlet boundary Γ_D.

The element load vector can be expressed by

$$\vec{t}_{loc} = -\frac{1}{2|T_i|} \int_{T_i} f \, d\mathbf{x} \begin{pmatrix} |e_1| & & \\ & |e_2| & \\ & & |e_3| \end{pmatrix} \vec{\zeta}$$

where $\zeta \in \mathbf{R}^3$ and $\zeta_j := \langle \mathbf{p}, \boldsymbol{\nu}_j \rangle - \epsilon^{-1} \int_{T_i} e^{-\Psi} \langle \mathbf{x}, \boldsymbol{\nu}_j \rangle \, d\mathbf{x}$ with $\mathbf{p} \in e_j$. A brief calculation shows that S_{loc} and $\vec{t}_{loc}$ can be read as the stiffness matrix and the load vector of the following local variational problem:

Find $p_h \in \mathcal{M}_h(T_i)$ such that

$$\epsilon^{-1} \left\langle \int_{\partial T_i} m_h \boldsymbol{\nu} \, d\Gamma, \int_{\partial T_i} e^{-\Psi} p_h \boldsymbol{\nu} \, d\Gamma \right\rangle = -\frac{1}{2|T_i|} \int_{\Omega} f \, d\mathbf{x} \cdot \pi(m_h) \,, \quad m_h \in \mathcal{M}_h(T_i) \quad (21)$$

where

$$\pi(m_h) = \int_{\partial T_i} m_h \left\langle \mathbf{x} - \epsilon^{-1} \int_{T_i} e^{-\Psi(\mathbf{x}')} \mathbf{x}' \, d\mathbf{x}', \boldsymbol{\nu} \right\rangle d\Gamma \,.$$

In practice the rationale behind solving (1) often is to determine the current $\mathbf{j}$ passing through a certain section of the boundary, for example Ohmic contacts. Unfortunately, we just discarded $\mathbf{j}_h$ in the process of static condensation. Therefore, we face the task of retrieving $\mathbf{j}_h$ from the values of the multiplier p_h. First we observe that this again can be done for each element T_i separately. Tedious tinkering with the full 7×7 local stiffness matrix is rewarded by a surprising result: The total flux through the edge of an element is readily available as a component of the residual of the equation $S_{loc} \vec{p}_{h,loc} = \vec{t}_{loc}$. In this context $\vec{p}_h$ covers three components of $\vec{p}$ which are linked to the edges of T_i. Thus we can obtain the desired fluxes at virtually no extra cost. This should be taken as an additional incentive to search for error estimators based on the flux. It seems wise to do so anyway, considering the pivotal role of the flux in mixed discretization schemes.

4.2. An Equivalent Nonconforming Petrov–Galerkin Ansatz. In section 3, setting out at the mixed hybrid problem, we unexpectedly arrived at a scheme that looked like a modified standard formulation (2). As well the genuine mixed discretization given before can be recast to resemble a variant of (2), though a good deal of twisting is needed. Our presentation partly follows that of Reusken in [12] Again we pick a single element T_i with edges $\{e_1, e_2, e_3\}$, outward normal unit vectors $\{\boldsymbol{\nu}_1, \boldsymbol{\nu}_2, \boldsymbol{\nu}_3\}$ and midpoints $\{\mathbf{m}_1, \mathbf{m}_2, \mathbf{m}_3\}$ of the edges. Furthermore, $R = 0$ is assumed.

Let $\{\phi_{i,h}^1, \phi_{i,h}^2, \phi_{i,h}^3\}$ be the local canonical basis of the Crouzeix–Raviart space $\mathcal{CR}_h$ of linear nonconforming functions given by

$$\phi_{i,h}^k(\mathbf{x}) = \frac{|e_k|}{|T_i|} \langle \mathbf{x} - \mathbf{m}_l, \boldsymbol{\nu}_k \rangle \,, \quad 1 \leq k \neq l \leq 3 \quad \mathbf{x} \in T_i$$

We additionally require continuity at the midpoints of interelement boundaries and in this case these functions (for all the T_i combined) span the space of test functions. Following the Petrov–Galerkin principle we use the scaled functions

$$\tilde{\phi}_{i,h}^k := \left(\frac{1}{|e_k|} \int_{e_k} e^{-\Psi} \, d\Gamma \right) \phi_{i,h}^k \,, \quad 1 \leq k \leq 3 \quad (22)$$

as a local basis of the ansatz space $\tilde{C\mathcal{R}}_h$, from which a global basis is constructed as before.

LEMMA 4.1. *Let* $\{\mu_{i,h}^1, \mu_{i,h}^2, \mu_{i,h}^3\}$ *be the set of local basis functions of* $\mathcal{M}_h$ *belonging to the edges of the element* T_i. *Then*

$$\int\limits_{T_i} \left(\frac{1}{|T_i|} \int\limits_{T_i} e^{\Psi(\mathbf{x}')}\, d\mathbf{x}' \right)^{-1} \left\langle \nabla \tilde{\phi}_{i,h}^k, \nabla \phi_{i,h}^l \right\rangle d\mathbf{x} = \left(\int\limits_{T_i} e^{-\Psi}\, d\mathbf{x} \right)^{-1} \left\langle \int\limits_{\partial T_i} \mu_{i,h}^l \boldsymbol{\nu}\, d\Gamma, \int\limits_{\partial T_i} e^{-\Psi} \mu_{i,h}^k \boldsymbol{\nu}\, d\Gamma \right\rangle$$

holds for $1 \leq k, l \leq 3$. *The functional* $\pi : \mathcal{M}_h(T_i) \mapsto \mathbf{R}$, *defined in (21), satisfies*

$$\pi(\mu_{i,h}^k) = -\int\limits_{T_i} \left(e^{-\Psi(\mathbf{x})} \left(\frac{1}{|T_i|} \int\limits_{T_i} e^{-\Psi(\mathbf{x}')}\, d\mathbf{x}' \right)^{-1} - 3 \right) \phi_{i,h}^k\, d\mathbf{x}, \quad 1 \leq k \leq 3$$

These relations give ample hint about how to alter a_{std} and f_{std} of (2) in such a way that the new forms can describe the variational problem (21):

$$
\begin{aligned}
a_{mod}(u, v) &= \sum_i \int\limits_{T_i} \left(\frac{1}{|T_i|} \int\limits_{T_i} e^{-\Psi}\, d\mathbf{x}' \right)^{-1} \langle \nabla u, \nabla v \rangle\, d\mathbf{x} \\
f_{mod}(v) &= \sum_i \frac{1}{2|T_i|} \int\limits_{T_i} f\, d\mathbf{x} \cdot \int\limits_{T_i} \left(e^{-\Psi} \left(\frac{1}{|T_i|} \int\limits_{T_i} e^{-\Psi}\, d\mathbf{x}' \right)^{-1} - 3 \right) v\, d\mathbf{x}
\end{aligned}
\tag{23}
$$

With these tailored expressions we finally have:

THEOREM 4.2. *The stiffness matrix arising from the discrete variational problem*

$$\text{Find } u_h \in \tilde{C\mathcal{R}}_{h|\Gamma_D} \text{ such that } a_{mod}(u_h, v_h) = f_{mod}(v_h), \quad v_h \in C\mathcal{R}_{h|\Gamma_D}$$

is equal to the stiffness matrix S *of (20) if we build the test space from the canonical nodal basis of the lowest order Crouzeix–Raviart space and the ansatz space from their scaled counterparts (22).*

The conspicuous term

$$\left(\frac{1}{|T_i|} \int\limits_{T_i} e^{-\Psi}\, d\mathbf{x} \right)^{-1}$$

in (23) is the definition of the harmonic average of the function e^{Ψ} over T_i. To this expression an important class of discretization schemes for the continuity equations (1), namely the *inverse averaging type methods*, owes its name. We consider it satisfactory that mixed methods belong to this prominent club that also includes the widely used Scharfetter–Gummel scheme.

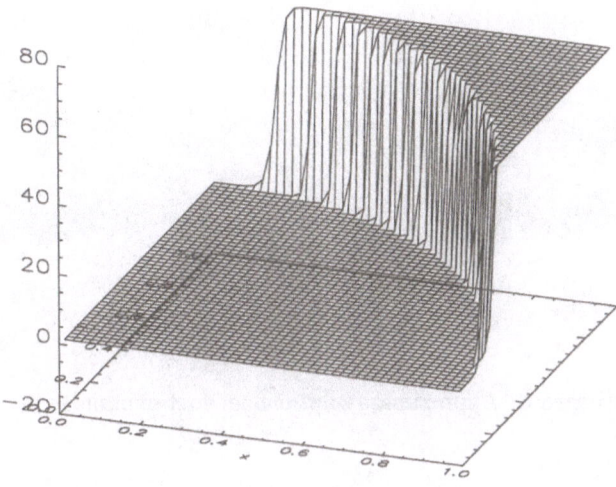

Figure 2: Exact solution for first experiment

5. Numerical experiments. Though general convergence estimates for standard mixed discretizations are available (cf. [1], [6]) they give little insight into the actual performance of the methods in practice. Too many details of their behaviour and qualitative phenomena still defy a theoretical explanation, let alone prediction. So we have to rely on numerical experiments to probe the quality of the method when applied to the continuity equation (1). In semiconductor device simulation the particular challenge of (1) is posed by the layer behaviour of the potential Ψ: Steep gradients occur near pn–junctions whereas Ψ is only slightly varying elsewhere. In order to model these conditions we used the "radial step potential" $\Psi(\mathbf{x}) = \sigma_a(r), r = \sqrt{x_1^2 + x_2^2}, \mathbf{x} = (x_1, x_2) \in \Omega, \sigma_a(r) = 1/(1 + e^{-a(r-1)})$. The parameter a governs the slope of the step: the larger a the sharper the drop. All calculations were carried out on $\Omega =]0, 1[^2$. If necessary, linear interpolation of Ψ and f was used and all linear systems have been solved exactly.

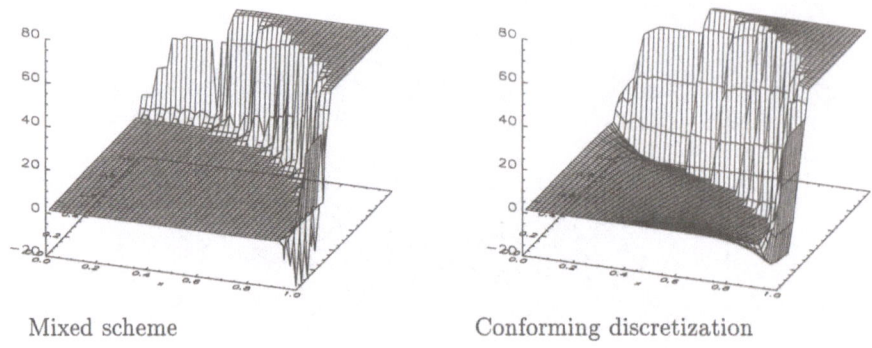

Mixed scheme   Conforming discretization

Figure 3: Approximate solutions for first experiment

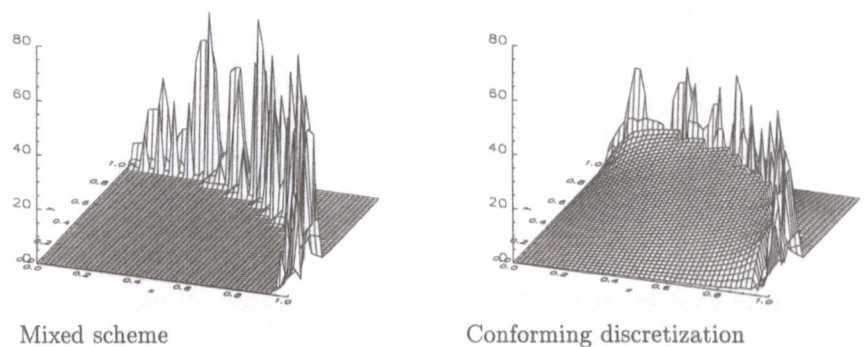

Mixed scheme   Conforming discretization

Figure 4: Absolute errors of approximate solutions of first experiment

For a first experiment Ω was triangulated by a rectangular grid of mesh width $h = \frac{1}{16}$ with each cell being further subdivided into two triangles. The radial step potential with $a = 500$ and vanishing right hand side were employed, along with boundary values that yield e^{Ψ} as exact solution. In physical parlance this situation is referred to as thermal equilibrium. We solved the problem by our mixed method and compared the solution and the errors with those obtained by using standard conforming linear elements. The figures 2 through 5 give a graphical representation of the results.

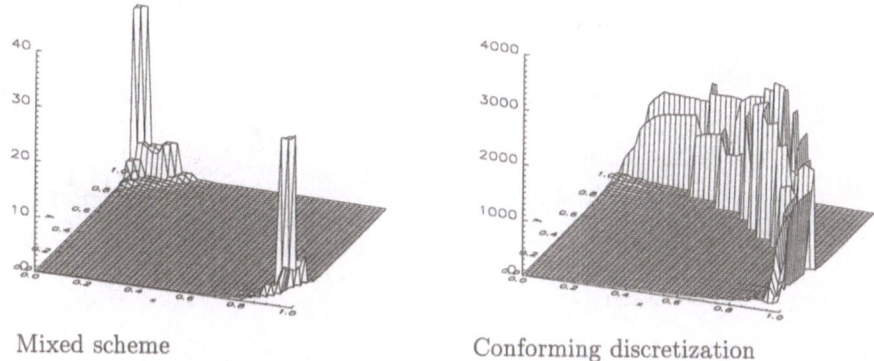

Mixed scheme   Conforming discretization

Figure 5: Absolute errors of the flux for first experiment

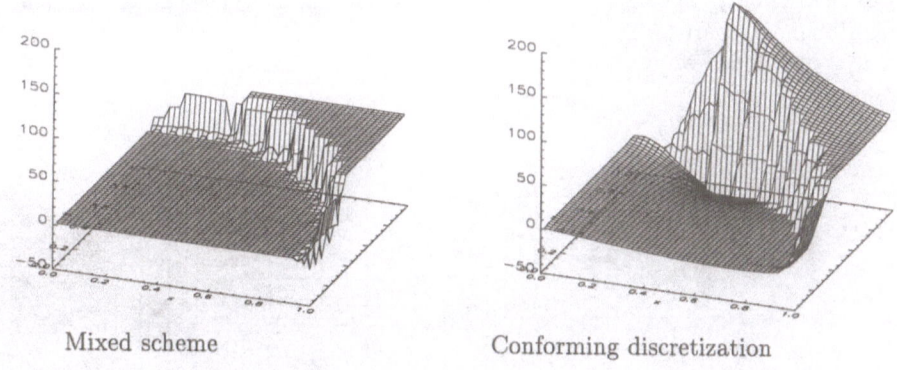

Mixed scheme   Conforming discretization

Figure 6: Approximate solution for 2nd experiment. (Two Neumann sides)

We see that the mixed discretization benefits from the very smooth, actually vanishing flux in this experiment. It copes with the internal layer of the solution for n far better than the conforming method and flatly outstrips the latter with respect to reproducing the zero flux. It should be noted that presumably not the method itself but post-processing has to be blamed for the spurious spikes that pop up close to the step. Perhaps this taint can be removed by a more refined post-processing that includes bubble functions as investigated in [1]. The mixed method may perform strikingly better in this setting, but that does the method only small credit, because the comparison is not quite fair, as the conforming discretization are prone to failure in the presence of dominating convection.

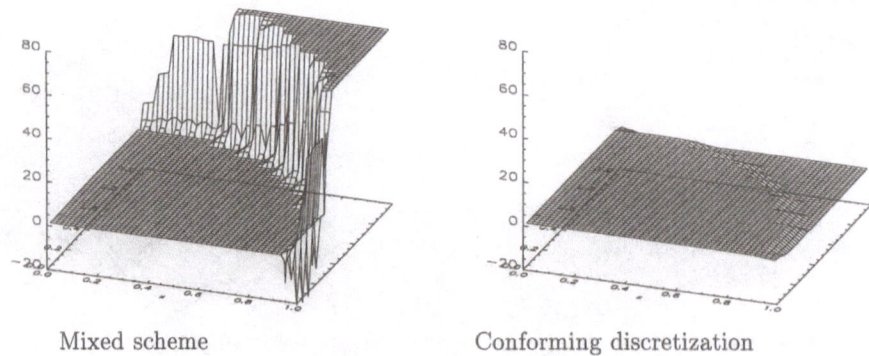

Mixed scheme   Conforming discretization

Figure 7: Approximate solution for 2nd experiment. (Three Neumann sides)

The next experiment was designed to investigate the impact of boundary conditions. The setting remained unchanged except that Neumann boundary conditions were imposed on one or more sides of Ω. The figures 6 and 7 show how both methods respond to additional Neumann boundary parts.

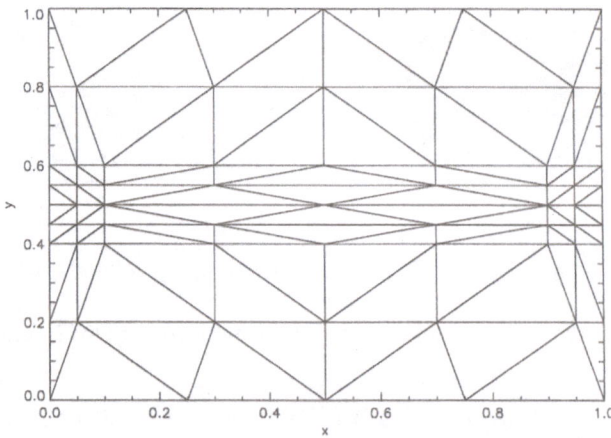

Figure 8: Triangulation with obtuse elements

The mixed solution turns out to be scarcely affected by changing boundary conditions. Conversely, the conforming approximation quickly deteriorates when the size of the Neumann boundary portions increases. This is hardly surprising, because at Neumann boundaries the flux is fixed and so is one unknown of the mixed ansatz. This accounts for the stabilizing effect of large Neumann parts on the mixed approximation.

Flux oriented discretizations are notorious for their vulnerability to "badly" shaped elements that have an obtuse angle. To expose this shortcoming of the mixed method we built a triangulation with obtuse elements in one part of the domain (See Figure 8), which was then regularly refined twice. This (third) experiment was conducted with a linear potential $\Psi(\mathbf{x}) = \langle \mathbf{c}, \mathbf{x} \rangle, \mathbf{c} \in \mathbf{R}^2$. The boundary values and right–hand side were adjusted to yield a polynomial solution. The distribution of different kinds of errors for three different directions $\mathbf{c}$ of the convection is displayed in figures 9, 10 and 11. Figure 12 shows how the presence of wretched elements leads to distortions in the calculated flux for a particular choice of $\mathbf{c}$. Besides we measured how errors depend on the direction of $\mathbf{c}$. $\mathbf{c}$ had length 100 and in very small steps was fully rotated around the origin (α denotes the angle relative to the abscissa.). The result is plotted in figure 13.

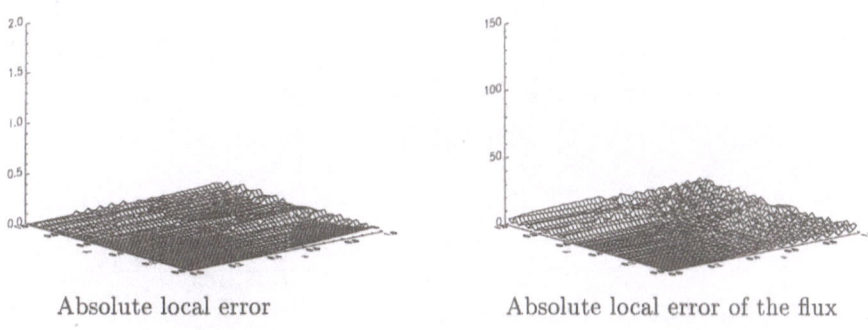

Absolute local error   Absolute local error of the flux

Figure 9: Results of experiment 3 with $\mathbf{c} = \binom{100}{0}$

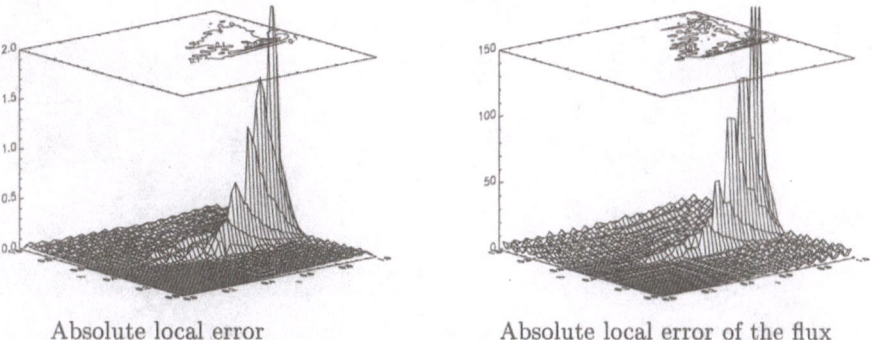

Absolute local error   Absolute local error of the flux

Figure 10: Results of experiment 3 with $\mathbf{c} = \binom{70}{70}$

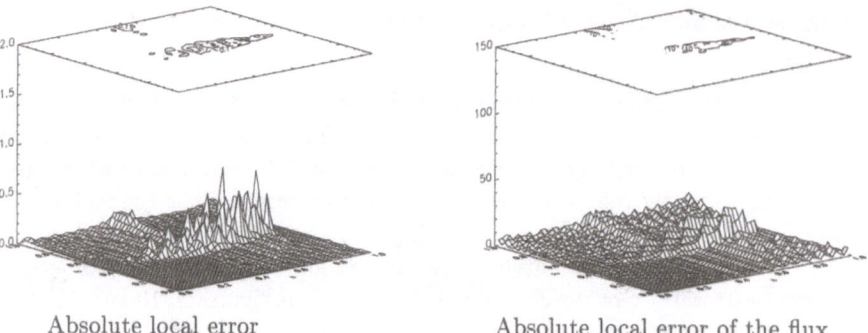

Absolute local error   Absolute local error of the flux

Figure 11: Results of experiment 3 with $\mathbf{c} = \left(\begin{smallmatrix} 0 \\ 100 \end{smallmatrix}\right)$

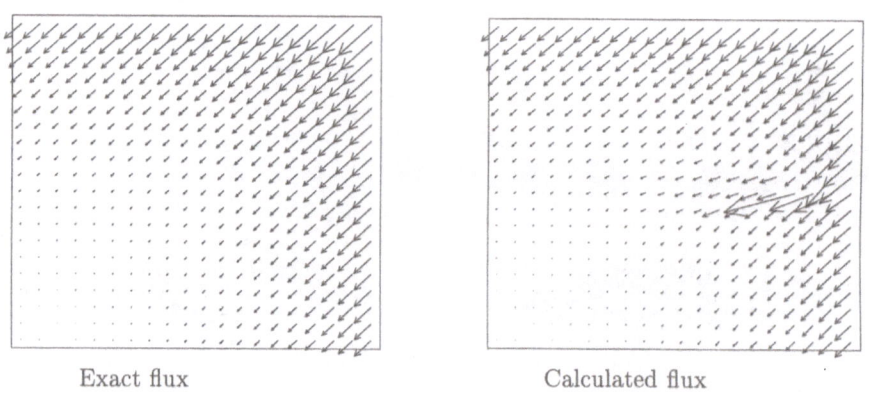

Exact flux   Calculated flux

Figure 12: Exact and calculated flux for $\mathbf{c} = \left(\begin{smallmatrix} 70 \\ 70 \end{smallmatrix}\right)$

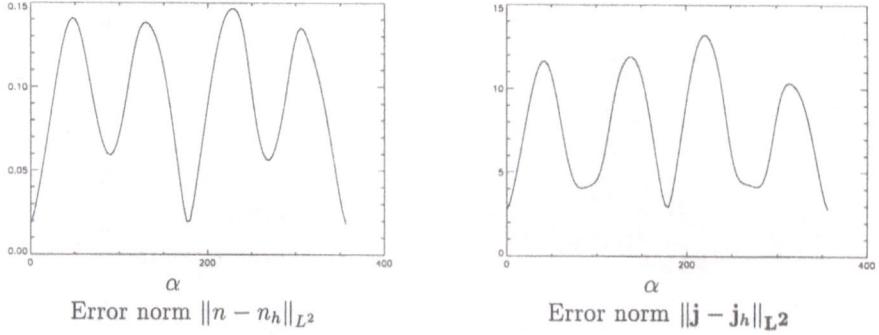

Error norm $\|n - n_h\|_{L^2}$ Error norm $\|\mathbf{j} - \mathbf{j}_h\|_{\mathbf{L}^2}$

Figure 13: L^2-Norm of the error depending on α for experiment 3

The results teach that for some values of c obtuse angles might cause disastrous instabil-ity. The plots also reveal that in these cases large errors spread to parts of the domain where the triangulation is not "marred", that is, a sort of pollution effect is present. Apparently other directions of c do no harm; we see an enigmatic relationship between error and direction of convection: Maybe a clever orientation of the obtuse elements can help to steer clear of the instability trap. Nevertheless, the results send a daunting message as far as an adaptive strategy is concerned: While for plane problems clever re-finement strategies for triangulations exist that avoid obtuse elements, in 3–D simulation they inevitably occur. The viability of the mixed method in practice hinges on whether instability can be managed.

REFERENCES

[1] D. N. ARNOLD, F BREZZI, *Mixed and Nonconforming Finite Element Methods: Implementation, Post-Processing and Error Estimates*, Math Modelling Numer Anal , 19, 7–35 (1985)

[2] B. R. BALIGA, S V. PATANKAR, *A New Finite Element Formulation for Convection-Diffusion Problems*, Numerical Heat Transfer, 3, 393–409 (1980)

[3] R.. BANK, J. BÜRGLER, W. FICHTNER, R. K SMITH, *Some Upwinding Techniques for Finite Element Approximations of Convection-Diffusion Equations*, Num. Math, 58, 185–202 (1990)

[4] P.. BJØRSTAD, O. B. WIDLUND, *Iterative Methods for the Solution of Elliptic Problems on Regions Partitioned into Substructures*, SIAM J. Numer Anal., 23, 1097–1120

[5] F. BREZZI, *On the Existence, Uniqueness and Approximation of Saddle Point Problems Arising from Lagrangian Multipliers*, R A I R.O Anal. Numer , 8, 129–151

[6] F. BREZZI, M. FORTIN, *Mixed and Hybrid Finite Element Methods*, Springer– Verlag, New York (1991)

[7] F BREZZI, L D MARINI, P PIETRA, *Two Dimensional Exponential Fitting and Application to Drift-Diffusion Models*, SIAM J. Numer. Anal , 26, 1347–1355 (1989)

[8] ——, *Numerical Simulation of Semiconductor Devices*, Comp Math. Appl Mech Eng , 75, 493–514 (1989)

[9] I EKELAND, R. TEMAM, *Convex Analysis and Variational Problems*, North–Holland, Amsterdam (1978)

[10] T J R HUGHES, A. BROOKS, *Streamline-Upwind Petrov-Galerkin Formulations for Convective Dominated Flows with particular Emphasis on the Incompressible Navier Stokes Equations*, Comput. Methods Appl Mech. Eng , 32, 199–259 (1982)

[11] P. A RAVIART, J. M THOMAS, *A Mixed Finite Element Method for Second Order Elliptic Problems*, Lecture Notes in Mathematics, 606, Springer–Verlag, Berlin (1977)

[12] A REUSKEN, *Multigrid Applied to Mixed Finite Element Schemes for Current Continuity Equations*, Preprint Technical University Einhoven (1990)

[13] VEUBEKE, B. FRAEIJS DE, *Displacement and Equilibrium Models in the Finite Element Method* in "Stress Analysis" (O. C Zienkiewicz, G Hollister, eds), John Wiley and Sons, New York (1965)

International Series of Numerical Mathematics, Vol. 117, © 1994 Birkhäuser Verlag Basel   219

A PIECEWISE LINEAR PETROV-GALERKIN ANALYSIS OF THE BOX-METHOD

T. KERKHOVEN
UNIVERSITY OF ILLINOIS
URBANA, IL 61801
U.S.A.

Abstract. For the type of boundary value problems which figure in the drift-diffusion semiconductor model, the perpendicular-bisecting box-method discretization yields M-matrices and maximum stability on Delaunay triangulations, whereas the piecewise linear finite element discretization need not. In two dimensions this discrepancy is due to variability of coefficients in the drift-diffusion equations. In three dimensions the difference is, furthermore, of geometric origin. Some results for both dimensionalities are presented and discussed.

Key words: Finite Elements, box-method, stability.

Acknowledgements: While working on this paper the author was supported by the National Science Foundation under grant ECS-91-20641.

1. Introduction. Under the assumption of Einstein's relations and at constant temperature T, the steady state drift-diffusion semiconductor model in terms of the electron density n, the hole density p, and the dimensionless electrostatic potential $u = (q\phi)/(k_B T)$ is given by

$$-\nabla \cdot [\epsilon_r \nabla u] + n - p - k_1 = 0, \tag{1.1}$$

$$\nabla \cdot k_B T \mu_n [n \nabla u - \nabla n] = -qR, \tag{1.2}$$

$$-\nabla \cdot k_B T \mu_p [p \nabla u + \nabla p] = -qR. \tag{1.3}$$

Here q is the size of the charge of the electron, μ_n and μ_p are the respective conduction electron and hole mobilities, and k_B is Boltzmann's constant. Discretization of the boundary value problems (BVPs) in usually proceeds by some version of the "box-method," see e.g. [3]. Computational semiconductor models are desired to provide extrema stability in a priori bounds on the extrema of the solution, properly defined contact currents which add up to 0 to reflect charge conservation, and acceptable accuracy. These properties of the discretized model depend on the choice of variables for discretization, on the discretization procedure, and on the mesh geometry. The discretization variables are usually chosen to be the electrostatic potential u and, either the densities n and p, or the Slotboom variables ν and ω, or the quasi-Fermi levels v and w. The box-method is the most prevalent discretization procedure, including a suitable expression for the current flux Φ_{jk} through box face f_{jk} between boxes B_j and B_k (often obtained by the one dimensional technique of Scharfetter and Gummel [10]). Both irregular triangular and nonuniform rectangular meshes are common. Possibly, the variables of the discretized equations are transformed to a different set to improve computational properties.

In the current continuity system, a change from the densities n and p in (1.1–1.3) to the Slotboom variables $\nu = n\exp(-u)$ and $\omega = p\exp(u)$ yields the system

$$-\nabla\cdot[\epsilon\nabla u] + [\nu\exp(u) - \omega\exp(-u) - k_1] = 0, \tag{1.4}$$

$$-\nabla\cdot[\mu_n\exp(u)\nabla\nu] = -\frac{q}{k_B T}R, \tag{1.5}$$

$$-\nabla\cdot[\mu_p\exp(-u)\nabla\omega] = -\frac{q}{k_B T}R. \tag{1.6}$$

The transformation to the system (1.4–1.6) in terms of ν and ω generates a self-adjoint formulation of the current continuity equations, without first order convective terms, and may be viewed as the application of an exponential "upwinding" technique. The pde's in (1.4–1.6) are of the general form

$$-\nabla\cdot[a(x,u)\nabla u] + f(x,u) = g(x), \tag{1.7}$$

with $\partial f/\partial u \geq 0$, on an N dimensional domain G, and subject to Dirichlet boundary conditions on the part $\Sigma_D G$ of the boundary of G, and to Neumann boundary conditions on the part $\Sigma_N G$. The second order term in (1.7) can be used to define the bilinear form

$$a(v,w) \equiv \int_G a(x)\nabla v\cdot\nabla w\,dx. \tag{1.8}$$

Employing the inner product (1.8), the BVP (1.7) is rewritten and further specified below.

DEFINITION 1.1. *Let $a(\cdot,\cdot)$ be defined by (1.8). For $u \in H^1$, let $F(u)$ denote the continuous linear functional on H^1_{0,Σ_D} defined by,*

$$F(u)(v) = \int_G f(x,u)v\,dx, \tag{1.9}$$

where f is monotonically increasing in its second argument, and $\partial f(\cdot,s)/\partial s \leq C$. Let $\bar{u} \in C^2(\bar{G})$, and let $g \in L_2$ be prescribed.

Then, the BVP for the function $u \in \bar{u} + H^1_{0,\Sigma_D}$, is defined by the gradient relation,

$$a(u,v) + F(u)(v) = \langle g,v\rangle, \quad \forall v \in H^1_{0,\Sigma_D}. \tag{1.10}$$

A standard extrema principle argument yields, for the solution u to (1.10), the a priori bounds,

$$\min\{\inf_{\Sigma_D G} u, f^{-1}(x,g_{\min})\} \leq u(x) \leq \max\{\sup_{\Sigma_D G} u, f^{-1}(x,g_{\max})\}. \tag{1.11}$$

A suitable discretization technique, combined with an appropriate mesh-geometry, transfers in the drift-diffusion system (1.4–1.6) to the discretized model the a priori extrema stability (1.11) of the Slotboom variables or the electrostatic potential.

In §2.1 a discrete extrema principle for a standard "box-method" discretization of (1.10) is related exclusively to the mesh-geometry, independently of the variable coefficient $a(x)$. Therefore, one set of sufficient conditions on the mesh geometry yields extrema stability of standard "box-method" discretizations for the second order part of (1.10), irrespective of the variable coefficient $a(x) > 0$. Moreover, the "box-method" discretization allows a straightforward interpretation of current fluxes Φ_{jk} for the discretized model.

On a simplicial mesh, both the box-method and a finite element discretization of (1.10) are expressed in terms of the geometric properties of the N-dimensional simplex S_r stated below.

DEFINITION 1.2. *Let S_r be an N-dimensional simplex. Then,*

- V_{S_r} *is the volume,*
- $\vec{x}_i$ *is a vertex,*
- e_{ij} *is the edge connecting vertices $\vec{x}_i$ and $\vec{x}_j$,*
- F_k *is the face opposite to the vertex k, with measure $|F_k|$. In the global simplicial mesh the notation F_{rs} is also employed, indicating the face in between the simplices S_s and S_r,*
- h_i *is the normal distance of $\vec{x}_i$ to F_i,*
- γ_{ij} *is the angle between the inward normal vectors to the faces F_i and F_j.*

In (1.7) the expression $a(x)\nabla u$ corresponds to a current density like object. A box method discretization relates fluxes Φ_{jk} of this current density, through box faces f_{jk} between adjacent boxes B_j and B_k, to the integral over source and drain terms inside a box B_j.

In [2], Bank and Rose have shown that, in two dimensions, the perpendicular bisector box-method discretization of the Laplacean generates the stress-matrix of the standard piecewise linear finite element method. This observation is used as a starting point for the presented finite element convergence theory for box-method discretizations of the drift-diffusion model. Both in the box-method, and in the finite element method for a piecewise linear approximation U_h, one component u_j of the solution vector $\mathbf{u}$ corresponds to every vertex $\vec{x}_j$ in the mesh. In the sequel, the notation $U_{\mathbf{u}} = \sum_j u_j \phi_{pl,j}(\vec{x})$ will be employed for the piecewise linear finite element function associated with the vector of nodal values $\mathbf{u}$. More generally, with the nodal vector $\mathbf{v}$ will also be associated the nodal piecewise polynomial function $V_{pp,\mathbf{v}} = \sum_j v_j \psi_j(\vec{x})$. These more general functions $\psi_j(\vec{x})$ are required to satisfy the nodality condition, $\psi_j(\vec{x}_l) = \delta_{jl}$, and for all $\vec{x}$, the positivity condition, $\psi_j(\vec{x}) \geq 0$.

As mentioned above, the standard box-method discretization (2.2) of (1.10) sets the sum of expressions for current fluxes Φ_{jk} through box faces f_{jk} equal to the integral over the source terms inside the box B_j. This implies that the box-method is flux-oriented, rather than element oriented. Therefore, in the box-method, the expression for the flux Φ_{jk} includes an approximation of the average of the variable coefficient $a(x)$ over the face f_{jk}. In the bilinear form (1.8), on the other hand, element avarages are taken. Furthermore, the finite element approxima- tion to the nonlinear term $F(u)(v)$ in (1.10) differs from its box-method discretization in(2.2). Therefore, in Definition 2.4 a relaxed version of the BVP in Definition 1.1 is introduced, with piecewise constant coefficient $\tilde{a}(x)$ and piecewise linear boundary conditions. Lemma 2.1 from [8] asserts that, to piecewise linear accuracy, the BVP in Definition 1.1, with coefficient $a(x)$ and boundary conditions $\bar{u}$, is a perturbation of the relaxed BVP in Definition 2.4.

In the BVP in Definition 2.4, the variable coefficient $\tilde{a}(x)$ is constant on single elements S_r. Therefore, on a single element S_r, the bilinear form (1.8), corresponding to the intermediate BVP in Definition 2.4, is similar to the Laplacean, and an analysis which generalizes the procedure in [2], becomes applicable.

Similarly, in Lemma 2.3 from [8] the box-method equations (2.2) are examined as an approx- imation to an intermediate, piecewise constant $\tilde{a}(x)$, discretization (2.12), in which the nonlinear term $F(u)(v)$ in (1.10) is discretized by a suitable Petrov-Galerkin method. Sufficient general- ity is allowed to include the common Scharfetter-Gummel flux approximation. Taken together, the results in Lemmas 2.1 and 2.3 allow the correspondence of finite element and box-method discretizations of (1.8) to be established for the Laplacean on a single simplicial element S_r.

This correspondence is detailed in §3, where the results for the Laplacean in [2] are gen- eralized both to dimensions of more than two, and to general Petrov-Galerkin methods. First, Lemma 3.2 from [8] asserts, for arbitrary dimension N, that for a piecewise linear approximation $U_h(\vec{x})$, the Petrov-Galerkin element matrix corresponding to the Laplacean, is not modified by substitution for piecewise constant box test functions $\phi_{B_j}(\vec{x})$, test functions $\psi_j(\vec{x})$ which assume on element-faces F_s equal averages as the box-functions ϕ_{B_j}. Subsequently, again for aribrary dimension N, by Lemma 3.3 from [8], for element stress-matrices for the Laplacean, it is shown that the box method stress-matrix $A_{B,\tilde{a}}$, defined in (2.13), is equal to the stress-matrix $A_{PG,\tilde{a}}$, generated by Petrov-Galerkin equations for a piecewise linear approximation U_h with piecewise constant box test functions $\phi_{B_j}(\vec{x})$.

Together, Lemma's 3.2 and 3.3 imply that, for any test functions $\psi_j(\vec{x})$, which assume equal element-face averages as the box test functions ϕ_{B_j}, the intermediate box-method equations in (2.12) are equal to the Petrov-Galerkin discretization (2.4) of the relaxed BVP in Definition 2.4. Therefore, error estimates for the solution U_h to the Petrov-Galerkin equations (2.4), with piecewise constant $\tilde{a}(\vec{x})$, are valid for the piecewise linear approximation U_u, corresponding to the solution vector u to the relaxed box-method equations (2.12).

In two dimensions the perpendicular bisecting box-method is indeed equal to the piecewise linear finite element stress-matrix, but different in higher dimensions. For this reason a finite element approximation theory, based on the Petrov-Galerkin method for a piecewise linear approximation with more general test functions [1], is constructed in [8]. The order of convergence for the relaxed Galerkin and Petrov-Galerkin discretizations of the single BVPs follows from the precise interpolation space in which the solution to these BVPs lies.

2. Maximum Principles in Computations. As mentioned above, discretizations of the drift-diffusion semiconductor model are usually developed by the box-method. On a domain G, discretization by the box method (and discretization by the Petrov-Galerkin method for a piecewise linear approximation U_h, to be discussed later on) commences with the definition of a set of vertices $\vec{x}_j$, connected by edges e_{jk}, partitioning the interior of G into simplices S_r, as in Definition 1.2. For the finite element discretizations this definition determines the entire mesh. For the box method, the definition of boxes B_j has to be added. First, the widely employed box discretization is developed.

2.1. Box Method Discretization. In the box-method discretization, one algebraic equation corresponds to every vertex $\vec{x}_i$ in the mesh. Charge conservation for the discretized equations is obtained by associating with every vertex $\vec{x}_i$, a dual box B_i such that the boxes B_i form a complete covering of the domain G without overlap, and that the sum of the fluxes Φ_{jk} from a given box B_j to all its neighboring boxes B_k is equal to the sum of the source terms inside the box B_k. Hence, discrete charge conservation is obtained with respect to a suitably chosen grid of boxes that is dual to the finite element grid. Stated formally,

DEFINITION 2.1. *Let S_r be an N-dimensional simplex as in definition 1.2. Then, on the simplex S_r the mesh of dual boxes is defined such that,*

- *To every vertex $\vec{x}_j$ in the simplicial mesh corresponds exactly one box B_j in the mesh of boxes.*
- *To every simplex S_r in the simplicial grid corresponds exactly one vertex $\vec{c}_r$ in the mesh of boxes.*
- *To every edge e_{jk} in the simplicial grid corresponds exactly one (not necessarily planar) face f_{jk} common to the two boxes B_j and B_k.*
- *Every element face F_{rs} in the simplicial mesh is divided in exactly N connected parts $p_{rs,j}$, where $p_{rs,j}$ contains the vertex $\vec{x}_j$ in the simplicial mesh.*
- *$A_{B,ij}$ is the ij-th component of the discretized self-adjoint operator.*

In the text below, usually the face f_{jk} is a single $N-1$ dimensional plane. This is, however, *not* required by the definition above, and f_{jk} may be an arbitrary surface.

The definition of the box method is completed by the rule which defines the fluxes Φ_{jk} through the faces f_{jk} between adjacent boxes. The box method discretization, based on perpendicular bisectors, of the equation (1.7) is usually defined with respect to the mesh of boxes B_j as follows. The faces f_{jk} of the boxes B_j are assumed to be normal to their corresponding edges e_{jk} in the simplicial mesh. The flux Φ_{jk} through face f_{jk} from box B_j into box B_k is defined by

$$\Phi_{jk} = a_{e_{jk}}(|f_{jk}|/|e_{jk}|)[u_j - u_k], \qquad (2.1)$$

where $a_{e_{jk}}$ approximates the average of $a(\vec{x})$ over inter-box face f_{jk}. As in [3], eqn. (32), the nonlinear term $f(\vec{x}, u)$ in (1.7) gives rise to a contribution on the diagonal, multiplied by the surface areas of the boxes B_j. To box B_j then corresponds the discretized equation

$$\sum_{\vec{x}_k} a_{e_{jk}}(|f_{jk}|/|e_{jk}|)[u_j - u_k] + |B_j|f(\vec{x}_j, u_j) = |B_j|g(\vec{x}_j). \qquad (2.2)$$

The solution **u** to (2.2) satisfies a discrete extrema principle if the off-diagonal elements $A_{B,jk} = -a_{e_{jk}}(|f_{jk}|/|e_{jk}|)$ of the discretized BVP are not positive. These off-diagonal elements $A_{B,jk}$ are equal to those for the discretized Laplacean times suitable average values over box faces f_{jk} of the variable coefficient $a(\vec{x})$. In fact, if the solution vector **u** assumes a maximum at an internal vertex $\vec{x}_j$, then $\sum_{\vec{x}_k} a_{e_{jk}}(|f_{jk}|/|e_{jk}|)[u_j - u_k] \geq 0$. This observation, combined with the corresponding one for the minimum, implies by a standard discrete extrema principle argument that,

$$\min\{\inf_{\Sigma_D G} u, f^{-1}(\vec{x}_j, g_{\min})\} \leq u_j \leq \max\{\sup_{\Sigma_D G} u, f^{-1}(\vec{x}_j, g_{\max})\}, \qquad (2.3)$$

analogously to the bounds for the solution u to (1.10).

The off-diagonal stress-matrix component $A_{B,jk}$ is not positive if the surface area $|f_{jk}|$ in the expression (2.1) for the flux Φ_{jk} is not negative. Therefore, the sign of the off-diagonal components of the discretized equations is determined by the geometry of the mesh, and not affected by variability of $a(\vec{x}) > 0$. All faces f_{jk} in the mesh have a positive area $|f_{jk}|$ if the domain G is triangulated by the Delaunay triangulation. The positive sign of the areas $|f_{jk}|$ follows from Delaunay's empty sphere condition [5], which states that the circumscribed sphere of no simplex S_r in the mesh contains vertices $\vec{x}_j$ which are foreign to S_r. For instance, if in two dimensions the two angles γ_1 and γ_2 opposite the same edge e_{jk} add up to more than π, then the circumcribed circles of the triangles contain foreign vertices and $|f_{jk}| < 0$. Hence, the occurrence of such faces with "negative" measure or size can be avoided by the choice of a suitable mesh geometry.

2.2. Piecewise Linear Finite Element Discretization. On a Delaunay triangulation, the box-method discretization discussed in the previous section provides extrema stability of the solution, and well defined current fluxes Φ_{jk} through the faces f_{jk} between boxes. Piecewise linear finite elements, on the other hand, provide a fully developed convergence theory on triangular non-uniform meshes which allows the incorporation of the mixed Dirichlet-Neumann boundary conditions in drift-diffusion models. Moreover, because of monotonicity, piecewise linear finite elements assume their extrema at the vertices of the finite element mesh. Therefore, a discrete extrema principle for the nodal values implies immediately a priori L_∞ bounds on the finite element approximation U_h which solves (2.4). The finite element discretization of (1.10) is given by

$$\langle a(\vec{x})\nabla U_h, \nabla \psi_i \rangle + F(U_h)(\psi_i) = \langle g(\vec{x}), \psi_i \rangle. \tag{2.4}$$

Here, U_h is in the subspace of piecewise linear finite element functions S_h, and the ψ_i are appropriate test functions which may be of a more general nature. The extra flexibility provided by the Petrov-Galerkin framework in the choice of test functions may be employed to improve a priori extrema stability. The boundary conditions are interpolated with the piecewise linear interpolant $\bar{u}_I$ so that $U_h \in \bar{u}_I + S_h$, where the members of S_h vanish on the Dirichlet boundary Σ_D of the polyhedral domain G. The functions of S_h are continuous and are linear in each simplex, S_r. As usual, $h = \max_r\{diam\ S_r\}$.

In the discretization of the drift-diffusion model (1.4–1.6), the electrostatic potential function u and the Slotboom variables ν and ω are approximated by piecewise linear finite element functions. The Petrov-Galerkin finite element equations for the approximation U_h to the electrostatic potential u are given by

$$\langle \epsilon(\vec{x})\nabla U_h, \nabla \psi_i \rangle + \langle \exp(U_h - v) - \exp(w - U_h), \psi_i \rangle = \langle k_1, \psi_i \rangle. \tag{2.5}$$

The quasi-Fermi levels v, w in (2.5) may be piecewise linear or may correspond to piecewise linear Slotboom variables ν and ω. Simlarly, the Petrov-Galerkin discretization of the current continuity equations for piecewise linear Slotboom variables ν and ω are given by

$$\langle \mu_n \exp(U_h)\nabla V_h, \nabla \psi_i \rangle = \langle \frac{q}{k_B T}R, \psi_i \rangle, \quad \text{for} \quad i = 1, \cdots M, \tag{2.6}$$

and

$$\langle \mu_p \exp(-U_h)\nabla W_h, \nabla \psi_i \rangle = \langle \frac{q}{k_B T}R, \psi_i \rangle, \quad \text{for} \quad i = 1, \cdots M. \tag{2.7}$$

In these latter equations $V_h \in \bar{V}_I + S_h$, $W_h \in \bar{W}_I + S_h$, where $\bar{V}$ and $\bar{W}$ suitably interpolate the boundary conditions.

Again, the BVP for (1.10) is considered on an N dimensional domain G. The entries of the stress-matrix for the piecewise linear finite element discretization of (1.10) on a simplicial mesh are expressed in terms of the geometric properties of the (N-dimensional) simplex S_r (in this context the simplicial finite element S_r) as stated below.

DEFINITION 2.2. *Let S_r be an N-dimensional simplex as in Definition 1.2. Let the piecewise linear nodal basis function ϕ_l satisfy $\phi_j(\vec{x}_l) = \delta_{jl}$. Then,*

-

$$A_{ij} \equiv \int_S a(\vec{x}) \nabla \phi_i \cdot \nabla \phi_j dx,$$

is the ij-th entry of the <u>element</u> stiffness matrix,

- $\langle a(\vec{x}) \rangle \equiv \int_S a(\vec{x}) dx / V_S$, *the average of $a(\vec{x})$ over the element S_r,*

- $A_{pl,ij}$ *is the ij-th component of the assembled discretized self-adjoint operator.*

It follows through elementary calculations, as shown in the appendix of [9], that

$$A_{pl,ij} \equiv \int_S a(\vec{x}) \nabla \phi_i \cdot \nabla \phi_j dx = \langle a(\vec{x}) \rangle \cos(\gamma_{ij}) \frac{1}{h_i h_j} V,$$

or

$$A_{pl,ij} = \langle a(\vec{x}) \rangle \cos(\gamma_{ij}) \frac{|F_i||F_j|}{N^2 V}.$$

The entries A_{jk} of the global stress-matrix in the finite element method are formed as sums of contributions from elemental stress-matrices for elements S_r over the "fan" of all elements S_r of which the edge e_{jk} is part. Hence,

$$A_{pl,jk} = \sum_{S_r \text{ containing } e_{jk}} \langle a(\vec{x}) \rangle_{S_{r_*}} \cos(\gamma_{ij}) \frac{|F_i||F_j|}{N^2 V}.$$

This implies that both in the box-method, and in the piecewise linear finite element method, the sparsity structure of the stress-matrix is determined by the positioning of the edges e_{jk} which connect the vertices $\vec{x}_j$ in the simplicial mesh. For both discretizations, the entry $A_{B,jk}$, or $A_{pl,jk}$, in the stress matrix can only be nonzero if the vertices $\vec{x}_j$ and $\vec{x}_k$ are connected by an edge e_{jk}. The two discretizations differ, however, in the arithmetic expression for the numerical value of the off-diagonal entry. In the box method, $A_{B,jk}$ corresponds directly to a flux Φ_{jk} through the face f_{jk} between the boxes B_j and B_k, and is set equal to a geometrical factor times a suitable approximation to the average value $\langle a(\vec{x}) \rangle_{f_{jk}}$ of the coefficient $a(\vec{x})$ over the face f_{jk}.

In other words, the box method is "edge-oriented", while the finite element method is "element-oriented". As a result, in the piecewise linear finite element method, the sign of the off diagonal elements A_{jk} in the stress-matrix depends not only on the geometry of the mesh, but on the variation of $a(\vec{x})$ as well. For instance, in two dimensions the expression for the off diagonal entry $A_{pl,jk}$ reduces to

$$A_{pl,jk} = \frac{1}{2}[\langle a(\vec{x}) \rangle_{T_1} \cot(\gamma_1) + \langle a(\vec{x}) \rangle_{T_2} \cot(\gamma_2)],$$

where the T_i are the two triangles adjacent to edge jk, and the γ_i are the two angles opposite to the edge jk. If arbitrary variation of $a(\vec{x}) > 0$ is allowed, the condition that all angles in the mesh are smaller than $\pi/2$ still yields non positive off diagonal elements in the discretization of the self-adjoint second order operator. Unlike the condition $\gamma_1 + \gamma_2 \leq \pi$, this single angle condition is not satisfied by the Delaunay triangulation and cannot be satisfied for arbitrary sets of vertices $\vec{x}_i$. For instance, it cannot be satisfied if three of the four internal angles γ_i of a quadrilateral satisfy $\gamma_i > \pi/2$. This demonstrates that, for variable $a(\vec{x}) > 0$, in two dimensions a priori extrema stability is more readily obtained by the box method with perpendicular bisectors than by piecewise linear finite elements.

As mentioned before, Lemma 3.2 implies that in three dimensions the stress-matrices for the perpendicular bisecting box method and the piecewise linear finite element method differ, even for constant $a(\vec{x}) > 0$. In particular, the piecewise linear finite element method need not generate non-positive off-diagonal elements in the stress-matrix on a Delaunay tetrahedryzation. In fact, it is not clear under what conditions the piecewise linear finite element method does yield an M-matrix as stress matrix in three dimensions [6].

It was mentioned before that the solution u to (1.10) satisfies the a priori L_∞ bounds (1.11). The solution vector u to the box-method equations (2.2) satisfies the analogous a priori extrema bounds (2.3). The system of equations generated by the piecewise linear finite element discretization of (1.10), on the other hand, contains off-diagonal components corresponding to the nonlinear term $f(\vec{x}, u)$. As a result, a discrete extrema principle prevails only if the second order term in (1.10) generates off-diagonal components sufficiently negative to dominate these positive off-diagonal contributions. Regularity conditions on the geometry of the mesh, as stated in Assumption 2.1 below, combined with a restriction on the meshwidth h, imply the desired a priori L_∞ bounds. Assumption 2.1 defines a mesh which generates off-diagonal elements of the stress-matrix which are bounded away from zero sufficiently much that the Laplacean may dominate the mass matrix. A suitable meshwidth restriction then produces the discrete extrema principle.

ASSUMPTION 2.1. *In N dimensions, where $N \geq 2$, let there be a $\rho > 0$ such that every off-diagonal element $A_{pl,jk}$ in the matrix satisfies,*

$$A_{pl,jk} = \sum_{S_r \text{ adjacent } jk} \langle a(\vec{x}) \rangle_{S_r} \cos(\gamma_{jk}^{(S_r)}) \frac{V^{(S_r)}}{h_j h_k} \leq -(\rho/h_{max}^2) \sum_{S_r \text{ adjacent } jk} V^{(S_r)}.$$

In Theorem **3.2** of [9], it is demonstrated that in N dimensions the piecewise linear finite element discretization (2.5) reproduces "reduced" bounds which are equal to the bounds (1.11) on u for the original system, provided that the grid satisfies Assumption 2.1, above, and the meshwidth restriction

$$h_{max} \leq \sqrt{(N+1)(N+2)\frac{\rho}{D}}, \tag{2.8}$$

where D is the Lipschitz constant of the function $f(\vec{x}, u)$ in (1.7) within the L_∞ bounds (1.11).

Assumption 2.1 and (2.8) taken together provide sufficient conditions for L_∞ stability. Furhermore, the following one dimensional example demonstrates that these conditions come close to being necessary as well. The piecewise linear finite element discretization of,

$$- u_{xx} + Au = 0, \tag{2.9}$$

on a uniform grid is given by (for $i = 1, \cdots, N$),

$$- u_{i-1} + 2u_i - u_{i+1} + (h^2/6)A[u_{i-1} + 4u_i + u_{i+1}] = 0. \tag{2.10}$$

For positive A, the solution to (2.9) is given by $u(\vec{x}) = \alpha \exp(\sqrt{A}x) + \beta \exp(-\sqrt{A}x)$, and satisfies a maximum principle. The difference equation (2.10) is solved by the exponential $u_i = a_1 \gamma_1^i + a_2 \gamma_2^i$ where the γ_k are the roots of the characteristic equation

$$1 * \gamma^2 - \frac{2 + (2/3)Ah^2}{1 - (1/6)Ah^2}\gamma + 1 = 0.$$

It follows straightforwardly that the product $\gamma_1 * \gamma_2 = 1$, and that the discriminant is positive. Therefore, the sign of the roots is positive if and only if their sum is positive. Hence, $\gamma_i > 0$, for $i = 1, 2$, only if $[2 + (2/3)Ah^2]/[1 - (1/6)Ah^2] > 0$. This implies that $h < \sqrt{6/A}$ is a necessary condition for a discrete maximum principle. If the meshwidth h exceeds this limit then both γ_1 and γ_2 are negative, and the solution oscillates between negative and positive values at successive gridpoints.

At a device length of 1μ and a suitable doping density of $10^{26}m^{-3}$, the use of quasi-Fermi levels as coordinates gives rise in Poisson's equation for the potential to a term $\langle f(\vec{x}, U), \phi \rangle$ equivalent to a value of $A = 6.13 * 10^5$. Hence, the off-diagonal elements of the Laplacean dominate the off-diagonal elements generated by the nonlinear term $F(u)(v)$ in (1.10) only for a meshwidth $h \leq 3.13 * 10^{-3}$, relative to the device size l_0. Even for a two dimensional computation, this highly restrictive upper bound on the meshwidth h results in the excessive requirement of more than 10^5 meshpoints.

2.3. Relaxation and Maximum Stability. The box-method equations (2.2) are defined in terms of function values at vertices $\vec{x}_j$, boxes B_j, and fluxes Φ_{jk} throught box faces f_{jk}. As mentioned before. variability of the coefficient $a(\vec{x})$ in (1.10), combined with the definition of matrix elements in terms of fluxes through box faces, causes discrepancies between the box-method and finite element methods. To examine these discrepancies, a relaxation of the BVP of

Definition 1.1, in which for $a(\vec{x})$ and $g(\vec{x})$ are substituted the piecewise constant functions $\tilde{a}(\vec{x})$ and $\tilde{g}(\vec{x})$, respectively, and for the boundary condition $\bar{u}$ the piecewise linear interpolant $\bar{u}_I$, is introduced next. Here, $\tilde{a}(\vec{x})$ and $\tilde{g}(\vec{x})$ are constant on every simplex S_r and set equal to "patch averages" $\langle a(\vec{x})\rangle_{P_s}$ and $\langle g(\vec{x})\rangle_{P_s}$, for patches P_s as in,

DEFINITION 2.3. *On a domain G, let the S_r be simplicial finite elements as in Definition 1.2. Then, the "patches" P_s on the domain G are defined as a subdivision of the simplicial finite element mesh, such that $P_s = \cup_{r\in I}S_r$ is a connected region of adjacent S_r. The averages of $a(\vec{x})$ and $g(\vec{x})$ may or may not be taken over identical patches P_s.* Moreover, the boundary conditions are approximated by a piecewise linear interpolant as in

DEFINITION 2.4. *Let $a_{\tilde{a}}(\cdot,\cdot)$ be equal to $a(\cdot,\cdot)$ in (1.8), except that for $a(\vec{x})$ is substituted the piecewise constant function $\tilde{a}(\vec{x})$. Let $F(u)$, $\bar{u} \in C^2(\bar{G})$, and $g \in L_2$ be as in Definition 1.1. On the boundary ΣG, let $\bar{u}_I$ interpolate the boundary condition $\bar{u}$. Let $\tilde{g}$ be a piecewise constant interpolant of g.*

Then, the BVP for the function $u_p \in \bar{u}_I + H^1_{0,\Sigma_D}$, is defined by the gradient relation,

$$a_{\tilde{a}}(u_p, v) + F(u_p)(v) = \langle \tilde{g}, v\rangle, \quad \forall v \in H^1_{0,\Sigma_D}. \tag{2.11}$$

Lemma 2.1 below yields a bound on the perturbation in the solution u to (1.10) in terms of the perturbations to $a(\vec{x})$, g, and the boundary data. In this lemma, it is shown that the solution u_p to this perturbed BVP approximates the solution u to (1.10) to piecewise linear accuracy.

Furthermore, an "element-based" box-method (2.12) is introduced, which differs from the box-method (2.2) in the averaging process of the variable coefficient $a(\vec{x})$ in (1.10), and in the discretization of the lower order terms in (1.10). In the box-method (2.2), the flux ϕ_{jk} between the adjacent vertices $\vec{x}_j$ and $\vec{x}_k$ is set equal to (2.1). The average $\langle a(\vec{x})\rangle_{f_{ij}}$ of the variable coefficient $a(\vec{x})$ over the face f_{ij} is often approximated in terms of the value of $a(\vec{x})$ at the vertices $\vec{x}_i$ and $\vec{x}_j$. Usually, this involves the Scharfetter-Gummel discretization [10]. The system (2.12) employs an alternative definition of fluxes between adjacent boxes B_j, and a Petrov-Galerkin discretization for the lower order terms, rather than the "lumping" techniques which characterize box-methods. The equations are given by

$$\sum_{j=1}^{M} A_{B,\tilde{a},ij} u_j + F(U_{\mathbf{u}})(\psi_i) = \langle \tilde{g}(\vec{x}), \psi_i\rangle. \tag{2.12}$$

In the element based box-method inner product, instead of face averages of the variable coefficient $\tilde{a}(\vec{x})$, element averages are taken, as in

$$\mathbf{u}^t A_{B,\bar{a}} \mathbf{v} = \sum_{S_r} \tilde{a}_r \sum_{e_{jk}} (|f_{jk}|/|e_{jk}|)(u_j - u_k)(v_j - v_k). \tag{2.13}$$

The difference between the inner products defined by the relaxed box-method inner product (2.12), and the standard box-method inner product (2.2), will be estimated using Lemma 2.3. First, the perturbation to the solution u, brought about by the transition to the BVP in Definition 2.4, from the BVP in 1.1, is examined.

LEMMA 2.1. *Let u solve the BVP defined in 1.1, in which the boundary conditions are described by $\bar{u} \in C^2(\bar{G})$, and the right hand side by $g \in L_2$. Let u_p solve the BVP defined in 2.4, in which $\tilde{a}(\vec{x})$ replaces $a(\vec{x})$ in 1.1 In which, on the Dirchlet part of the boundary $\Sigma_D G$, the function $\bar{u}_I$ interpolates the boundary condition $\bar{u}$, and where $\tilde{g}$ is a piecewise constant interpolant of g. Then, there exist constants c and d, independent of the meshwidth h, such that,*

$$\sqrt{a(u - u_p, u - u_p)} \leq \sqrt{a(\bar{u} - \bar{u}_I, \bar{u} - \bar{u}_I)} + c\|\bar{u} - \bar{u}_I\|_{H^1} +$$

$$d\|g - \tilde{g}\| + \|\sqrt{\frac{a(\vec{x})}{\tilde{a}(\vec{x})}} - \sqrt{\frac{\tilde{a}(\vec{x})}{a(\vec{x})}}\|_{L_q}\|\sqrt{\tilde{a}(\vec{x})}\nabla u_p\|_{L_r}, \tag{2.14}$$

where $(1/q) + (1/r) = (1/2)$.

In Poisson's equation for the potential, $a(\vec{x})$ is equal to the dielectric constant of the material and, therefore, piecewise continuous. In the current continuity equations this factor is equal to $\mu_n \exp(u)$ or $\mu_p \exp(-u)$, where u is the electrostatic potential in units of the thermal potential $U_T \equiv (k_B T)/q$. For the latter case, the last term in (2.14) is estimated in the next lemma. By a careful approach, a factor $\exp[(1/2)(u_{max} - u_{min})]$ in the bound is avoided. First, some regularity of the shape of the patches P_s is introduced.

ASSUMPTION 2.2. *For the finite element mesh on the N-dimensional domain G, there exists a single constant $c_{min} > 0$, such that the measure $|P_s|$ of all patches P_s is bounded from below in terms of their diameter h_{P_s} by*

$$|P_s| \geq c_{min} h_{P_s}^N. \tag{2.15}$$

Moreover, on G, the patch diameters h_{P_s} are bounded uniformly by $h_{P_s} \leq h_P$. Notice that Assumption 2.2 does not preclude a large variation of element measures $|S_r|$ in the finite element mesh.

LEMMA 2.2. *Let the domain G be partitioned into patches P_s, and let $\tilde{a}(\vec{x})$ be a piecewise constant function, equal to "patch averages" $\langle a(\vec{x}) \rangle_{P_s}$, as mentioned above. Let $a(\vec{x}) = \mu_n \exp[u(\vec{x})]$, where $\mu_n(\vec{x})$ is constant on each element S_r, and let $q \geq p$.*

Then,

$$\|\sqrt{\frac{a(\bar{x})}{\tilde{a}(\bar{x})}} - \sqrt{\frac{\tilde{a}(\bar{x})}{a(\bar{x})}}\|_{L_q} \leq \tag{2.16}$$

$$2\cosh[(C/c_{\min}^{(1/s)})h_{P_s}^{1-(N/s)}\|\nabla U_h\|_{s,G}](C/c_{\min})h_{P_s}^{1+(N/q)-(N/p)}|\nabla U_h|_{p,G}.$$

In [8] it is argued that for reasonable device geometries in 2 and 3 dimensions the value of r in Lemma 2.1 and p in Lemma 2.2 for a mixed Dirichlet-Neumann BVP (1.10) should satisfy $\|\nabla u\| \in L_t$ for $t < 4$. Technically, if the solution is sufficiently smooth, this result falls short (by the amount ϵ) of realizing full piecewise linear accuracy in terms of the meshwidth h.

In Lemma 2.1 the difference between the BVP in Definition 1.1, and the BVP with piecewise constant $\tilde{a}(\bar{x})$ and $\tilde{g}(\bar{x})$ and piecewise linear boundary conditions $\bar{u}_I$ in Definition 2.4, was examined as a perturbation to the BVP in Definition 1.1. As mentioned above, this perturbation introduces a piecewise linear approximation error. Therefore, henceforth the BVP in Definition 1.1 will be replaced by the perturbed BVP in Definition 2.4.

As mentioned before, box-methods are edge (flux) and box (volume) based, rather than element based. An intermediate, element-based box-method was introduced in (2.12). The difference between a usual box-method (2.2) and this intermediate discretization (2.12) is estimated in Lemma 2.3 below. There, this discrepancy due to the different fluxes is bounded in terms of the maximum variation of $a(\bar{x}) = \mu_n \exp(u)$ over patches P_s. The expression of $A_{B,jk}$ in the box-method, in §2.1, as a geometrical factor times a face-average factor $a_{e_{jk}}$, does yield the desired L_∞ stability. In the error analysis presented here, this deviation from weighted element averages $\tilde{a}_r$ does, however, reduce the order of accuracy of the finite element approximation. A standard error analysis for numerical integration is applied to the "lumping" of first order terms. Similarly, in the bilinear finite element scheme proposed in [7] $\int \rho \phi dx$ is integrated numerically by the trapezoidal rule in both dimensions. This too results in a "lumping" of the mass matrix for Poisson's equation. As mentioned before in §2.2, "lumping" significantly improves extrema stability of the discretized equations.

For all patches P_s, adjacent to edge e_{kl}, the "flux-averages" $\exp(U)_{kl}$ over box faces f_{kl} are assumed to satisfy the condition,

$$\exp(u_{\min,P_s}) \leq \exp(U)_{kl} \leq \exp(u_{\max,P_s}). \tag{2.17}$$

It is easily shown that the coefficient,

$$\frac{(1/2)(u_k - u_l)}{\sinh[(1/2)(u_k - u_l)]}\exp[(1/2)(u_k + u_l)], \tag{2.18}$$

in the box-method Scharfetter-Gummel discretization [10] satisfies the assumption (2.17). The areas $|B_j|$ of the boxes B_j into which element S_r is partitioned in Definition 2.1 satisfy,

$$\sum_j |B_j| = |S_r|. \tag{2.19}$$

LEMMA 2.3. *Let A_B be the matrix which corresponds to the box-method inner product for (2.2). Let the expression for the "flux average" $\exp(U)_{kl}$ satisfy (2.17), and let the box areas $|B_j|$ satisfy (2.19).*

For the vector $\mathbf{v} \in \mathbf{R}^M$, let $F_v(\mathbf{v})$ denote the continuous linear functional on $\mathbf{R}^M$ defined by

$$F_v(\mathbf{v})(w) = \sum_i |B_i| f(\bar{x}_i, v_i) w_i. \tag{2.20}$$

Let $\langle f, g \rangle_v$ denote the continuous linear functional on $\mathbf{R}^M \times \mathbf{R}^M$ defined by

$$\langle f, g \rangle_v = \sum_i |B_i| f(\bar{x}_i) g(\bar{x}_i). \tag{2.21}$$

Then,

$$|\mathbf{v}^t A_B \mathbf{w} + F_v(\mathbf{v})(\mathbf{w}) - \mathbf{v}^t A_{B,\bar{a}} \mathbf{w} - F(V_{\mathbf{v}})(W_{pp,\mathbf{w}})| \leq \tag{2.22}$$

$$2 \sinh[(C/c_{\min}^{(1/s)}) h_P^{1-(N/s)} \|\nabla U_h\|_{s,G}] \sqrt{\mathbf{v}^t A_B \mathbf{v}} \sqrt{\mathbf{w}^t A_{B,\bar{a}} \mathbf{w}} +$$

$$d \|\nabla V_{\mathbf{v}}\| \|\nabla W_{pp,\mathbf{w}}\| h^1.$$

Lemma 2.1 implies that the transition from BVP 1.1 to BVP 2.4 with piecewise constant $\bar{a}(\bar{x})$, piecewise constant $\bar{g}(\bar{x})$, and piecewise linear $\bar{u}_I$ introduces only a piecewise linear approximation error.

The proof of Lemma 2.3 yields that the face-averages in the box-method discretization (2.2) of BVP 1.1 approximate the element-averages in the element based box-method discretization (2.12) of BVP 2.4 up to an error term with an exponent of the meshwidth h equal to only $1 - (N/s)$. Assuming again that $s < 4$ in drift-diffusion simulations subject to mixed Neumann-Dirichlet boundary conditions, this approximation results an order of accuracy less than $1/2$ in two dimensions, and less than $1/4$ in three dimensions.

3. Piecewise linear finite elements and the box method. With respect to the mesh of boxes B_j, the box method is defined in (2.2) in §2.1. The lemma from [8] stated below examines some linear algebraic properties the bilinear form on $\mathbf{R}^M \times \mathbf{R}^M$, which is defined by the perpendicular bisectors box-method.

LEMMA 3.1. *The box-method in Lemma 3.3 defines a symmetric positive definite bilinear form* $\mathbf{u}^t A_B \mathbf{v}$.

The lemma above implies that the expression $\sqrt{\mathbf{u}^t A_B \mathbf{u}}$ is a norm on $\mathbf{R}^M$.

By the relaxation results in §2.3, the BVP in Definition 1.1 is an approximation, of piecewise linear accuracy, to the relaxed BVP in Definition 2.4, in which the variable coefficient $\tilde{a}(\tilde{x})$ is constant on patches P_s, as specified in Definition 2.3. The accuracy of the expression of the current fluxes Φ_{kl} in the classical box-method (2.2) as products of geometrical factors times face averages of the variable coefficient $a(x)$, was estimated in Lemma 2.3. Let the test functions $\psi_j(\tilde{x})$ in (2.12) and in (2.4) be chosen equal. Then, the term $F(U_h)(\psi_i)$ in the Petrov-Galerkin discretization (2.4) of the relaxed BVP in Definition 2.4, and the term $F(U_u)(\psi_i)$ in the relaxed box-method (2.12) are equal, and the only differences between the two BVPs are in the discretization of the second order terms and the right hand sides.

At this point, notice that because $\tilde{a}(\tilde{x})$ and $\tilde{g}(\tilde{x})$ are constant on patches P_s, to set the discretization (2.4) of (2.11) equal to (2.12), it suffices to examine only Poisson's equation for constant right hand side on an element by element basis. This is done in Lemmas 3.2 and 3.3 from [8], below. In fact, the elemental finite element equations for Poisson's equation with a piecewise constant right hand side, are shown to be invariant under a change of test function ϕ, provided that the averages of ϕ over faces F_s and over finite element volumes S_r remain unchanged.

LEMMA 3.2. *On the N-dimensional simplicial element S_r, let $\tilde{g}(\tilde{x})$ be a constant function. Let U_h be a linear function, and let the functions $\phi^{(1)}$ and $\phi^{(2)}$ satisfy over all element faces F_s,*

$$\int_{F_s} \phi^{(1)} dx = \int_{F_s} \phi^{(2)} dx \quad \forall F_s, \tag{3.1}$$

and over the finite element volume S_r.

$$\int_{S_r} \phi^{(1)} dx = \int_{S_r} \phi^{(2)} dx, \tag{3.2}$$

Then,

$$\int_G \nabla U_h, \nabla \phi^{(1)} dx = \int_G \nabla U_h, \nabla \phi^{(2)} dx, \tag{3.3}$$

and,

$$\int_G \tilde{g} \phi^{(1)} dx = \int_G \tilde{g} \phi^{(2)} dx, \tag{3.4}$$

Notice that because U_h is linear on S_r, this lemma can be applied to box-method equations.

Lemma 3.2 implies that, for piecewise constant $\bar{a}(\vec{x})$, the stress-matrix is not affected by the shape of the box B_j, but only by the measure of the parts $p_{s,j}$ of the intersections of B_j with the faces F_s. For the two dimensional case this observation was made first in [2]. In this same publication Bank and Rose showed, in two dimensions, the results in Lemma 3.3 below. This lemma asserts that the stress-matrix of the intermediate box method based on perpendicular bisectors, (2.12), is equal to the Petrov-Galerkin method stress-matrix $A_{PG,\bar{a}}$ for a piecewise linear approximation U_h, with piecewise constant test functions ϕ_{B_j}, where the box-faces f_{jk} are normal to the corresponding edges e_{jk}.

LEMMA 3.3. *For all edges e_{jk} of the element S_r, let f_{jk} be the face between the boxes B_j and B_k in Definition 2.1, and let f_{jk} be normal to e_{jk}. Let the piecewise constant test functions ϕ_{B_j} be equal to 1 inside the box dual to vertex $\vec{x}_j$, and 0 elsewhere.*

Then, the intermediate box-method stress-matrix, $A_{B,\bar{a}}$, as defined in (2.12), is equal to the stress-matrix $A_{PG,\bar{a}}$ generated by the Petrov-Galerkin method for a piecewise linear approximation U_h, and box functions ϕ_{B_j}.

In the two-dimensional analysis in [2], Bank and Rose developed a finite element error analysis based on the Petrov-Galerkin framework with the piecewise constant test functions. Alternatively, the box-method equations can be considered as approximate Galerkin or Petrov-Galerkin equations for a piecewise linear approximation with conforming, piecewise linear, or piecewise polynomial test functions. The transition from the piecewise constant test functions to these continuous test functions is then facilitated by Lemma 3.2. In [8] Lemma 3.3 is applied to the relaxed, piecewise constant $\bar{a}(\vec{x})$, discretization (2.12). This allows writing the stress-matrix $A_{B,\bar{a}}$ as the Petrov-Galerkin stress-matrix $A_{PG,\bar{a}}$, introduced in the lemma above. Next, Lemma 3.2 allows the substitution for piecewise constant box test functions ϕ_{B_j} of less than H^1 regularity, test functions $\psi_j \in H^1$. Subsequently, a variational principle, or a more general inf-sup condition for Petrov-Galerkin equations with an inner product which is continuous on $H^1 \times H^1$ so that standard finite element error analysis becomes applicable, is employed. This approach circumvents the introduction of the Assumption (2.4) in [2] to accomodate nonconforming test functions.

Corollary 3.4 below establishes equivalence of the piecewise linear finite element discretization with suitable box-methods. Application of Lemma 3.2 to the Laplacean in N dimensions demonstrates that the boxes B_j are required to equidivide the faces F_{rs} in the finite element mesh.

COROLLARY 3.4. *In N dimensions, let the boxes B_j from Definition 2.1 be defined with respect to a simplicial mesh as in Definition 1.2. Let the boxes B_j equidivide all faces F_{rs} in* *the finite element mesh Then, the piecewise linear finite element discretization of the Laplacean method on N-simplices (2 4) is equal to the box method discretization (2 2)*

It follows from Corollary 3 4 that in more than two dimensions the box method based on perpendicular bisecting faces f_{jk} of edges e_{jk}, is in general *not* equivalent to the piecewise linear finite element method For instance, in three dimensions, boxes B_j based on perpendicular bisecting planes, partition the triangular faces F_{rs} into three parts of equal surface area, only if the boxes B_j pass through both the midpoints of the edges e_{jk} in the finite element grid, and the centroids of the inter element faces F_{rs} This occurs only if all faces F_{rs} are equilateral In other words, all tetrahedra S_r in the mesh have to be equilateral However, three dimensional space cannot be tetrahedrized with equilateral tetrahedra without leaving holes Hence, the box method based on perpendicular bisecting planes does not coincide with the usual piecewise linear finite element method in three dimensions

It follows immediately that in two dimensions, for piecewise constant $\bar{a}(\bar{x})$ and $\bar{g}(\bar{x})$, and piecewise linear boundary conditions, the classical box method based on a dual mesh of perpendicular bisectors is equivalent to the piecewise linear finite element method The perturbations due to more general $a(\bar{x})$, $g(\bar{x})$, and boundary conditions which are covered in Lemma 2 1 of §2 3 are included without effort The use of the current fluxes (2 1) in the classical box-method introduces a perturbation which can be accounted for by use of Strang's first lemma (see e g Theorem **4.1.1** in [4]) and Lemma 2.3 of §2 3

REFERENCES

[1] I Babuska and A K Aziz *The Mathematical Foundations of the Finite Element Method with Applications to Partial Differential Equations* Academic Press, New York, San Fransisco London, 1972

[2] Randolph E Bank and Donald J Rose Some Error Estimates for the Box Method *SIAM J on Numer Anal* 24 777–787 1987

[3] Randolph E Bank Donald J Rose and Wolfgang Fichtner Numerical Methods for Semiconductor Device Simulation *SIAM J on Scient and Statist Comp* , 4(3) 416–435, September 1983

[4] Philippe G Ciarlet *The Finite Element Method for Elliptic Problems* North Holland Amsterdam, New York Oxford 1978

[5] B Delaunay Sur La Sphere Vide *Izvestia Akademii Nauk SSSR Otdelenie Matematicheskii i Estestvennyka Nauk* 7(6) 792–800 1932

[6] Herbert Edelsbrunner Spatial Triangulations with Dihedral Angle Conditions In *Proc Int Workshop on Discrete Algorithms and Complexity*, pages 83–89 1989

[7] Jorg Machel and Siegfried Selberherr A Novel Finite Element Approach to device Modeling *IEEE TRans on Electr Dev* ED 30(9) 1083–1092, 1983

[8] Thomas Kerkhoven Petrov Galerkin Finite Element Approximation And The Box Method Technical report, University of Illinois, in preparation

[9] Thomas Kerkhoven and Joseph W Jerome L_∞ Stability of Finite Element Approximations to Elliptic Gradient Equations *Numerische Mathematik*, 57 561–575, 1990

[10] D Scharfetter and H K Gummel Large signal analysis of a silicon read diode oscillator *IEEE Trans Electron Devices* ED 20 64–77, 1969

International Series of Numerical Mathematics, Vol. 117, © 1994 Birkhäuser Verlag Basel

STABILITY ANALYSIS OF THERMOCAPILLARY
CONVECTION IN SEMICONDUCTOR CRYSTAL GROWTH

HANS D. MITTELMANN

Department of Mathematics

Arizona State University

Tempe, AZ 85287–1804

K.-T. CHANG, D. F. JANKOWSKI   G. P. NEITZEL

Department of Mechanical   *The George W. Woodruff School*

and Aerospace Engineering   *of Mechanical Engineering*

Arizona State University   *Georgia Institute of Technology*

Tempe, AZ 85287–6106   *Atlanta, GA 30332–0405*

Abstract. Some of the most challenging eigenvalue problems arise in the stability analysis of solutions to parameter-dependent nonlinear partial differential equations. Linearized stability analysis requires the computation of a certain purely imaginary eigenvalue pair of a very large, sparse complex matrix pencil. A computational strategy, the core of which is a method of inverse iteration type with preconditioned conjugate gradients, is used to solve this problem for the stability of thermocapillary convection. This convection arises in the float-zone model of crystal growth governed by the Boussinesq equations. The results obtained complete the stability picture augmenting the energy stability results (Mittelmann et al. 1992), here a real eigenvalue of a Hermitian eigenvalue problem had to be determined, and recent experimental results.

1. Introduction

The stability properties of thermocapillary convection as they occur in models of the float-zone crystal-growth process have been the subject of intense scrutiny for several years. It is believed that the instability of this form of convection, driven by temperature-induced surface-tension gradients, is responsible for the appearance of undesirable striations (Eyer et al. 1985) in material grown by the float-zone method. In terrestrial applications of the float-zone process, thermocapillarity must compete with buoyancy as a convection-driving force. The float-zone process, however, has been discussed as one which could benefit, from the standpoint of producing larger crystals of material, by being utilized in a microgravity environment. In this setting, since the effect of buoyancy will be significantly reduced, thermocapillary convection takes on a greater significance. If indeed it is the appearance of oscillatory thermocapillary convection which causes material striations, then a knowledge of the conditions under which this will occur, particularly in space, is crucial to the growth of material of the highest quality.

The *half-zone* model of a float-zone melt was created as a vehicle for performing terrestrial experiments in a geometry with a cylindrical-free surface for which the dominant convection-producing mechanism was thermocapillarity. It consists of a pair of fixed, coaxial, vertically oriented cylinders with a bridge of liquid suspended between them. The upper cylinder is heated to a higher temperature than the lower one, so that the density within the liquid bridge is stably stratified due to buoyancy. In an experiment, the temperature difference between the cylindrical rods is increased and the value at which steady, axisymmetric convection disappears is recorded and converted to a dimensionless *onset Marangoni number*.

There have been several sets of experiments of this variety performed; perhaps the best examples are those of Preisser, Schwabe and Scharmann (1983) and Velten, Schwabe and Scharmann (1991). The first of these employed a sodium nitrate ($NaNO_3$) bridge, determining onset Marangoni numbers for a material with *Prandtl number*, $Pr = 7$, using flow visualization to detect the appearance of oscillatory modes. Unfortunately, the Prandtl numbers associated with *electronic materials* grown via the float-zone method are $O(10^{-2})$. The second set of experiments used, in addition, potassium chloride (KCl), which has $Pr = 1$, to provide results which would help bridge the gap between model and reality. These experiments were further concerned with quantitatively determining the spatial structure of the oscillating flow.

Several theoretical and numerical investigations of thermocapillary convection in half-zone models have also been conducted. The early analyses of Xu and Davis idealized the half-zone to have an infinite aspect ratio and employed linear-stability theory to determine sufficient conditions for instability to infinitesimal disturbances. The results obtained were more than an order of magnitude *below* the experimentally determined Marangoni numbers at which oscillatory convection was noted, and it was speculated by Xu and Davis that this was due, in part, to the assumption of infinite aspect ratio. An infinitely long zone is susceptible to a much broader class of disturbances, notably long-wavelength ones, and hence should be less stable.

Motivated by the above experimental and theoretical work, Shen, Neitzel, Jankowski and Mittelmann (1990) undertook an energy-stability analysis of a half-zone of $O(1)$ aspect ratio. Energy theory provides a sufficient condition for stability to disturbances of arbitrary amplitude, and it is such a guarantee of *oscillation-free convection* which is desired by the crystal grower. Their results, computed primarily for $Pr = 1$, compared favorably with the experimental results of (Preisser et al. 1993). However, their analysis was based on the simplifying assumption of permitting only axisymmetric disturbances, while the oscillations observed by Preisser et al. were often *non*-axisymmetric and for a material with significantly larger Prandtl number. Attempts to compute results for larger Prandtl numbers than unity to provide a more direct comparison with the experimental results for $NaNO_3$ met with severe computational difficulties and would likely (based on trends observed for $Pr = 1.6$) have provided sufficient conditions for stability significantly higher than the critical values obtained experimentally.

The results of Shen et al. (1990) indicated, at a minimum, the need to permit non-axisymmetric disturbances to perturb the numerically determined basic state. Such computations were undertaken by Neitzel, Law, Jankowski and Mittelmann (1991). Fortunately, experiments performed by Velten et al. (1991) for KCl provided results for unit Prandtl number which eliminated the need to attempt the more difficult calculations for larger Prandtl numbers. These newer energy-theory results of Neitzel et al. are in excellent agreement with the experimentally determined onset Marangoni numbers in order of magnitude, although the azimuthal structure emerging from the energy theory does not (and should not necessarily) agree with that observed experimentally.

The present work seeks to complete the stability picture for the finite half-zone with a non-deformable free surface by calculating linear-stability limits for this basic state. The degree of closeness of the linear- and energy-theory results provides a bound for the region of parameter space *possibly* subject to subcritical instability. Since hysteresis has been observed in some of the experimental results, such a bound is of interest. The basic state whose linear-stability properties are to be determined is two-dimensional and thus is known only numerically. This two-dimensionality further requires that the equations governing the stability problem remain *partial* differential equations. The approach to the determination of these limits utilizes a pair of staggered grids, as employed so successfully to discretize the integrals in the maximum problem of energy-stability theory, to discretize the linearized disturbance equations on the radial/axial plane. Fourier analysis is used to account for the azimuthal structure. The following sections describe the analysis and present the linear-theory results obtained. These are compared with the experimental results of Velten et al. (1991), the energy- and linear-theory results of Neitzel et al. (1991) and Xu and Davis (1984), respectively, and finally with some recent direct numerical simulations of oscillatory thermocapillary convection in a model half-zone by Rupp, Müller and Neumann (1989).

2. Linear-stability analysis

The basic state of thermocapillary convection in a half-zone with non-deformable free surface is identical to that employed and described in Shen et al. and Neitzel et al. (1991). The axisymmetric Boussinesq equations are discretized in stream-function/vorticity form using finite differences on an equally spaced grid in $r - z$ plane. The computed basic state consists of a single toroidal thermocapillary eddy with flow on the free surface directed toward the bottom, cold cylinder. Isotherms become increasingly deformed from their conductive profiles as the Prandtl number of the melt is increased. Details of the numerical procedure and a discussion of the results obtained may be found in Shen et al. and Neitzel et al. (1991).

The stability analysis of the basic-state velocity, $U(x)$, temperature, $T(x)$, and pressure, $P(x)$, fields begins in the usual fashion by assuming there exists a solution to the Boussinesq equations of the form

$$q(x,t) = Q(x) + q'(x,t)$$

where q refers to any flow quantity (i.e., velocity, temperature or pressure), a capital letter denotes the basic state and a prime is used to denote a disturbance. Substitution of the solution into the Boussinesq equations and linearization in disturbance quantities leads to the *linearized disturbance equations* (dropping primes):

$$Re[u_t + Uu_r + uU_r + Wu_z + wU_z] = -p_r + \left[\frac{1}{r}(ru)_r\right]_r + \frac{1}{r^2}u_{\varphi\varphi} + u_{zz} - \frac{2}{r^2}v_\varphi, \quad (2.1)$$

$$Re\left[v_t + Uv_r + Wv_z + \frac{Uv}{r}\right] = -\frac{1}{r}p_\varphi + \left[\frac{1}{r}(rv)_r\right]_r + \frac{1}{r^2}v_{\varphi\varphi} + v_{zz} + \frac{2}{r^2}u_\varphi, \quad (2.2)$$

$$Re[w_t + Uw_r + uW_r + Ww_z + wW_z]$$

$$= -p_z + \frac{Gr}{Re}\theta + \left[\frac{1}{r}(rw)_r\right]_r + \frac{1}{r^2}w_{\varphi\varphi} + w_{zz}, \quad (2.3)$$

$$Ma[\theta_t + U\theta_r + uT_r + W\theta_z + wT_z] = \frac{1}{r}[r\theta_r]_r + \frac{1}{r^2}\theta_{\varphi\varphi} + \theta_{zz} \quad (2.4)$$

$$(ru)_r + v_\varphi + (rw)_z = 0. \quad (2.5)$$

In equations (2.1)–(2.5) we have scaled velocities by $\gamma\Delta T/\mu$, pressure by $\gamma\Delta T/R$, temperature by ΔT and time by $R\mu/(\gamma\Delta T)$, where $\Delta T = T_H - T_C$, γ is the (positive) rate of decrease of surface tension with respect to temperature, and μ is the coefficient of dynamic viscosity of the liquid in the zone. The disturbance temperature is denoted by θ.

The dimensionless parameters appearing in (2.1)–(2.5) are

$$\text{Reynolds number} \quad Re = \frac{\gamma R\Delta T}{\mu\nu},$$

$$\text{Grashof number} \quad Gr = \frac{g\alpha R^3\Delta T}{\nu^2},$$

$$\text{Marangoni number} \quad Ma = \frac{\gamma R\Delta T}{\mu\kappa},$$

where $\nu = \mu/\rho$ is the kinematic viscosity, ρ is the mean density, α is the coefficient of volumetric expansion and g is the gravitational acceleration. The Prandtl number, Pr, is given by $Pr = Ma/Re$.

The boundary conditions which complete the specification of the problem are

$$u = v = w = \theta = 0, \quad z = 0, \Gamma, \tag{2.6a-b}$$

$$u = w_r + \theta_z = v_r - v/r + \theta_\varphi = \theta_r + Bi\theta = 0, \quad r = 1, \tag{2.7a-d}$$

in addition to the requirement that all flow quantities remain bounded at $r = 0$. The quantity $\Gamma = H/R$ is the dimensionless *aspect ratio* of the zone. The additional parameter appearing in the free-surface heat-transfer condition (2.7d) is the Biot number, defined as

$$Bi = hR/k,$$

where h is a heat-transfer coefficient and k is the thermal conductivity of the liquid.

We make use of Floquet theory and normal modes to assume that all flow quantities may be decomposed as

$$q(r, \varphi, z, t) = q^*(r, z) \exp(\sigma t + im\varphi) \tag{2.8}$$

where $\sigma = \sigma_R + i\sigma_I$ is the complex growth rate and m is restricted to be an integer. Marginal stability corresponds to the condition $\sigma_R = 0$. The form (2.8) is now substituted into equations (2.1–5) and the boundary conditions. The following system results.

$$u_{rr} + \frac{1}{r}u_r + u_{zz} - \frac{m^2+1}{r^2}u - \operatorname{Re}\left[\sigma u + Uu_r + uU_r + Wu_z + wU_z\right] - \frac{2im}{r^2}v - p_r = 0, \tag{2.9}$$

$$v_{rr} + \frac{1}{r}v_r + v_{zz} - \frac{m^2+1}{r^2}v - \operatorname{Re}\left[\sigma v + Uv_r + Wv_z + \frac{Uv}{r}\right] + \frac{2im}{r^2}u - \frac{im}{r}p = 0, \tag{2.10}$$

$$w_{rr} + \frac{1}{r}w_r + w_{zz} - \frac{m^2}{r^2}w - \operatorname{Re}\left[\sigma w + Uw_r + uW_r + wW_z + Ww_z\right] - p_z + \frac{Gr}{Re}\theta = 0, \tag{2.11}$$

$$\theta_{rr} + \frac{1}{r}\theta_r + \theta_{zz} - \frac{m^2}{r^2}\theta - Ma[\sigma\theta + U\theta_r + uT_r + wT_z + W\theta_z] = 0, \tag{2.12}$$

$$u_r + \frac{u}{r} + \frac{im}{r}v + w_z = 0. \tag{2.13}$$

It has to be augmented with the boundary conditions (2.6)–(2.7) and *one* of the following sets.

$$u = v = w_r = \theta_r = 0, \quad m = 0,$$

$$u + iv = u_r = w = \theta = 0, \quad m = 1, \tag{2.14a-d}$$

$$u = v = w = \theta = 0, \quad m \geq 2.$$

A discrete version of this problem is a complex, generalized eigenvalue problem of the form

$$Ax = \sigma Bx, \tag{2.15}$$

where x is the vector of unknown velocity, temperature and pressure values at the nodes of the appropriate grid.

The corresponding eigenvalue problem from the energy-theory analysis of this basic state was, at worst, complex-Hermitian (in addition to being indefinite and sparse), whereas (2.15) has no such symmetry. An additional complication, which also existed as part of the energy-stability analysis is the fact that the basic-state velocity and temperature fields depend upon the stability parameter, Ma (equivalently, Re). For the energy-theory calculations, this required an additional level of iteration to obtain the energy limit, Ma_E. Since we formulate the problem with σ as the eigenvalue, our procedure is to fix the Prandtl, Grashof and Biot numbers as well as the azimuthal wavenumber m and calculate the eigenvalue of system (2.15) with largest *real* part, call it σ^*, for various values of the Marangoni number. The Marangoni number Ma^* corresponding to $\sigma_R^* = 0$ is the value above which infinitesimal disturbances of azimuthal wavenumber m will grow. The linear-stability limit, $Ma_L(Pr, Gr, Bi)$, is therefore given by

$$Ma_L(Pr, Gr, Bi) = \min_m Ma^*(Pr, Gr, Bi; m). \qquad (2.16)$$

3. The numerical method

As was shown above, the determination of the linear stability bounds Ma_L requires the solution of the following eigenvalue problem

$$Ax = \sigma Bx \qquad (3.1a)$$

where A, B are $N \times N$ matrices, $N = k + l$, which are partitioned as

$$A = \begin{pmatrix} A_{11} & A_{12} \\ A_{21} & A_{22} \end{pmatrix}, \quad B = \begin{pmatrix} B_{11} & 0 \\ 0 & 0 \end{pmatrix} \qquad (3.1b)$$

$A_{11}, B_{11} \in C^{k,k}$, $A_{22} \in C^{l,l}$, here k is the number of velocity/temperature unknowns and l the number of pressure unknowns. The matrices are derived from linearizing the boundary value problem for the Boussinesq equations at the solution (basic state) for specific values of the parameters (Pr, Gr, Bi). A is complex and non-Hermitian while B_{11} is taken as a multiple of the identity matrix. If the real parts of all eigenvalues are negative, the basic state is stable, and for increasing Ma that value Ma^* has to be found where for the first time an eigenvalue, to be called the critical eigenvalue in the following, crosses the imaginary axis. Here, it is expected that this corresponds to a simple Hopf bifurcation point (Hopf 1942), that is, there will be exactly one complex-conjugate eigenvalue pair with nonzero imaginary part which crosses the imaginary axis with nonzero speed.

There is a sizable literature on the numerical computation of Hopf bifurcation points; instead of attempting to give a necessarily rather incomplete listing of works, we refer to a recent collection of articles on bifurcation problems (Mittelmann & Roose 1990) in which several papers address this issue.

Generalized eigenvalue problems of the form (3.1), on the other hand, in particular if they arise from applications as is the case here, have been studied thoroughly, and many contributions have been made to their numerical solution. Again, we confine references to just one recent survey work (Kerner 1989) which also includes an extensive bibliography.

Frequently, Hopf bifurcation points are detected during a continuation process. For this, in general, all eigenvalues of (3.1) are computed, a procedure which is prohibitively expensive for large problems ($N > 10^4$). After the detection of a Hopf point its precise location may be determined through characterizing extended systems, see (Spence et al. 1990), (Dedier et al. 1990) and the references therein. This technique has proven to be useful for problems of moderate size while it is of limited use for very large problems.

From the literature dealing with the computation of Hopf points in flow problems (Christodoulou & Scriven 1988) shall be quoted exemplarily. In this work a relatively sophisticated combination of numerical techniques was applied to a nontrivial hydrodynamic stability problem. There is one feature common to the numerical approaches in the above and many other works, namely, the direct solution of the occurring linear systems of equations. This still limits the size of the problems. The dimensions of the eigenvalue problems solved are a few thousand. On the other hand, in Shen et al. the iterative solution of these problems permitted (in-core) solution of problems of dimension $10^4 - 10^5$. From the experience with the related but different computations of the energy stability bounds in Shen et al., there was reason to hope that, also for the determination of the linearized stability limits, a method of inverse iteration type would yield the desired results.

If it was known in advance with which imaginary part β the critical eigenvalue crosses the imaginary axis, then inverse iteration with shift $i\beta$ would allow detection of this Hopf point at least when the computation is started in the stable range, not too far from the critical parameter value and continuation in this parameter is employed. It is clear that, in order to safeguard the obtained results, computations have to be done with different values of β or, alternatively, generalizations of the numerical algorithm, as, for example, the Arnoldi method, have to be used to compute more than one eigenvalue at a time.

Let s denote the shift which is anticipated to, usually, be purely imaginary with, say, positive imaginary part β. The following form of inverse iteration was successfully applied to the present problem.

$$(A - sB)\delta x_k = (\sigma_k B - A)x_k, \tag{3.2a}$$

$$\sigma_{k+1} = \frac{(x_k + \delta x_k)_i}{\sigma_k [x_k]_i}, \tag{3.2b}$$

$$x_{k+1} = \frac{x_k + \delta x_k}{[x_k + \delta x_k]_i}, \tag{3.2c}$$

where $[x]_i$ denotes the component of the largest modulus of a vector $x \in C^n$. Here, x_0 was initially, i.e., for the first Marangoni number used, chosen as a random vector and σ_0 as 1. For a fixed shift convergence of both the eigenvalue and eigenvector approximations will in general be linear with a factor

$$\left| \frac{s - \sigma^*}{s - \sigma_n} \right| < 1 \tag{3.3}$$

where σ^*, σ_n are the nearest and the next nearest eigenvalue to s. A faster convergence for the σ_k could be obtained by also iterating with approximate left eigenvectors and using the generalized Rayleigh quotient, cf., e.g., Kerner (1989).

The essential computational work involved in the proposed method is the solution of the linear system in (3.2a). The matrix $C = A - sB$ is complex and non-Hermitian. First, an equivalent real system of order $2N$ was formed for the matrix

$$\overline{C} = \begin{pmatrix} Re(C) & -Im(C) \\ Im(C) & Re(C) \end{pmatrix}.$$

This system was then solved using the conjugate gradient method for the normal equations (Paige & Saunders 1982). For this the matrix $\overline{C}$ was first scaled by multiplying the matrices A_{ij} in (3.1b) with appropriate powers of the discretization parameters in order to balance $\overline{C}$. Then, additionally, a diagonal scaling was used such that the columns of $\overline{C}$ had unit norm. Different preconditionings may well lead to a more efficient solution method; several, however, were tried without yielding a substantial reduction in work. These included outer-inner iterations utilizing the partitioning in (3.1b) as, for example, in (Bank, Welfert & Yserentant 1990) for the Stokes problem. Here, however, $A_{11} - sB_{11}$ is neither Hermitian nor definite. An incomplete LU decomposition was used to precondition the inner iterations. Nevertheless, a higher efficiency is to be expected from an adaptive multi-level approach such as hierarchical bases (Bank et al.). Instead, the finite difference discretization given above was used on a fixed grid producing results that should be comparable to the energy stability results of Shen, et al.

Instead of the normal equations, the system with $\overline{C}$ could have been solved directly by suitable generalizations of conjugate gradients, as biconjugate gradients, biconjugate gradients squared, GMRES, etc. No preconditioners could easily be found which made these methods sufficiently efficient for the cases to be solved. While $\overline{C}$ will in general not be exactly singular if $i\beta \neq Im(\sigma^*)$, it will be nearly singular. It has also to be noted that the values of $Im(\sigma^*)$ were in the range $10^{-2} - 10^{-1}$.

4. Numerical results and Comparisons

Algorithm (3.2) was applied to the eigenvalue problem (3.1) for aspect ratio $\Gamma = 1$ and the parameter values $Pr = 1$, $Gr = 0$ for which experimental results are given in Velten et al. For coarse discretizations on square grids, up until about $\delta r = \delta z = \frac{1}{20}$, the results were checked with a complex QZ routine. Then, the approximate values of Ma^* and σ^* were known including the trend for decreasing gridsize. It was not difficult to find suitable guesses for the shift s and a Marangoni number somewhat smaller than the expected Ma^*. Continuation in Ma with a bisection or secant method on $Re(\sigma^*)$ was used to determine Ma^*.

The critical eigenvalues computed were in the range .02i–.03i. The complex linear systems (3.2a) had total dimension N where $l = (1 + \frac{1}{\delta r})(1 + \frac{1}{\delta z})$, $k = \frac{4}{\delta r \delta z} + \frac{fac}{\delta z}$ in (3.1b) and $3 \le fac \le 5$ depending on m. The number of cg iterations was less than N but $O(N)$ indicating the need for a better preconditioner. While a more complete discussion of results is given in Neitzel et al. (1992) it should be noted that the least-stable mode from energy theory, here $m = 1$, is not the same as the most-unstable mode from the linear theory. In fact, the linear bound is obtained for $m = 2$

$$Ma_E \approx 1685, \qquad Ma_L \approx 2484.$$

For this bound $\delta r = \delta z = 1/59$ resulted in an order $N = 17,709$ for the complex eigenvalue problem (3.1). Some of the calculated results are given in Table 1.

Γ	Gr	Bi	Ma_L
1.0	0	0	2484
1.2	0	0	1820
1.4	0	0	1674
1.0	0	0.30	3141
1.2	0	0.30	1916
1.4	0	0.30	1640

Table 1. Linear-stability results for half-zone basic states with $Pr = 1$.
For all cases presented, the minimizing azimuthal wavenumber is $m = 2$.

Data from Table 1 for the case $Gr = 0$, $Bi = 0.3$ are plotted in Figure 1, along with experimental data of Velten et al. for KCl ($Pr = 1$) and energy-stability results of Neitzel et al., which were computed for the same parameters. Apparent from the figure is the fact that the linear-stability results lie above the energy-theory ones, as they should since they provide sufficient conditions for instability and stability, respectively. There is also a reasonably large region between Ma_E and Ma_L in which subcritical instability may be

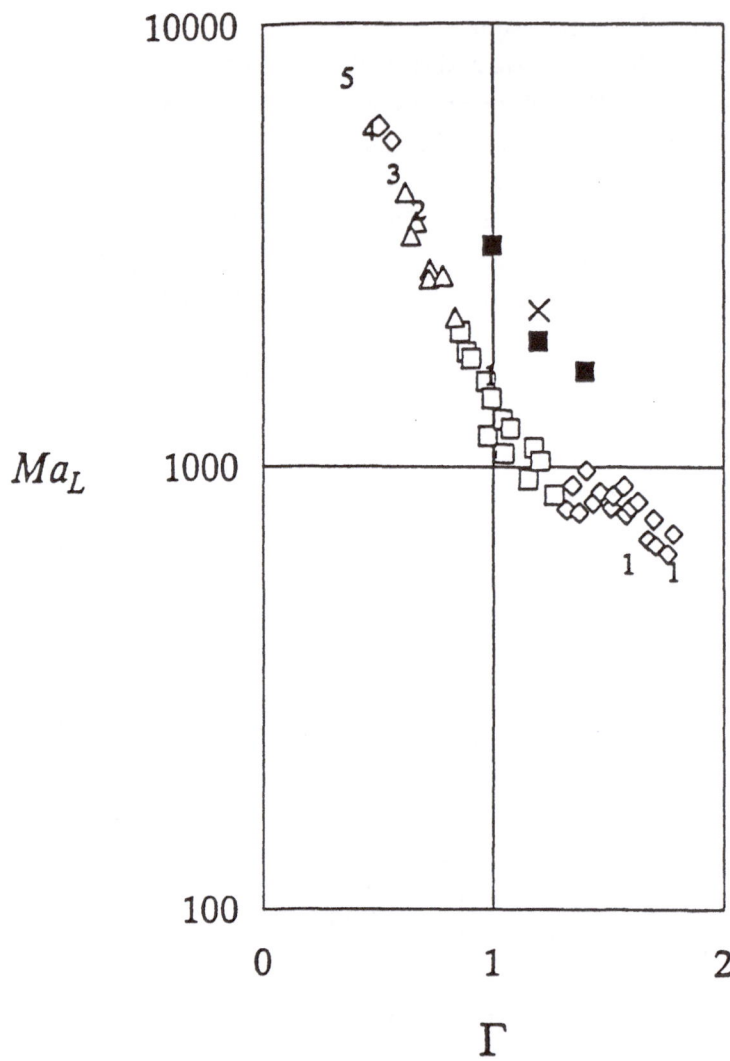

Figure 1. Comparison of linear-stability results (solid symbols) for $Gr =$ 0, $Bi = 0$ with those of energy stability theory (numerals corresponding to azimuthal wavenumber m), laboratory experiment (open symbols) and numerical experiment (single $\times$). Azimuthal wavenumber for linear theory and experiment is indicated by symbol type: triangle, $m = 3$; square $m = 2$; diamond, mixed mode.

possible for all aspect ratios for which results are available. If the level of the experimental data may be believed, i.e., if accurate material properties were used to compute Marangoni numbers corresponding to the conditions under which oscillatory flow was observed, then the existence of experimental points in this region may imply that the instability is indeed subcritical. Except for the case of $\Gamma = 1.4$, where the experimental data indicate the presence of a mixed-mode structure, the $m = 2$ azimuthal structure of the computed values of Ma_L is in agreement with that determined in the laboratory.

A comparison with the earlier linear-stability results of Xu and Davis (1984) shows that the stability enhancement predicted by them for zones of finite, $O(1)$ aspect ratio is realized. Some direct comparisons of their results with those of energy theory may be found in the work of Neitzel et al. (1991), but it is sufficient to state here that, for cases comparable to those considered in the present work, the linear limits calculated by Xu and Davis are $O(10^2)$.

As a final comparison, Rupp, Müller and Neumann (1989) have performed time-dependent, three-dimensional numerical simulations of the flow in a model half-zone similar to that considered in the present paper, with an adiabatic, non-deformable free surface. For a zone of aspect ratio $\Gamma = 1.2$, the Marangoni number at which oscillations were observed in their calculations is (converted to the present scaling) $Ma = 2250$, and this is shown plotted as a single × in Fig. 1. This compares with the value from Table 1 of 1828 and an experimental value of approximately 1030. Consequently, the numerical computations do not appear to be capable of reproducing oscillatory flow at either the experimentally determined Marangoni number or the linear-stability limit, Ma_L. Rupp et al. speculate that non-axisymmetric heating in the experiments may be partially responsible for the discrepancy between the experimental data and their results, but this will not explain the difference between their calculated result and the sufficient condition for instability provided by linear theory for a similar basic state.

Acknowledgments

This work was supported by the US Air Force Office of Scientific Research under grant AFOSR-90-0080 (HDM) and by the Microgravity Science and Applications Division of NASA under grant NAG-3-1221.

REFERENCES

BANK, R. E., WELFERT, B. D. & YSERENTANT, H. (1990) A class of iterative methods for solving saddle point problems. *Numer. Math.* **56**, 645-666.

CHRISTODOULOU, K. N. & SCRIVEN, L. E. (1988) Finding leading modes of a viscous free surface flow: An asymmetric generalized eigenproblem. *J. Sci. Comp.* **3**, 355-405.

DEDIER, B., ROOSE, D. & VAN ROMPAY, P. (1990) Interaction between fold and Hopf curves leads to new bifurcation phenomena. 171-186 in MITTELMANN & ROOSE.

EYER, A., LEISTE, H. & NITSCHE, R. (1985) Floating zone growth of silicon under microgravity in a sounding rocket, *J. Crystal Growth* **71**, 173–182.

HOPF, E. (1942) Abzweigung einer periodischen Lösung von einer stationären Lösung eines Differentialsystems. *Bericht der Math.-Phys. Klasse der Sächsischen Akademie der Wissenschaften zu Leipzig* **94**.

KERNER, W. (1989) Large-scale complex eigenvalue problems. *J. Comp. Phys.* **85** 1-85.

MITTELMANN, H. D., LAW, C., JANKOWSKI, D. F. & NEITZEL, G. P. (1992) A large sparse and indefinite generalized eigenvalue problem from fluid mechanics, SIAM J. Sci. Stat. Comp. **13**, 411-424.

MITTELMANN, H. D. & ROOSE, D. (eds.) (1990), Continuation techniques and bifurcation problems. *J. Comp. Appl. Math.* **26** and ISNM 92, Birkhäuser-Verlag, Basel.

NEITZEL, G. P., CHANG, K.-T., JANKOWSKI, D. F. & MITTELMANN, H. D. (1992) Linear-stability theory of thermocapillary convection in a model of the float-zone crystal growth process. Submitted to Phys. Fluids A.

NEITZEL, G. P., LAW, C. C., JANKOWSKI, D. F. & MITTELMANN, H. D. (1991) Energy stability of thermocapillary convection in a model of the float-zone crystal-growth process. Part 2. Non-axisymmetric disturbances, *Phys. Fluids A* **3**, 2841-2846.

PAIGE, C. C. & SAUNDERS, M. A. (1982) LSQR: An algorithm for sparse linear equations and sparse least squares. *ACM Trans. Math. Software* **8**, 43-71.

PREISSER, F., SCHWABE, P. & SCHARMANN, A. (1983) Steady and oscillatory thermocapillary convection in liquid columns with free cylindrical surface. *J. Fluid Mech.* **126**, 545-567.

RUPP, R., MÜLLER, G. & NEUMANN, G. (1989) Three-dimensional time dependent modelling of the Marangoni convection in zone melting configurations for GaAs, *J. Crystal Growth* **97**, 34–41.

SHEN, Y., NEITZEL, G. P., JANKOWSKI, D. F. & MITTELMANN, H. D. (1990) Energy stability of thermocapillary convection in a model of the float-zone crystal-growth process. *J. Fluid Mech.* **217**, 639-660.

SPENCE, A.. CLIFFE, K.A. & JEPSON, A. D. (1990) A note on the calculation of paths of Hopf bifurcations. 125-131 in MITTELMANN & ROOSE.

VELTEN, R., SCHWABE, D. & SCHARMANN, A. (1991) The periodic instability of thermocapillary convection in cylindrical liquid bridges, *Phys. Fluids A* **3**, 267-279.

XU, J.-J. & DAVIS, S. H. (1984) Convective thermocapillary instabilities in liquid bridges, *Phys. Fluids* **27**, 1102–1107.

International Series of Numerical Mathematics, Vol. 117, © 1994 Birkhäuser Verlag Basel

THE METHOD OF BALIGA-PATANKAR AND 3-D DEVICE SIMULATION

F. MONTRONE

Technische Universität München, Mathematisches Institut, Arcisstraße 21,

D-80333 München, Germany

The carrier transport in a semiconductor device is described by the classical drift-diffusion equations, which can be discretized by the method of Baliga-Patankar [6]. However, the computation of the minority charge densities is sensitive with respect to round-off errors, if the Baliga-Patankar discretization scheme is used. This will be shown by a 2-D counterexample.

A new, stable discretization scheme will be proposed which preserves the advantage of the Baliga-Patankar discretization scheme: No restrictions on the angles of the finite-elements in the mesh need to be imposed.

AMS Subject Classification: 65N30, 65N50.

Key words: drift-diffusion equation, Baliga-Patankar scheme, continuity equation, box method, finite volume method, 3-D device simulation, tetrahedron, adaptive refinement.

1   Notation

The stationary electrical behavior of a semiconductor device is characterized by the following scaled system of partial differential equations for the electrostatic potential Ψ, the electron density η, and the hole density ρ (cf. [3–5]):

$$
\begin{aligned}
\lambda^2 \Delta \Psi &= \eta - \rho - C & \text{(Poisson equation)} \\
\operatorname{div} J_\eta &= R & \\
\operatorname{div} J_\rho &= -R & \text{(Continuity equations)}
\end{aligned}
\tag{1}
$$

Here, λ is the constant, scaled Debye length, C models the doping profile as a function of the point x in the device region Ω, J_η and J_ρ denote the electron and hole current density vector, respectively, and $R = R(\Psi, \eta, \rho)$ describes the carrier recombination-generation rate.

The carrier transport is modeled by the classical drift-diffusion approximation

$$J_\eta = (\nabla\eta - \eta\nabla\Psi)\mu_\eta \,,$$
$$J_\rho = -(\nabla\rho + \rho\nabla\Psi)\mu_\rho \,,$$

(2)

where μ_η and μ_ρ denote the electron and hole mobility, respectively.

Concerning the boundary of the device, boundary conditions of Dirichlet-type prevail at metal contacts, while boundary conditions of Neumann-type are imposed at insulating boundary parts.

The Gummel iteration is applied to (1). A finite element method with a linear ansatz function for Ψ is used to solve the Poisson equation. Hence, $\nabla\Psi$ becomes a constant vector in every finite element.

2   The Baliga-Patankar Discretization

The Baliga-Patankar method [1] is a special finite volume or box method.

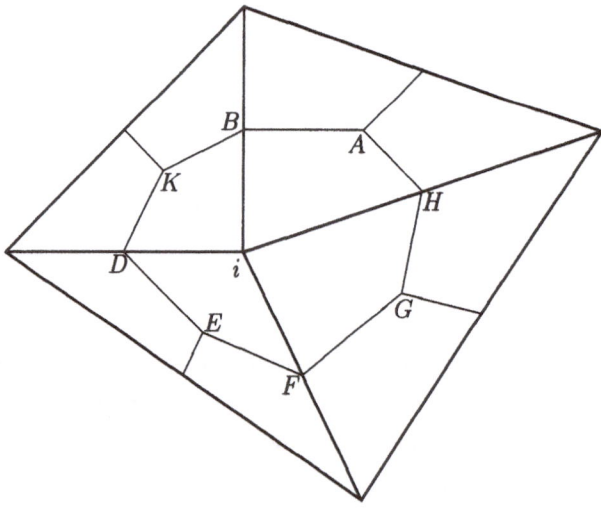

Figure 1

Finite elements are triangles in the 2-D case and tetrahedra in the 3-D case. For each node i of the N nodes of the triangulation, a control volume V_i is constructed by joining the center of gravity of each triangle with the midpoints of the corresponding sides. In the 3-D case the centers of gravity of the four faces of a tetrahedron are additionally joined with the center of gravity of the tetrahedron. In Figure 1, V_i is defined by the polygon $ABKDEFGHA$. In contrast to the construction by perpendicular bisectors, this construction allows obtuse angles of the finite elements.

The continuity equation $\mathrm{div}J_\eta = R$ is integrated over these control volumes V_i, and the left-hand side, $\int_{V_i} \mathrm{div}J_\eta\,dx$, is transformed into a surface-integral of the outward normal component of J_η:

$$\int_{\partial V_i} (J_\eta \circ n)\,ds = \int_{V_i} R\,dx \qquad (i = 1, ..., N). \tag{3}$$

In the 2-D case, the left-hand side of (3), which describes the current passing through the surface of the control volume V_i, is expressed in each triangle by means of the ansatz function

$$\eta = \alpha \frac{1}{\|\nabla\Psi\|}\left[\exp\big(\|\nabla\Psi\|(X - X_f)\big) - 1\right] + \beta Y + \gamma\,, \tag{4}$$

where α, β, and γ are constant coefficients. (X, Y) is a local coordinate system defined in each finite element in such a way that $\nabla\Psi = (\|\nabla\Psi\|, 0)^\top$ holds in this local coordinate system.
X_f can be defined by $X_f := \max\{X_i \mid i \text{ is a vertex of the finite element}\}$.

This ansatz has the following advantage: The discretization of the left-hand side in (3) results in a linear expression with respect to the coefficients α, β, and γ.

As an additional requirement, the continuity of the electron density at the nodes is claimed.

The resulting system of equations has the form

$$
\begin{pmatrix}
\square & & & & & | & | & | \\
& \square & & & & | & | & | \\
& & \square & & & | & | & | \\
& & & \ddots & & & \vdots & \\
& & & & \square & | & | & | \\
\hline
= & = & = & \cdots & = & &
\end{pmatrix}
\begin{pmatrix}
\alpha \\ \beta \\ \gamma \\ \vdots \\ \delta \\ \hline \vec{\eta}
\end{pmatrix}
=
\begin{pmatrix}
0 \\ 0 \\ 0 \\ \vdots \\ 0 \\ \hline \vec{f}(\vec{\eta})
\end{pmatrix}. \tag{5}
$$

α, β, γ, δ denote the coefficients of the ansatz function and $\vec{\eta}$ the nodal values of η. $\vec{f}$ is defined by its components $f_i(\vec{\eta}) := \int_{V_i} R\,dx$, and $\square$ describes a 3×3 matrix.

The structure of the sparse matrix in (5) is symmetric. The continuity of η at the nodes and the conservation of the current are described by the upper and lower part of the equations (5), respectively.

The 3-D case with 5,000 nodes, for example, would lead to 5,000 equations in the lower part of (5) and to about 100,000 equations for the coefficients.

A condensation of the matrix yields

$$M\vec{\eta} = \vec{f}(\vec{\eta}) \tag{6}$$

with a matrix M, which is still sparse.

3   Sensitivity

A 2-D example will show that the computation of the minority-charge densities by the method of Baliga-Patankar is sensitive with respect to round-off errors:

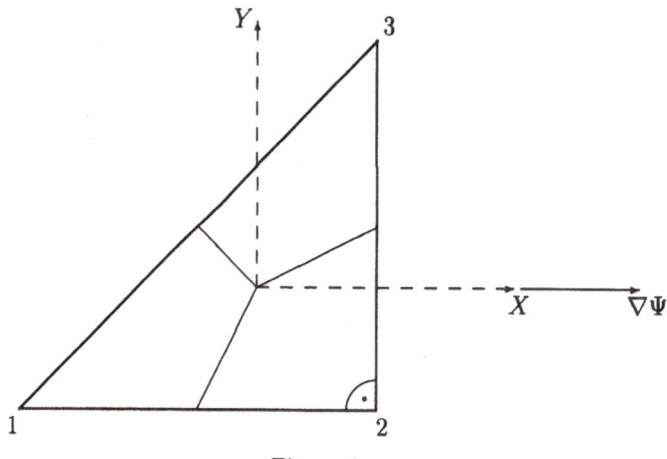

Figure 2

The device of this example has the shape of a triangle as shown in Figure 2.

It is assumed that there is no externally applied potential. This leads to a vanishing recombination-generation term R.

The nodes 1, 2, and 3 are supposed to have the local coordinates $(-\frac{2}{3}, -\frac{1}{3})$, $(\frac{1}{3}, -\frac{1}{3})$, and $(\frac{1}{3}, \frac{2}{3})$, so the edges [1,2] and [2,3] are normalized to the length 1.

The scaled nodal values of the electrostatic potential Ψ, provided by the built-in potential, shall be given by $\Psi_1 = -3$ and $\Psi_2 = \Psi_3 = 17$. Thus, $\|\nabla\Psi\| = 20$ holds.

Homogeneous boundary conditions of Neumann-type are assumed at the edges [1,2] and [1,3]: in other words, the outward current density component vanishes.

Let the boundary data for η on the edge [2,3] be given by the Dirichlet-type boundary condition $\eta_2 = \eta_3 = 3$.

The exact solution for the only unknown, η_1, is given by $3e^{-20} \approx 6 \cdot 10^{-9}$.

Now the Baliga-Patankar method is applied to this specific example:
The original ansatz function (4) with $X_f := X_3 = \frac{1}{3}$ is slightly changed into

$$\eta = \alpha \frac{1}{\|\nabla\Psi\|} \exp\big(\|\nabla\Psi\|(X - X_3)\big) + \beta Y + \tilde{\gamma}, \tag{7}$$

where $\tilde{\gamma} := \gamma - \dfrac{\alpha}{\|\nabla\Psi\|}$. This is possible, since $\|\nabla\Psi\| \gg 0$.

Equation (5) now reads as follows:

$$\left(\begin{array}{ccc|ccc} \frac{1}{20} & 1 & -\frac{1}{3} & -1 & 0 & 0 \\ \frac{1}{20}e^{-20} & 1 & -\frac{1}{3} & 0 & -1 & 0 \\ \frac{1}{20} & 1 & \frac{2}{3} & 0 & 0 & -1 \\ \hline 0 & -60 & 5 & 0 & 0 & 0 \\ 0 & 0 & 0 & 1 & 0 & 0 \\ 0 & 0 & 0 & 0 & 0 & 1 \end{array}\right) \left(\begin{array}{c} \alpha \\ \tilde{\gamma} \\ \beta \\ \eta_2 \\ \eta_1 \\ \eta_3 \end{array}\right) = \left(\begin{array}{c} 0 \\ 0 \\ 0 \\ 0 \\ 3 \\ 3 \end{array}\right) \tag{8}$$

Condensation leads to (no matter, whether (4) or (7) is used):

$$\left(\begin{array}{ccc} 15 + \dfrac{60\exp(-20)}{1 - \exp(-20)} & \dfrac{-60}{1 - \exp(-20)} & -15 \\ 1 & 0 & 0 \\ 0 & 0 & 1 \end{array}\right) \left(\begin{array}{c} \eta_2 \\ \eta_1 \\ \eta_3 \end{array}\right) = \left(\begin{array}{c} 0 \\ 3 \\ 3 \end{array}\right) \tag{9}$$

This system immediately shows the effect of round-off errors: $\eta_1 = 3e^{-20}$ can only be obtained through the expression

$$\left\{-3 \cdot 15 + \left[3 \cdot \left(15 + \dfrac{60\exp(-20)}{1 - \exp(-20)}\right)\right]\right\} \cdot \dfrac{1 - \exp(-20)}{60}, \tag{10}$$

where the number of digits canceled is given by $\log_{10}\eta_2 - \log_{10}\eta_1 \approx 9$.

Since the scaled η varies from 10^{-17} up to 10^3 in simulations, the number of digits canceled in this example would be even larger, if only $\|\nabla\Psi\|$ was chosen larger.

In contrast, the Scharfetter-Gummel scheme leads to a stable numerical computation of the solution through the expression

$$3\frac{1 - \exp(-20)}{\exp(20) - 1}.$$

The sensitivity of the Baliga-Patankar scheme is not caused by the process of matrix condensation. Even if the system (8) is not condensed, the solution process remains unstable. In this case $\beta = \frac{\partial \eta}{\partial Y}$ is computed as $\eta_3 - \eta_2$. The values for η_2 and η_3 will have absolute errors in the range of the machine precision eps, and therefore also $\frac{1}{3}\beta$. This leads to an absolute error ε_{abs} of η_1 in the range of eps and thus, to a relative error ε_{rel} of η_1 with $|\varepsilon_{rel}| = |\varepsilon_{abs}/\eta_1| \gg$ eps. If the original ansatz function (4) is used, no improvement can be observed, independent of the choice of X_f.

A computation which is *locally* performed with a very high machine precision of e.g. $\text{eps}^{\text{high}} = 10^{-20}$ in the region of a *pn*-junction would not be a remedy, either:
The absolute error $\varepsilon_{\text{large}}$ of the large η-components in the *n*-region would approximately be of the size of η_{large} eps , since computation in the *n*-region is done with machine precision eps. But $\varepsilon_{\text{large}}$ will pass through the *pn*-junction with a weighting-factor of size one, which results in an absolute error of the same size for the small η-components.

An application of the Scharfetter-Gummel scheme leads to weighting factors which are quite different. A system of equations is obtained which can be written as (6). The components m_{ik} in the matrix M have quite different size: Roughly speaking, if m_{kl} is multiplied by $\eta_l \gg \eta_k$, the products $m_{kl}\eta_l$ and $m_{kk}\eta_k$ are approximately of the same magnitude. The Scharfetter-Gummel discretization, however, does not allow obtuse angles in the triangulation.

4   A Special Discretization Scheme

The proposed discretization scheme is now constructed in the 2-D case:

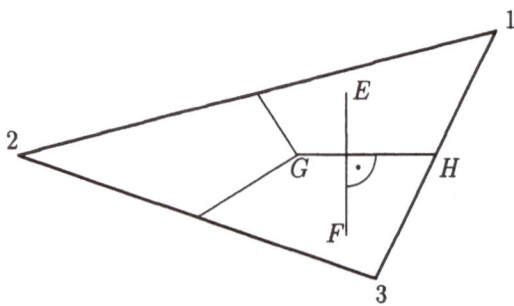

Figure 3

Let the triangle $\Delta 123$ be of arbitrary shape and construct the Baliga-Patankar control volumes as illustrated in Figure 3. An analytical expression for $\int\limits_{GH} (J_\eta \circ n)\, ds$, the current passing through $GH \subset \partial V_1$, can be obtained in the following way: Choose two arbitrary points E and F within the triangle $\Delta 123$ such that $EF \perp GH$. The equation $J_\eta = \mu_\eta(\nabla\eta - \eta\nabla\Psi)$ is integrated along the straight line EF using the Scharfetter-Gummel assumptions of a constant current density and electron mobility in the finite element $\Delta 123$. The integration yields

$$J_\eta\big|_{GH} \circ n = \frac{\mu_\eta}{\|E - F\|}\Big[\mathcal{B}\big(\Psi(E) - \Psi(F)\big)\eta(E) - \mathcal{B}\big(\Psi(F) - \Psi(E)\big)\eta(F)\Big], \quad (11)$$

where $\mathcal{B}(v) := \dfrac{v}{\exp(v) - 1}$ denotes the Bernoulli function.

Since E and F are not nodes of the triangulation, $\eta(E)$ and $\eta(F)$ must be expressed in terms of the nodal values of η:

In every finite element an exponential behavior of η is described by the ansatz function

$$\eta = \gamma e^{\alpha x + \beta y} \qquad (12)$$

with coefficients α, β, γ. Defining the functions φ_i, $i = 1, 2, 3$, to be the barycentric coordinates of the triangle $\Delta 123$, one obtains

$$\eta(E) = \eta_1^{\varphi_1(E)}\eta_2^{\varphi_2(E)}\eta_3^{\varphi_3(E)}.$$

The abbreviation $\delta\Psi := \sum\limits_{i=1}^{3}\big[\varphi_i(E) - \varphi_i(F)\big]\Psi_i$ is used which leads to

$$\int\limits_{GH} (J_\eta \circ n)\, ds = \mu_\eta \frac{\|G - H\|}{\|E - F\|}\Big[\mathcal{B}(\delta\Psi)\prod_{i=1}^{3}\eta_i^{\varphi_i(E)} - \mathcal{B}(-\delta\Psi)\prod_{i=1}^{3}\eta_i^{\varphi_i(F)}\Big]. \qquad (13)$$

Hence, for the left-hand side in (3), $\int\limits_{\partial V_i}(J_\eta \circ n)\, ds$, an analytical expression $F_i(\vec\eta)$ is obtained which is nonlinear with respect to η_k, $k = 1, ..., N$. Within the resulting system of nonlinear equations

$$\vec{F}(\vec\eta) = \vec{f}(\vec\eta), \qquad (14)$$

where $\vec{F} := (F_1, F_2, ..., F_N)^\top$, however, the small $\vec\eta$-components can be computed in a numerically stable way as well as the larger ones.

Concerning numerical stability, a maximum distance between E and F is recommended. Therefore, these points should lie on the edges.

Remark

With $E \in [1,2]$, for example, it is possible to express $\eta(E)$ in terms which are linear with respect to η_1 and η_2. The resulting system of equations, however, suffers from analogous shortcomings as described above for the Baliga-Patankar system of equations. This is shown by the following example:

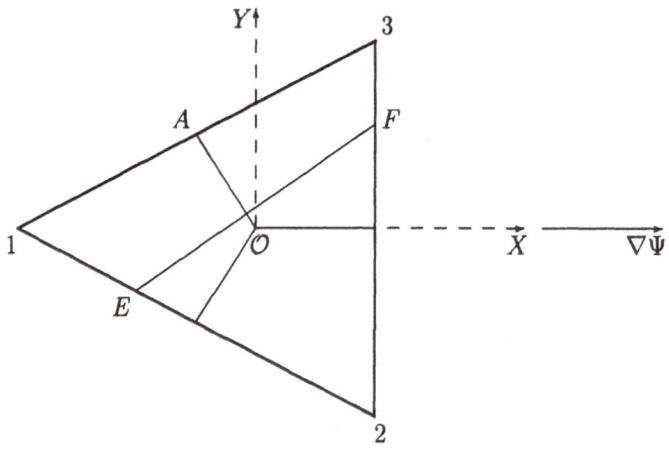

Figure 4

The device is shaped like the triangle illustrated in Figure 4 with nodes 1, 2, and 3, which have the coordinates $(-\frac{2}{3}, 0)$, $(\frac{1}{3}, -\frac{1}{2})$, and $(\frac{1}{3}, \frac{1}{2})$. The same boundary conditions as in the previous example are assumed with slightly changed boundary data:

$$\Psi_1 = -3, \ \Psi_2 = \Psi_3 = 18, \quad \eta_2 = \eta_3 = 3.$$

The solution is given by $\eta = 3e^{21(X - 1/3)}$, so the result $\eta_1 = 3e^{-21}$ has to be computed.

For symmetry reasons $\int_{\partial V_1} (J_\eta \circ n)\, ds = 2 \int_{OA} (J_\eta \circ n)\, ds$ holds and hence, in this example, the continuity equation reads

$$\int_{OA} (J_\eta \circ n)\, ds = 0.$$

With $E := (-\frac{1}{3}, -\frac{1}{6})$, $F := (\frac{1}{3}, \frac{5}{18})$ inserted in (11), this yields

$$\mathcal{B}(-14)\eta(E) - \mathcal{B}(14)\eta(F) = 0. \tag{15}$$

The solution of the one-dimensional continuity equation on the edge [1,2] as the boundary value problem

$$\frac{d^2\tilde{\eta}(t)}{dt^2} - 21\frac{d\tilde{\eta}(t)}{dt} = 0 \quad \text{for } t \in [0,1] \quad \text{and} \quad \begin{array}{l} \tilde{\eta}(0) = \eta_1 \\ \tilde{\eta}(1) = \eta_2 \end{array}$$

leads to

$$\tilde{\eta}(t) = \frac{e^{21} - e^{21t}}{e^{21} - 1}\eta_1 + \frac{e^{21t} - 1}{e^{21} - 1}\eta_2 . \tag{16}$$

With $\eta(E) = \tilde{\eta}(\frac{1}{3})$ in (16) and $\eta(F) = \frac{2}{9}\eta_2 + \frac{7}{9}\eta_3$, equation (15) results in

$$\left(\frac{e^{21} - e^7}{e^{21} - 1}\mathcal{B}(-14), \frac{e^7 - 1}{e^{21} - 1}\mathcal{B}(-14) - \frac{2}{9}\mathcal{B}(14), -\frac{7}{9}\mathcal{B}(14) \right) \left(\eta_1, \eta_2, \eta_3 \right)^{\mathsf{T}} = 0 .$$

Using $\eta_2 = \eta_3 = 3$, the last equation gives

$$\eta_1 = 3\frac{e^{21} - 1}{e^{21} - e^7}\left[\frac{7}{9}\mathcal{B}(14) - \left(\frac{e^7 - 1}{e^{21} - 1}\mathcal{B}(-14) - \frac{2}{9}\mathcal{B}(14) \right) \right]\frac{1}{\mathcal{B}(-14)},$$

or, if multiplications are carried out

$$\eta_1 = 3e^{-21}\left\{ \frac{1}{1 - e^{-14}}\left[\left(\frac{7}{9}e^7 - \frac{7}{9}e^{-14} \right) - \left(\frac{7}{9}e^7 - 1 + \frac{2}{9}e^{-14} \right) \right] \right\}.$$

It can be seen that about three digits are canceled. Again, if $\|\nabla\Psi\|$ is enlarged or the angle $\sphericalangle 213$ is decreased, even more digits are canceled.

The following nonlinear equation is proposed for (15):

$$\mathcal{B}(-14)\eta_1^{2/3}\eta_2^{1/3} - \mathcal{B}(14)\eta_2^{2/9}\eta_3^{7/9} = 0 \tag{17}$$

Here, when computing η_1, no cancellation is observed.

Numerical Implementation

The proposed scheme has been implemented for the 3-D case:
The points E and F are chosen to lie on the surface of the finite element such that the center of gravity of the tetrahedron is situated on the straight line EF, and EF is perpendicular to the selected plane quadrangle of ∂V_i.

The values $\varphi_i(E)$ and $\varphi_i(F)$ do not change during the iteration processes and therefore, they need to be computed only once.

The equations (14) are solved by fixed-point iteration:

$$\vec{F}(\vec{\eta}^{\text{new}}) = \vec{g}(\vec{\eta}^{\text{old}}) \tag{18}$$

where $\vec{g}$ denotes the numerical approximation of $\vec{f}$ by $g_i(\vec{\eta}) := R(\Psi_i, \eta_i, \rho_i)\mathcal{A}_i$ with $\mathcal{A}_i := \int\limits_{V_i} 1 ds$.

$\vec{\eta}^{\text{new}}$ is computed by the Gauß-Seidel-Newton method:

$$F_i(\eta_1^{\text{old}}, \ldots, \eta_{i-1}^{\text{old}}, \eta_i^{\text{new}}, \eta_{i+1}^{\text{old}}, \ldots, \eta_N^{\text{old}}) - g_i(\vec{\eta}^{\text{old}}) = 0$$

is solved for $i = 1, \ldots, N$ by the Newton method, since $\dfrac{\partial F_i}{\partial \eta_i^{\text{new}}}$ can be easily computed in an analytical way.

5 Numerical Example

The proposed discretization scheme was applied to the following 3-D example, which is used in [2] to investigate punch through effects in depletion type trench cells. In the following, distances are measured in μm.

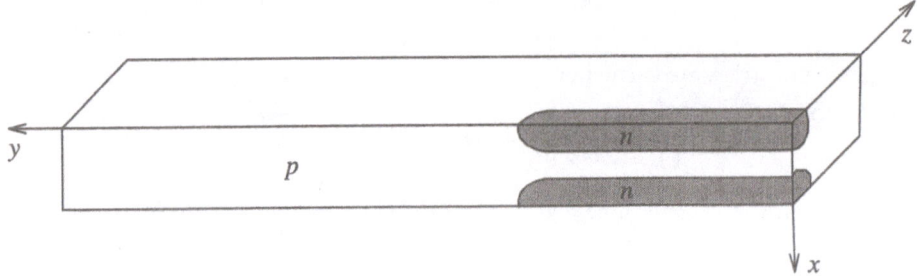

Figure 5

Figure 5 shows the geometry of the device. y is directed into the depth of the device.

The device region is $\Omega := \{(x, y, z) \mid 0 \leq x \leq 2.3,\ 0 \leq y \leq 15,\ 0 \leq z \leq 4\}$. There are three contacts $\Gamma_1 := \{(x, y, z) \mid 0 \leq x \leq 0.3,\ y = 0,\ 0 \leq z \leq 0.3\}$, $\Gamma_2 := \{(x, y, z) \mid 2 \leq x \leq 2.3,\ y = 0,\ 0 \leq z \leq 0.3\}$, and $\Gamma_3 := \{(x, y, z) \mid 0 \leq x \leq 2.3,\ y = 15,\ 0 \leq z \leq 4\}$. 8 Volt bias is applied at Γ_2, while at Γ_1 and Γ_3 no external potential is applied. The Shockley-Read-Hall and the Auger recombination-generation terms are implemented: $R := R_{\text{SRH}} + R_{\text{Au}}$.

The doping profile was kindly provided by W. Bergner (SIEMENS AG München).

An initial triangulation is constructed by subdividing the rectangular parallelepiped into smaller ones and then triangulating each small one by 5 tetrahedra. The subdivision is performed with the following values:

x-direction 0, 0.2, 0.4, 0.6, 0.72, 0.82, 0.92, 1.04, 1.2, 1.4, 1.6, 1.8, 2.0, 2.2, 2.3,

y-direction 0., 0.3, 0.9, 2.4, 3.3, 3.65, 3.8, 4.0, 4.2, 4.6, 5.0, 5.2, 5.35, 5.5, 5.65, 5.8, 6.05, 6.2, 6.45, 6.8, 7.4, 8.2, 9.5, 10.8, 13., 15, and

z-direction 0, 0.1, 0.15, 0.3, 0.5, 0.7, 0.9, 1, 1.25, 1.6, 2.1, 3, 4.

Based on the results of the simulation on this coarse grid, the grid is refined in the following way: every edge $[k, l]$ is refined, if $[k, l]$ is within $\{(x, y, z) \mid 1 \leq x \leq 2,$ $4.2 \leq y \leq 15,\ 0 \leq z \leq 0.4\}$ and fulfills $\eta_k > \eta_l 10^6$ or $\eta_k < \eta_l 10^{-6}$ or $\rho_k > \rho_l 10^6$ or $\rho_k < \rho_l 10^{-6}$; furthermore, every edge within $\{(x, y, z) \mid 0.72 \leq x \leq 1.8, 0 \leq y \leq 5.4,$ $0 \leq z \leq 0.3\}$ is refined, if it is not perpendicular to the xy-plane. In the remaining device region the coarse grid is sufficient. The resulting number of tetrahedra is given by 30581 and the number of nodes by $N = 6850$. An edge is refined by taking its midpoint as a new node. The implemented refinement algorithm does not refine any additional edge besides the selected ones.

To visualize the result of the simulation, an orthogonal 2-D grid is used with linear interpolation. The Figures 6, 7, and 8 show the computed Ψ, η and ρ in the plane $y = 0$. The depletion regions of η and ρ arise as expected, since the pn-junction is reverse biased. In Figures 9, 10, and 11, the computed Ψ, η and ρ are visualized in the plane $z = 0$. Here, the punch through effect can be observed in the depth of the device at $y \approx 4.6$. This observation is confirmed by Figures 12 and 13 which show the plane $y = 4.6$. The results correspond to the simulations published in [2].

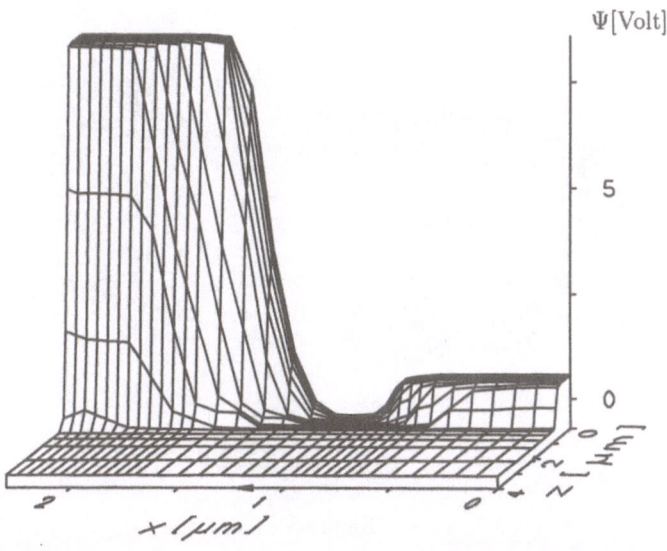

Figure 6

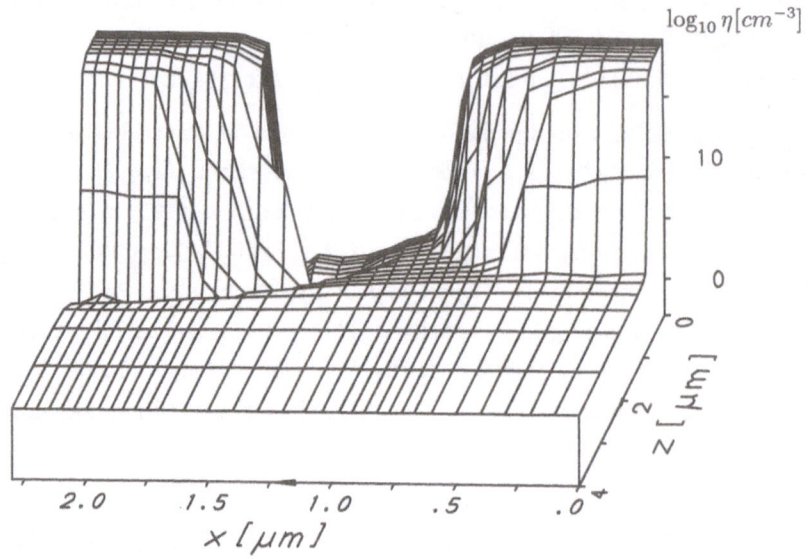

Figure 7

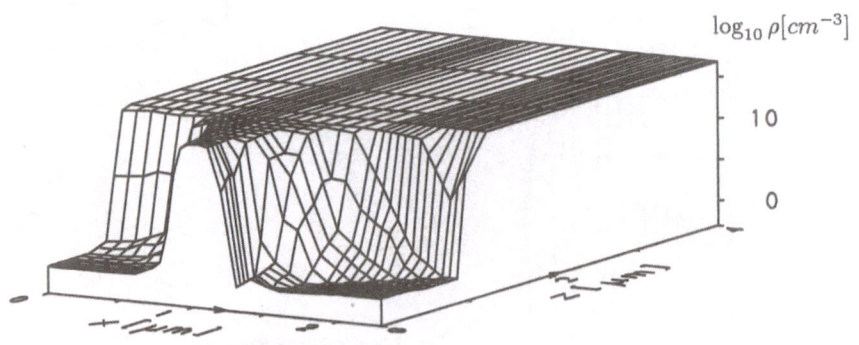

Figure 8

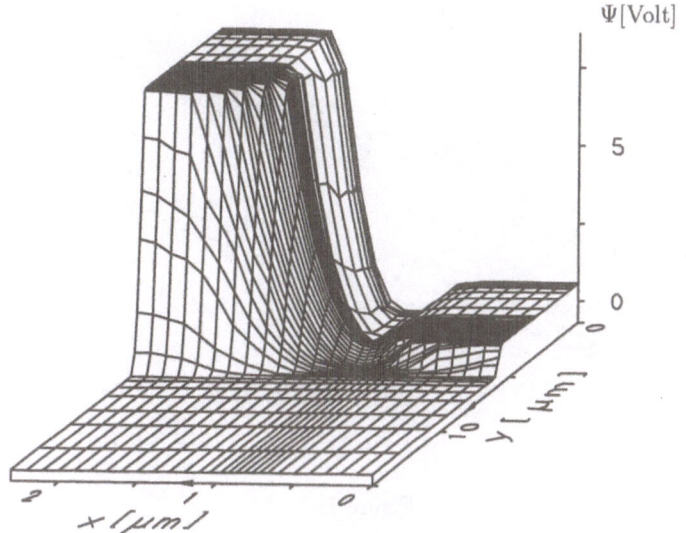

Figure 9

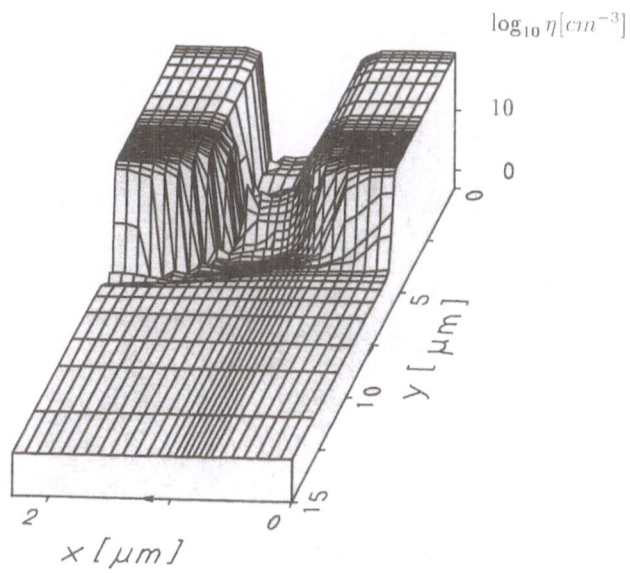

Figure 10

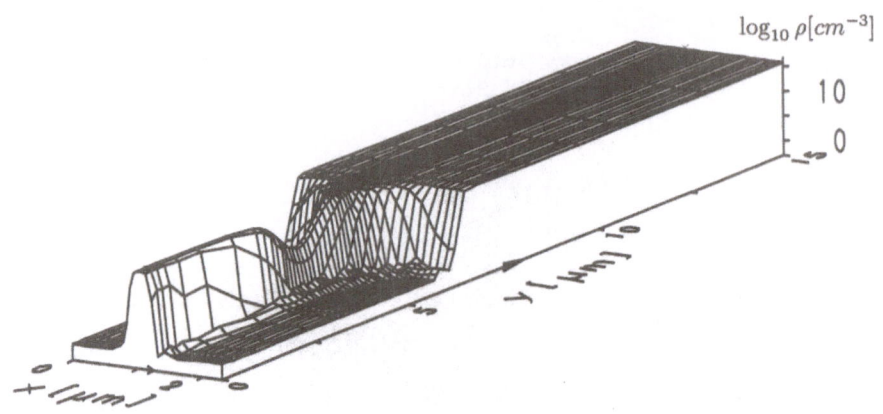

Figure 11

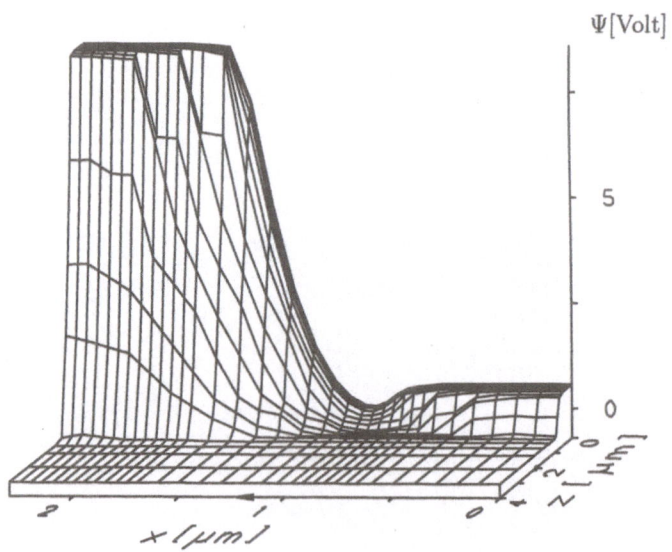

Figure 12

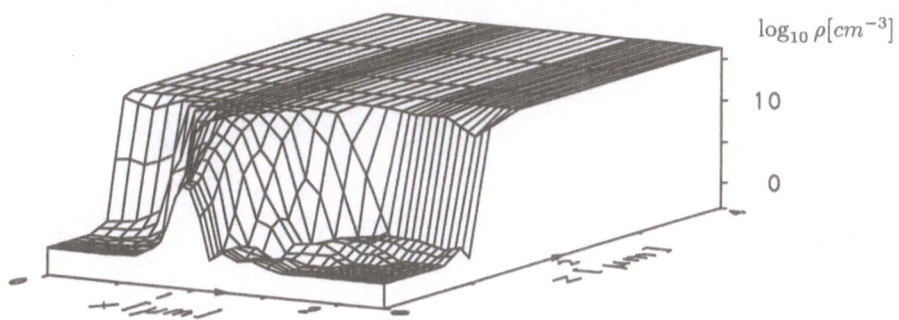

$\log_{10} \rho [cm^{-3}]$

10

0

Figure 13

Acknowledgment

The author acknowledges the support for this work given by Prof. Dr. Dr.h.c. R. Bulirsch. He is indebted to Prof. Dr. R.H.W. Hoppe, W. Bergner (SIEMENS AG München) and the members of the Numerical Analysis group at the TU München for stimulating discussions.

The author thanks the Bayerische Forschungsstiftung for the support in FORTWIHR: Bayerischer Forschungsverbund für Technisch-Wissenschaftliches Hochleistungsrechnen. This work is part of the project 4.4 "Numerical Simulation of Electric Circuits and Semiconductor Devices".

References

[1] B. R. BALIGA, S. V. PATANKAR: *A new finite-element formulation for convection-diffusion problems*, Numerical Heat Transfer, vol. 3, pp. 393–409, 1980.

[2] W. BERGNER, R. KIRCHER: *SITAR – An Efficient 3-D Simulator for Optimization of Nonplanar Trench Structures*, IEEE Transactions on computer-aided design, vol. 9, no. 11, pp. 1184–1188, Nov. 1990.

[3] R. BULIRSCH, A. GILG, K. MERTEN, K. STEGER: *Numerische Simulation in der Halbleiterindustrie*, Informatik Forsch. Entw. vol. 5, pp. 42–56, Springer-Verlag, 1990.

[4] R. KIRCHER, W. BERGNER: *Three Dimensional Simulation of Semiconductor Devices*, Birkhäuser, Basel-Boston-Berlin, 1991.

[5] P. A. MARKOWICH: *The Stationary Semiconductor Device Equations*, Springer, Wien-New York, 1986.

[6] N. SHIGYO, T. WADA, S. YASUDA: *Discretization Problem for Multidimensional Current Flow*, IEEE Transactions on computer-aided design, vol. 8, no. 10, pp. 1046–1050, Oct. 1989.

International Series of Numerical Mathematics, Vol. 117, © 1994 Birkhäuser Verlag Basel

A mass conserving moving grid method for dopant simulation

M.Paffrath

SIEMENS AG, ZFE BT SE 43
D-81739 München
Otto-Hahn-Ring 6

1   Introduction

One major problem in 2D process simulation is the computation of doping profiles on domains with moving boundaries. Consider a structure with two subdomains Ω_1 and Ω_2 of materials silicon (Si) and silicon dioxide SiO_2, respectively (cf. Fig. 1). During an oxidation step, oxygen or steam diffuses through the oxide and reacts with silicon at the $Si - SiO_2$ interface to form new oxide. One volume unit of silicon is transformed into about 2.3 volume units of oxide. This volume expansion causes the oxide subdomain to be lifted up.

The present physical models for oxide growth do not depend on the dopant concentration values and so allow a fully decoupled numerical approach to the oxidation-diffusion problem. Within our process simulator MIMAS-II, there are two different modules for oxide growth and dopant diffusion, respectively [10]. Suppose the oxidation module determines the structure at discrete times $0 = t_1^{ox} < t_2^{ox} < ... < t_n^{ox} = t_f$ and is one time step ahead of the diffusion module. Then in t_i^{ox} the diffusion module can get all the structural information (boundaries, interfaces, flow velocities in SiO_2) corresponding to

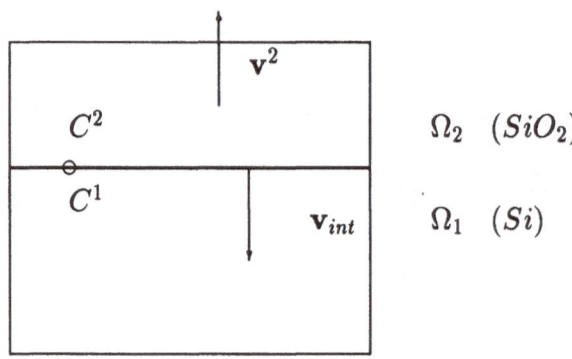

Figure 1: Computational domain

t_{i+1}^{ox} that is needed to compute the dopant concentrations for the next time interval $[t_i^{ox}, t_{i+1}^{ox}], i = 0, ..n-1$.

In this paper we will concentrate on the dopant diffusion part of the problem and outline the solution strategies implemented in MIMAS-II. The main features are a moving grid method combined with a special mass conserving finite element discretization. Finally numerical results are presented.

2   Drift-diffusion equations

In each subdomain Ω_k a diffusion equation of the general form

$$\frac{\partial C^k}{\partial t} = -\nabla \cdot \mathbf{F}^k \qquad k = 1, 2 \tag{1}$$

on some time interval $[0, t_f]$ has to be solved. C^k denotes the total dopant concentration, $\mathbf{F}^k$ the flux vector in Ω_k.

In the silicon subdomain Ω_1 a standard formulation for the flux is given by [16]:

$$\mathbf{F}^1 = -D^1 \nabla C^E \pm D^1 C^E \nabla \Psi \tag{2}$$

In the oxide subdomain Ω_2 a simplified diffusion model is assumed:

$$\mathbf{F}^2 = -D^2 \nabla C^2 + \mathbf{v}^2 C^2 \tag{3}$$

D^k denotes the diffusion coefficient in Ω_k, k=1,2, C^E the electrically active concentration of the dopant , Ψ the electrostatic potential in Ω_1.

The sign of the field term in (2) corresponds to the charge state of the dopant. The convective term $\mathbf{v}^2 C^2$ in (3) accounts for the fact that Ω_2 is moving with velocity $\mathbf{v}^2$ relative to the bulk of Ω_1 (cf. Fig. 1). Physical models with analytical formulas for D^k, C^E, Ψ are e.g. given in [7],[4],[6]. In mathematical terms, D^k, C^E, Ψ are nonlinear functions in C^k that have to be substituted in (2) and (3), respectively. In the case of several dopants, (1) is replaced by a system of partial differential equations.

At outer boundaries the diffusive part of the flux in normal direction is 0:

$$\begin{aligned}
\mathbf{F}^1 \cdot \mathbf{n} &= 0 \\
\mathbf{F}^2 - \mathbf{v}^2 C^2 &= 0
\end{aligned} \tag{4}$$

The diffusion equations in Ω_1 and Ω_2 are coupled by a segregation condition at interface Γ (cf. Fig. 1):

$$\begin{aligned}
(-\mathbf{F}^1 + \mathbf{v}_{int} C^1) \cdot \mathbf{n}^1 &= +F_{seg} \\
(-\mathbf{F}^2 + \mathbf{v}_{int} C^2) \cdot \mathbf{n}^2 &= -F_{seg}
\end{aligned} \tag{5}$$

with $\mathbf{v}_{int}$ denoting the interface velocity of Γ and $\mathbf{n}^k$ the outer normal to $\partial \Omega_k$, k=1,2. F_{seg} is given by some segregation model, a standard one [1] is:

$$F_{seg}(C^1, C^2) = h(C^2 - \frac{C^1}{m}) \tag{6}$$

h and m denote transport and segregation coefficients,respectively.

3 Discretization

Time and spatial discretization are separated by the well-known method of lines. Spatial discretization is accomplished by linear finite elements. The variational formulation of (1) reads: For $k = 1, 2$ find $C^k \in T^k$ such that for all functions $\varphi^k \in W^k$

$$\int_{\Omega_k} \frac{\partial C^k}{\partial t} \varphi^k = + \int_{\Omega_k} \mathbf{F}^k \cdot \nabla \varphi^k - \int_{\partial \Omega_k} \mathbf{F}^k \cdot \mathbf{n}^k \varphi^k \tag{7}$$

Our objective is to apply some kind of moving grid technique. Assume that the referential frame is moving with velocity $\mathbf{w}^k$ in Ω_k. Following [13], (7) is modified using Reynolds transport theorem and Greens theorem :

$$\frac{d}{dt}(\int_{\Omega_k} C^k \varphi^k) = \int_{\Omega_k} \frac{\partial}{\partial t}(C^k \varphi^k) + \int_{\partial \Omega_k} C^k \varphi^k \mathbf{w}^k \cdot \mathbf{n}^k \tag{8}$$

Substitution into (7) gives:

$$\frac{d}{dt}(\int_{\Omega_k} C^k \varphi^k) = + \int_{\Omega_k} \mathbf{F}^k \cdot \nabla \varphi^k + \int_{\Omega_k} C^k \frac{\partial}{\partial t} \varphi^k + \int_{\partial \Omega_k} (-\mathbf{F}^k + \mathbf{w}^k C^k) \cdot \mathbf{n}^k \varphi^k \quad (9)$$

At outer boundaries of Ω_1 holds:

$$\mathbf{w}^1 \cdot \mathbf{n}^1 = 0 \qquad (10)$$

and at outer boundaries of Ω_2:

$$\mathbf{v}^2 \cdot \mathbf{n}^2 - \mathbf{w}^2 \cdot \mathbf{n}^2 = 0 \qquad (11)$$

At the interface Γ $\mathbf{w}^1$ and $\mathbf{w}^2$ are equal to $\mathbf{v}_{int}$. From (5),(10),(11) then follows:

$$\frac{d}{dt}(\int_{\Omega_k} C^k \varphi^k) = + \int_{\Omega_k} \mathbf{F}^k \cdot \nabla \varphi^k + \int_{\Omega_k} C^k \frac{\partial}{\partial t} \varphi^k \pm \int_{\Gamma} F_{seg} \varphi^k \qquad (12)$$

The sign of the boundary integral in (12) is positive for Ω_1 and negative for Ω_2.

It should be noted that, in contrast to equation (7), (12) constitutes a system of differential equations. It is equivalent to a finite element formulation in a space-time-domain $(\Omega(t), 0 \le t \le t_f)$.

The finite element approximation reads: find $C_h^k \in T_h^k$ such that for all functions $\varphi_h^k \in W_h^k$

$$\frac{d}{dt}(\int_{\Omega_k} C_h^k \varphi_h^k) = + \int_{\Omega_k} \mathbf{F}^k \cdot \nabla \varphi_h^k + \int_{\Omega_k} C_h^k \frac{\partial}{\partial t} \varphi_h^k \pm \int_{\Gamma} F_{seg} \varphi_h^k \qquad (13)$$

For linear finite elements $\frac{\partial}{\partial t} \varphi_h^k$ can be substituted by an expression that is computationally easier available. The total time derivative of φ_h^k is 0 because φ_h^k does not change with t on the moving mesh. Therefore holds:

$$\frac{d}{dt}\varphi_h^k = \mathbf{w}_h^k \cdot \nabla \varphi_h^k + \frac{\partial}{\partial t}\varphi_h^k = 0 \qquad (14)$$

This leads to:

$$\frac{d}{dt}(\int_{\Omega_k} C_h^k \varphi_h^k) = + \int_{\Omega_k} \mathbf{F}^k \cdot \nabla \varphi_h^k - \int_{\Omega_k} C_h^k \mathbf{w}_h^k \cdot \nabla \varphi_h^k \pm \int_{\Gamma} F_{seg} \varphi_h^k \qquad (15)$$

Term $\mathbf{w}_h^k$ denotes the velocity points in the moving grid.

An essential requirement in dopant simulation is that the discretization should be conservative, i.e. the law of mass conservation is preserved. It is easy to verify that the formulation (15) satisfies this requirement. Choose $\varphi_h^k \equiv 1$ in (15) and sum up over k, then the left hand side gives

$$\sum_{k=1}^{2} \frac{d}{dt}(\int_{\Omega_k} C_h^k) = \frac{d}{dt}(\sum_{k=1}^{2} \int_{\Omega_k} C_h^k) = \frac{d}{dt}(\int_{\Omega} C_h) = 0 \qquad (16)$$

Because of a significant convective term $\mathbf{v}^2 - \mathbf{w}^2$ in Ω_2, for W_h^k the standard Galerkin test space $H^1(\Omega_2)$ will not be appropriate. Here a variant of the streamline upwind Petrov-Galerkin method is applied [5] . Test function φ_h^2 is composed of the standard Galerkin test function $\psi_h^2 \in H^1(\Omega_2)$ and a function d, given by

$$d = \frac{D_{num}}{\mid \mathbf{v}^2 - \mathbf{w}^2 \mid^2}(\mathbf{v}^2 - \mathbf{w}^2) \cdot \nabla \psi_h^2 \qquad (17)$$

where $D_{num} \geq 0$ is a parameter controlling the numerical dissipation.

For time integration of (15) the TR-BDF(2) method is applied [2].

4   Grid generation and adaption

In dopant simulation, three main grid strategies for the moving boundary problem may be distinguished:

1. using a stationary grid in all subdomains (e.g. [9]).

2. using a stationary grid in the silicon subdomain and a grid moving with material flow in the oxide subdomain (e.g. [8]).

3. using a coordinate transformation method, where the grid is moving with the $Si - SiO_2$ interface (e.g. [12],[17]).

In our opinion, the draw-back of the first and second strategy is that they need some heuristics for redistribution of dopant concentrations and grid refinement at the moving interface. The third strategy in its basis form has geometrical restrictions, but allows a more rigorous mathematical handling of segregation effects. Our method [11] is a variant of the third strategy. To overcome geometrical restrictions, the coordinate transformation method is generalized in the sense that the coordinate domain is not defined a priori but is adapted to the geometrical evolution of the physical structure. So the method becomes more flexible concerning large variations of geometry. As times for redefinition of the coordinate domain oxidation times t_i^{ox} are the natural choice.

Suppose that a grid $\mathcal{T}(t_i^{ox})$ of the diffusion module is given. The algorithm can be briefly stated as follows:

Algorithm 1

1. *Decide whether the grid should be adapted to doping profile. If yes, generate a new grid $\mathcal{T}(t_i^{ox})$.*

2. *Compute boundary and interface points of $\mathcal{T}(t_{i+1}^{ox})$ thus defining $\partial\Omega(t_{i+1}^{ox})$.*

3. *Map $\mathcal{T}(t_i^{ox})$ to $\mathcal{T}(t_{i+1}^{ox})$ by solving two Laplace equations*

$$\Delta \begin{pmatrix} \xi_1 \\ \xi_2 \end{pmatrix}(x) = 0 \qquad (18)$$

   for coordinates ξ_1, ξ_2 in $\Omega(t_{i+1}^{ox})$ with Dirichlet boundary conditions given by step 2.

4. *Determine $\mathcal{T}(t)$ for $t_i^{ox} < t < t_{i+1}^{ox}$ by linear interpolation:*

$$\mathcal{T}(t) = \mathcal{T}(t_i^{ox}) + \frac{t - t_i^{ox}}{t_{i+1}^{ox} - t_i^{ox}}(\mathcal{T}(t_{i+1}^{ox}) - \mathcal{T}(t_i^{ox})) \qquad (19)$$

In step 1, the new grid is generated according to a density function defining the required distances between nodes [14]. While in t_i^{ox} the grid is adapted to the doping profile, between times t_i^{ox} and t_{i+1}^{ox} the grid is only adapted to geometry.

We note that step 4 gives the velocity of the moving grid and so defines term $\mathbf{w}_h^k$ in (15).

5   Interpolation of concentrations

After remeshing (step 1 of algorithm 1), concentrations are linearly interpolated for the new mesh. For each point x of the new mesh, the old mesh is searched through for a triangle T containing x. It would be a very time consuming task to search through the complete list of triangles for each x. Therefore a hash search is introduced.

First define a rectangular background grid R with elements $R_{i,k}$ covering the

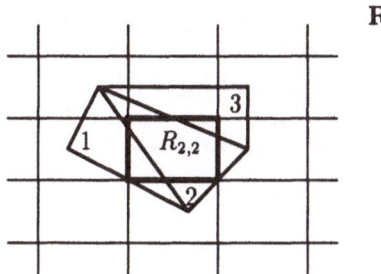

Figure 2: Hash search

structure. R has to be equidistant in each coordinate. In a preprocessing step set up a list $L_{i,k}$ of old triangles intersecting $R_{i,k}$ (for all i,k). In figure 2 e.g. the list for element $R_{2,2}$ is given by triangles 1,2,3. Then, in order to find T, determine R_{i_0,k_0} containing x (which takes only few operations) and search through L_{i_0,k_0}.

6   Example: Sloped-Wall SWAMI

In [3] the so-called sloped-wall S̲ide W̲A̲ll M̲asked I̲solation (SWAMI) process was presented as a technique to produce a fully recessed zero bird's beak and a nearly planar surface.

Simulation has to deal with extreme slopes in the oxide/silicon interface. Towards the end of processing even a turning point and a cusp emerge in the interface.

The initial structure is shown in Fig. 3. It consists of single-crystal (100) silicon as silicon substrate, a thin oxide layer for stress relief and a nitride mask. Fig. 3 also shows the concentration of implanted boron used as channel stopper. Contour lines are given for boron concentrations $10^{17\,5}$, 10^{18} and $10^{18\,5}$ (number of atoms per unit-volume), respectively. Fig. 4 shows the corresponding numerical grid.

After implantation a field oxide is grown at a temperature of 900 °C for 280 min under wet oxidation conditions. Fig. 5 shows the grid at time $t_1 = 6$ min. Up to t_1, the grid has only been adapted to moving boundaries, it is topologically equivalent to the initial grid (cf. algorithm 1). At t_1 a new grid is generated which better resolves the boron profile (Fig. 6). Fig. 7 shows the boron profile at the end of the process step.

References

[1] D. A. Antoniadis, M. Rodoni, and R. W. Dutton,
Impurity redistribution in SiO2-Si during oxidation: A numerical solution including interfacial fluxes.
J. Electrochem. Soc., Solid-State Sci. Techn. **126**(11), 1939–1945 (1979).

[2] R. Bank, W. M. Coughran, W. Fichtner, E. H. Grosse, D. J. Rose, and R. K. Smith,
Transient simulation of silicon devices and circuits.
IEEE-CAD **4**(4), 436–451 (1985).

[3] K. Y. Chiu, J. L. Moll, K. M. Cham, J. Lin, C. Lage, S. Angelos, and R. L. Tillman,
The sloped-wall SWAMI - A defect-free zero bird's-beak local oxidation process for scaled VLSI technology.
IEEE-ED, **30**(11), 1506–1511, November(1983).

[4] R. B. Fair,
Concentration profiles of diffused dopants in silicon. In: *Impurity doping processes in silicon.*, F. F. Y. Wang (ed.)
North-Holland Publishing Company (1978).

[5] T. J. R. Hughes and M. Mallet,
A new finite element formulation for computational fluid dynamics:III.The generalized streamline operator for multidimensional advective-diffusive systems.
Comput. Meths. Appl. Mech. Engrg. **58**, 305–328 (1986).

[6] M. R. Kump and R. W. Dutton,
The efficient simulation of coupled point defect and impurity diffusion.
IEEE Trans. Computer Aided Des. **7**(2), 191–204 (1988).

[7] F. Lau,
Modeling of polysilicon diffusion sources.
International Electron Devices meeting, Technical Digest, 737–740
(1990).

[8] M. E. Law, C. Rafferty, and R. W. Dutton,
*SUPREM IV.*In: *Process simulators for silicon VLSI and high speed
GaAs devices.*, J. D. Plummer et al. (ed.)
Stanford Electronics Lab., Stanford University, July (1987).

[9] K. Nishi, K. Sakamoto, S. Kuroda, J. Ueda, T. Miyoshi, and S. Ushio,
*A general-purpose two-dimensional process simulator-OPUS-for ar-
bitrary structures.*
IEEE-CAD, **8**(1), 23–32, January (1989).

[10] M. Paffrath, W. Jacobs, W. Klein, E. Rank, K. Steger, U. Weinert,
and U. Wever,
Concepts and algorithms in process simulation.
Surv.Math.Ind. **3**, 149–183 (1993).

[11] M. Paffrath and K. Steger,
*Method of temporary coordinate domains for moving boundary value
problems.*
IEEE-CAD **12**(6), 746–756 (1993).

[12] B. R. Penumalli,
*A comprehensive two-dimensional VLSI process simulation program,
BICEPS.*
IEEE-ED, **30**(9), 986–992, September (1983).

[13] E. J. Probert, O. Hassan, and K. Morgan,
*An adaptive finite element method for transient compressible flows
with moving boundaries.*
Int. J. Numer. Methods Eng. **32**, 751–765 (1991).

[14] M. Schweingruber and E. Rank,
Adaptive mesh generation for triangular or quadrilateral elements.
Proceedings of the first European conference on numerical methods
in engineering, Bruxelles (1992).

[15] M. Sharma and G. F. Carey,
*Semiconductor device simulation using adaptive refinement and flux
upwinding.*
IEEE-CAD **8**(6), 590–598 (1989).

[16] R. Tielert,
Two-dimensional numerical simulation of impurity redistribution in VLSI processes.
IEEE Trans. Electron Dev. **27** , 1479–1483 (1980).

[17] K. Wimmer, R. Bauer, S. Halama, G. Hobler, and S. Selberherr,
Transformation methods for nonplanar process simulation.
Proceedings SISDEP 4 (Zurich, Switzerland), 131–138, September (1991).

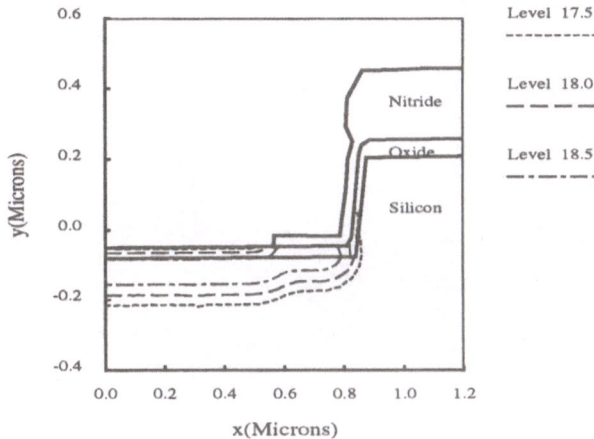

Figure 3: Boron concentration after implanation

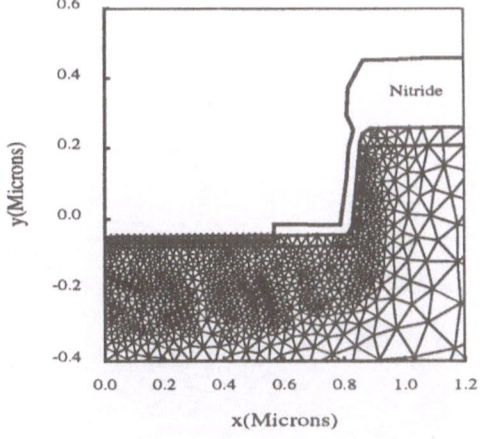

Figure 4: Grid after boron implantation

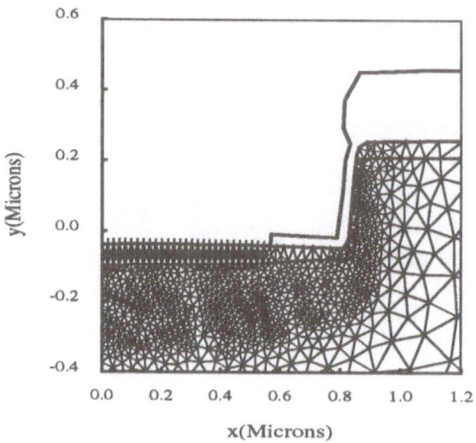

Figure 5: Grid before adaption at time t = 6 [min]

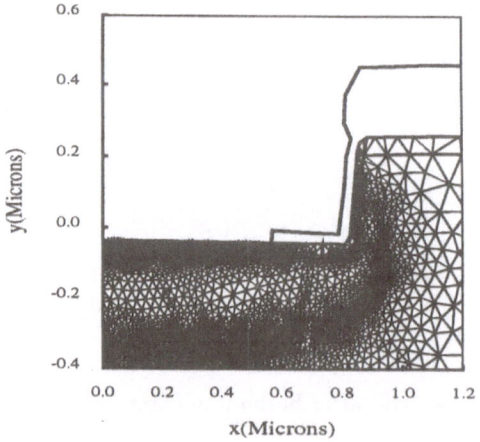

Figure 6: Grid after adaption at time t = 6 [min]

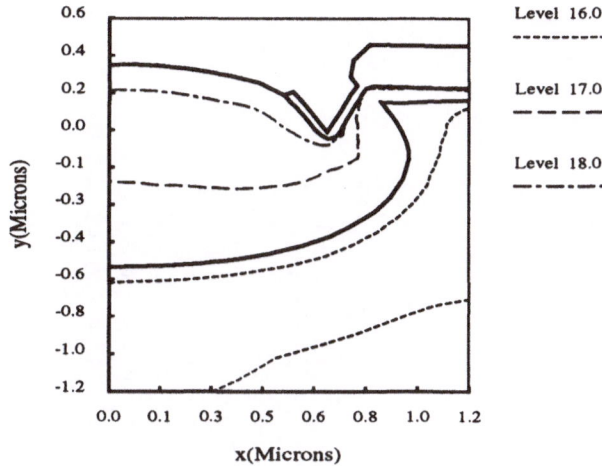

Figure 7: Boron concentration at time t = 280 [min]

International Series of Numerical Mathematics, Vol. 117, © 1994 Birkhäuser Verlag Basel

Numerical approaches to the kinetic semiconductor equations

J. Wick, Kaiserslautern

1 Introduction

The numerical computation of carrier transport effects in semiconducturs is mostly done with a Monte-Carlo simulation [5]. It does not refer to a mathematical model equation, since it models the motion of the carriers itself. Thus this is a very flexible technique. Its main disadvantage are the low accuracy and the effects of noise. Hence for reasonable results a large amount of carriers must be considered. This requires both large storage as many computer time. Thus the method becomes expensive.

Here we will present several suggestions to cut down these costs. The basis for all these methods are the mathematical description of the carrier transport. If classical interaction are considered only, we find for it a Boltzmann-like kinetic equation. If we omit collisions we obtain the Vlasov equation, which is well-known in stellar dynamics and plasma physics. In these cases the Monte-Carlo approach can also considered as a characteristic method. This gives rise to other concepts of convergence [10], [3]. Hence it is quite natural to investigate their possible generalizations to a kinetic equation with collisions. Here we will report these extensions to the kinetic semiconductor equations.

(i) Cottet and Raviart [3] gave a convergence proof of the Monte-Carlo method for the Vlasov equation as a characteristic method - or as a discretization with a Lagrangian grid - in certain function spaces. The generalization was considered by [6],[11].

(ii) The convergence proof of Neunzert [10] is based on the fact, that the Vlasov evolution preserves an initial discrete measure. This remains not true in the semiconductor case. Hence a rediscretization at every timestep is required [8], [7] [12].

(iii) A collision can also be modeled as a flux, which leads to a conservation law in divergence form. Then a discrete measure will remain discrete under the associate evolution. This is the basis of third method [9], [2].

In the sequel we present the equation under consideration and discuss the mentioned three numerical methods.

2 The kinetic equation

The function $f : [0, \infty) \times \Omega_x \times \Omega_k \to \mathbf{R}$ denotes the probability that at time $t \in [0, \infty)$ and at the position $x \in \Omega_x$ an electron has the state $k \in \Omega_k$, $\Omega_x \subset \mathbf{R}^3, \Omega_k \subset \mathbf{R}^3$. Hence we require

$$0 \leq f \leq 1. \tag{2.1}$$

The velocity of an electron is assumed to be a given function of its state

$$v = v(k). \tag{2.2}$$

The acceleration of an electron is governed by a force field K. This is composed by an external and a selfconsistent field. For simplicity we do not consider a selfconsistent field, which will require a coupling with Maxwell's equations.
The collisions are described with a scattering kernel S by the collision integral

$$P(f, f) := \int_{\Omega_{k'}} [S(k, k')f'(1 - f) - S(k', k)f(1 - f')]dk'. \tag{2.3}$$

$f = f(t, x, k), f' = f(t, x, k')$.

The factors $(1 - f), (1 - f')$ take into account Pauli's exclusion principle and they guarantee, that (2.1) holds for all t provided the initial condition has this property.
If we do not take into account the motion of holes, the evolution of the electrons is modeled by

$$\partial_t f + < v, \text{ grad}_v \ f > + < K, \text{ grad}_k \ f > = P(f, f). \tag{2.4}$$

(2.3) shows immediately the conservation of electrons. (2.4) is the supplement with the initial condition

$$f(0, x, k) = f^0(x, k) \in L^1, \quad \int_{\Omega_x} \int_{\Omega_k} f^0(x, k)dkdx = \overline{f^0}$$

and appropriate boundary conditions.
Quite often,

$$f << 1 \tag{2.5}$$

is assumed and the Pauli factors in (2.3) are neglected. This simplifies the collision integral into

$$Q(f) = \int_{\Omega_k'} [S(k, k')f' - S(k', k)f]dk'.$$

3   General aspects of the particle approximation

Our aim is to approximate the solution of

$$\partial_t f + \langle v, grad_x f \rangle + \langle K, grad_k f \rangle = P(f, f)$$

or

$$\partial_t f + \langle v, grad_x f \rangle + \langle K, grad_k f \rangle = Q(f)$$

supplemented with the initial condition

$$f(0, x, k) = f^0(x, k) \in L^1(\Omega_x \times \Omega_k \to \mathbf{R}_+); \ \int f^0 dx dk = \overline{f^0}$$

and appropriate boundary conditions by

$$\sum_{i=1}^{n} w_i f_i(t) \delta(x - x_i(t)) \delta(k - k_i(t)).$$

We will call $(x_i(t), k_i(t))$ the "position on the i-th particle" and $w_i f_i(t)$ its "weight". The "space-" or "x-positions" refers to the space coordinates of the particles.

In general, we will not find two particles with the same space position. Hence for the approximate computation of the collision term a modification is required. The usual technique is to divide Ω_x into cells. Then all particles in a cell are taken into account for the computation of the collision integral for all position in the cell. Because of this local homogenization in x-space and the numerical experience with the Vlasov equation we restrict our consideration here to the space-homogeneous case. Hence we deal with

$$\partial_t f + \langle K, \ grad_k \ f \rangle = P(f, f) \tag{3.1}$$

or its linearized form

$$\partial_t f + \langle K, \ grad_k \ f \rangle = Q(f) \tag{3.2}$$

In addition, we will assume $\Omega_k = \mathbf{R}^3$, in order to avoid the treatment of boundary problems.

4   The characteristic method

We find

$$\dot{k} = K \tag{4.1}$$

$$\dot{f} = Q(f) \tag{4.2}$$

as the associated system of characteristic equations to (3.2). K is supposed to be known, hence $k(t) = k(t, k^0)$ can be considered as a given function for every initial value $k^0 \in \mathbf{R}^3$.

Now, we construct the particle approximation. Let be I a finite or countable index set. We suppose w_i are the weights and k_i are the nodes of an integration formula over $\mathbf{R}^3$. Then we put

$$f^0(k) = \int_{\mathbf{R}^3} f^0(k')\delta(k - k')dk' \approx \sum_{i \in I} w_i f^0(k_i)\delta(k - k_i) =: \tilde{f}^0(k).$$

We want to approximate $f(t, k)$ by

$$\tilde{f}(t, k) := \sum_{i \in I} w_i f_i(t)\delta(k - k_i(t)),$$

where $k_i(t) := k(t, k_i)$ and $f_i(t)$ will be computed using (4.2). Its right hand side is an integral, hence we use the same weights as above and $\{k_i(t)\}$ as nodes for the numerical computation.

$$Q_i(t) = \tilde{Q}(f)(k_i(t)) := \sum_{j \in I} w_j (S(k_i(t), k_j(t))f_j(t) - S(k_j(t), k_i(t))f_i(t)) \qquad (4.3)$$

Hence

$$\frac{d}{dt}f_i(t) = Q_i(t), \quad f_i(0) = \tilde{f}^0(k_i) \qquad (4.4)$$

will be an approximation of (4.2).
From (4.4), it follows that

$$\frac{d}{dt}\sum_{i \in I} w_i f_i(t) = \sum_{i \in I}\sum_{j \in I} w_i w_j (S_{ij}(t)f_j(t) - S_{ji}(t)f_i(t)) = 0 \qquad (4.5)$$

This show the conservation property of the discretized equation.

For a real computation mostly a regular grid was used

$$k_i = (i_1\Delta k, i_2\Delta k, i_3\Delta k) \quad , i_j \in \mathbb{Z}$$

This requires the introduction of artificial boundaries to get a finite number of grid points [11]. One way to avoid this are integration formulas of Gauss-type [4]. This method has many advantages:

- It is easy to understand

- It is easy to obtain approximate values of $f(t, k)$ for all point (t, k) of interest

- Its convergence is proven.

There are also disadvantages:

- The effort for the computation is $O(n^2)$ as (4.3) shows

This restricts the number of particles which can be treated in reasonable time with regular grids. Especially 3D-computations will be of low accuracy. Gaussian formulas reduce the number of the required particles, nevertheless 3D-computations are very expensive. Another serious restriction is

- the decoupling effect:

In a real semiconductor both the electric field as the collision process changes the k-values. In the characteristic method the k-values are influenced by the electric field only. The collision process is reflected in the change of weights. This leads to some difficulties. To show this let us consider a simple 1D relaxation model with

$$ S(k, k') = f^0(k), \quad \int_{\mathbf{R}} f^0(k)dk = 1, \quad f^0 \in (L^1 \cap L^\infty)(\mathbf{R}). $$

Thus the equation becomes

$$ f_t + Kf_k = f^0 - f, \quad f(0, k) = f^0(k). \tag{4.6} $$

In addition, we assume K to be constant. Then we can solve (4.6) explicitly

$$ f(t, k) = e^{-t}f^0(k - Kt) + \int_0^t e^{-s}f^0(k - Ks)ds $$

and its stationary solution reads as

$$ f_\infty(k) = \int_0^\infty e^{-s}f^0(k - Ks)ds. $$

Since

$$ ||f(t, \cdot) - f_\infty||_\infty \leq e^{-t}||f^0||_\infty + ||\int_t^\infty e^{-s}f^0(\cdot - Ks)ds||_\infty $$

$$ \leq 2e^{-t}||f^0||_\infty $$

the stationary solution will be approached exponentielly.
Choosing

$$ f^0(k) = \begin{cases} 1 + k & -1 \leq k < 0 \\ 1 - k & 0 \leq k < 1 \\ 0 & \text{otherwise} \end{cases} \tag{4.7} $$

we obtain

$$f_\infty(k) = \begin{cases} 0 & -\infty < k \le -1 \\ 1 + k - K + K \exp\left(-(k+1)/K\right) & -1 < k \le 0 \\ 1 - k + K + K \exp\left(-(k+1)/K\right) - 2K \exp\left(-k/K\right) & 0 < k \le 1 \\ 2K \exp\left(-k/K\right)(\text{Cosh}\left(\frac{1}{K}\right) - 1) & 1 < k < \infty \end{cases}$$

$$\tag{4.8}$$

Its numerical relevant support is still finite.

Now, in the characteristic method for (4.6) all particles have constant speed K

$$k_\imath(t) = k_\imath^0 + Kt.$$

Starting with a finite number of particles, they leave the numerical relevant support in finite time. Then the only information we get from these particles is: The solution is (almost) zero at their location. For this 1D example we can avoid this if we take out particles with small weights on the right and put them back - with the same weights - on the left. But for higher dimensional problems the situation can become much worser. Thus the decoupling of transport and collision require additional and difficult precautions to get a reliable code.

A hint of numerical difficulties caused by the decoupling effect can be obtained by the conservation property (4.5). Hence it should be not used to speed up the code itself.

5   The Rediscretization method

As a first approach we consider (3.1) with a vanishing field K. Then the equation reads as

$$\partial_t f = P(f, f), \quad f(0, 1) = f^0 \in L^1(\mathbf{R}^3). \tag{5.1}$$

We suppose

$$0 \le f^0 \le 1 \quad \text{and} \quad \int_{\mathbf{R}^3} f^0(k)dk = \overline{f^0}. \tag{5.2}$$

As mentioned above (5.2) holds also for the solution of (5.1). Therefore $f(t, \cdot)$ can be considered as a density of a measure $\mu(t)$ with respect to the Lebesgue measure λ

$$d\mu(t) = f(t, \cdot)d\lambda.$$

Thus we can interpret (5.1) as an equation of time-depending measures. We do not write down the measure formulation of (5.1). But for brevity we call $\mu(t)$ a solution of (5.1) if $f(t, \cdot)$ is a solution.

In the weak topology any measure can be approximated by equi-weighted discrete measures. Thus we search for $k_j(t)$, such that

$$\frac{\overline{f^0}}{n} \sum_{j=1}^{n} \delta(k - k_j(t)) =: \mu_n(t)$$

is an approximation of $\mu(t)$.
For this purpose we discretize (5.1) by using

$$\partial_t f(i\Delta t, \cdot) = \frac{1}{\Delta t}(f((i+1)\Delta t, \cdot) - f(i\Delta t, \cdot)) + O(\Delta t)$$
$$= \frac{1}{\Delta t}(f^{i+1} - f^i) + O(\Delta t).$$

and consider

$$f^{i+1} \approx f^i + \Delta t P(f^i, f^i)$$

If μ_n^i is a discrete approximation of μ^i, the associated measure to f^i, the right hand side of (5.3) reads as

$$\mu_n^i(B) + \Delta t \int_B \int_{\mathbf{R}^3} S(k, k') \mu_n^i(dk') \lambda(dk) - \Delta t \int_B \int_{\mathbf{R}^3} S(k, k') \mu_n^i(dk') \mu_n^i(dk)$$
$$- \Delta t \int_B \int_{\mathbf{R}^3} S(k', k) \lambda(dk') \mu_n^i(dk) + \Delta t \int_B \int_{\mathbf{R}^3} S(k', k) \mu_n^i(dk') \mu_n^i(dk) =: \nu_n^{i+1}(B)$$

where $B \subset \mathbf{R}^3$ is an arbitrary Borel set.

In general, the first integral does not denote a discrete measure. Hence at every time step a discrete approximation of ν_n^{i+1} is required to find μ_n^{i+1}. Moock gave in this PhD-thesis one algorithm for these computation [7].

The extension to the case $K \neq 0$ is simple, we use the characteristic system (4.1). Hence the Rediscretization method contains three stages for one time step

(i) Compute ν_n^{i+1}

(ii) Rediscretize ν_n^{i+1} and get $\frac{\overline{f^0}}{n} \sum_{j=1}^{n} \delta(k - \kappa_j^{i+1})$

(iii) Solve $\dot{k} = K, k(i\Delta t) = \kappa_j^{i+1}$, get $k((i+1)\Delta t; \kappa_j^{i+1}) =: k_j^{i+1}$ and put

$$\mu_n^{i+1} = \frac{\overline{f^0}}{n} \sum_{j=1}^{n} \delta(k - k_j^{i+1}).$$

The 1D-version of this method for (3.1) can be found in [8]. Moock [7] has extended this in several dimensions and Pfau has [12] transfered it to (3.1) in 1D. A generalization to higher dimension should be obtained by a combination of the considerations in [7] and [12].

The application of the method to (4.6) showed a numerical stationary solution after an appropriate time.
Thus we find as advantages of the method

- Its convergence is shown for nearly all cases

- No artificial boundary conditions are required

For numerical experiments it seems that the cardinality of the approximating point set n can be chosen less than 100 also for 2D space-homogeneous problems [7]. Thus the method can be suitable for space-inhomogeneous problems, too.
But there are also disadvantages

- It is not so easy to understand

- Until now, no recipe is known for the computer of values of $f(t, k)$.

- The numerical effort in the rediscretization step is $O(n^2)$

Since the number of used particles is smaller, the last point in less serious as in the previous method.

6 The CRF-method

Another proposal using the measure theoretic interpretation is based on the following observation.

Consider a general conservation law in divergence form

$$\partial_t g + \text{ div } (gG) = O. \tag{6.1}$$

Assuming G as a C^1-vector field (6.1) can be written as

$$\partial_t g + < G, \text{ grad } g >= -g \text{ div } G.$$

Then

$$\dot{k} = G ; \quad \dot{g} = -g \text{ div } G \tag{6.2}$$

is the associated characteristic system. The formal solution of the last equation reads as

$$g = g^0 \exp\left(-\int_0^t \operatorname{div} G \, ds\right).$$

With T_{t0} we denote the solution operator of the equation of the characteristic ground curve $\dot{k} = G$. Now, the Jacobian of $T_{t0}k^0$ with respect to the initial vector k^0 can be represented with the trace of the Jacobian of G, more detailed

$$J(k) := \det\left(\frac{\partial(t_{t0}k^0)_i}{\partial k_j^0}\right) = \exp\left(\int_0^t \sum_{j=1}^n \frac{\partial G_j}{\partial k_j} ds\right) = \exp\left(\int_0^t \operatorname{div} G \, dx\right).$$

Hence we can get the solution of (6.1) knowing $T_{t0}k^0$ only.

$$g(t, T_{t0}k^0) = g^0(k^0)J^{-1}(k^0)$$

and by $T_{0t} := T_{t0}^{-1}$ we obtain

$$g(t, k) = g^0(T_{0t}k)J^{-1}(T_{0t}k).$$

Now, we assume $g \geq 0$ and $g(t, \cdot) \in L^1(\mathbf{R}^s), s \in \mathbb{N}$ Then we can associate a measure $\nu(t)$ to $g(t, \cdot)$ and the measure solution of (6.1) reads as

$$\nu(t) = \nu(0) \circ T_{0t}. \tag{6.3}$$

Whether G is divergencefree or not the characteristic ground curves are enough to get the solution of the conservation law (6.1). For its point approximation no computation of the Jacobian is required as (6.3) shows.

Now, we came back to the problem (3.1). At first we solve

$$\operatorname{div} \Psi = -P(f, f) \tag{6.4}$$

supplemented by the no-outflow condition

$$\lim_{|k^j| \to \infty} \Psi^j = 0 \quad j = 1, 2, 3 \tag{6.5}$$

This means the Collision are Redefined as a Flux and this explains the name of the method. The boundary conditions guarantee the conservation property of the equation. The associated "velocity field" F is given according to

$$\int_B \Psi dk = \int_B F \, f dk \tag{6.6}$$

for all Borel sets $B \subset \Omega$.
Define
$$I(k^i) := \{\kappa = (\kappa^1, \kappa^2, \kappa^3) \in \mathbf{R}^3, \; -\infty < \kappa^i < k^i\} \quad i = 1, 2, 3$$
and
$$\partial I(k^i) := \{\kappa \in \mathbf{R}^3, \; \kappa^i = k^i\} \quad i = 1, 2, 3.$$

¿From (6.4) and (6.5) we obtain using the Gaussian theorem

$$-\int_{I(k^i)} P(f,f)d\kappa = \int_{I(k^i)} \mathrm{div}\,\Psi d\kappa = \int_{\partial I(k^i)} \Psi^i \, dw^i \quad i = 1, 2, 3$$

where dw^i is the appropriate surface element. The integration w.r.t i-th coordinate yields

$$P^i(k^i) := \int_{I(k^i)} \Psi^i d\kappa = -\int_{I(k^i)} \left(\int_{I(k^i)}^{\kappa^i} P(f,f)d\bar\kappa^i \right) d\kappa \tag{6.7}$$

If we choose $B = I(k^i)$ in (6.6) we find

$$\int_{i(k^i)} F^i f d\kappa = P^i(k^i)$$

Replacing $f d\kappa = f\lambda(d\kappa)$ by a discrete approximation and using the trapezoidal rule we obtain a linear system of order n to compute the vales of P^i at the support of the discrete measure (cf. [9], [1]). The matrix of the system is diagonal dominant and independent of the particle positions. Hence only one LR- decomposition is required to solve the linear system for each time step.

The method was tested for the polar-optical scattering in GaAs with axial symmetry. This leads to a 2D-problem in k-space. For this case the scattering kernel itself is a singular distribution. The application of the two other method - and also the Monte-Carlo method - requires a mollification of the kernel. Because of the integration in the computation of P^i, this is not necessary here. The axial component becomes an explicitly computable expression. The radial component is given as an elliptic integral. Its computation takes the largest amount of the computation time. We found the mean velocity and the mean energy comparable with other computations [11]. But in the CFR-method we deal with 27 particles only [2].
It should be recognized, that this method has also several disadvantages

- Its not easy to explain

- There is no convergence proof

- The computation of P^i can be difficult

- Again the values of $f(t, k)$ can not be computed

- Its still of order $O(n^2)$

- The code shows a limitation of the time step.

Of course, the $O(n^2)$-effort is not to restrictive for the small particle numbers in use. The last property seems to reflect the CFL-condition, where n must be considered as a measure of the space discretization [1]. In other words, $n\Delta t$ must be controled.
The advantages are

- Its seems to work for small particle number

- A wide class of collision kernels can be treated without mollification

- The non-linear case is included

- No artificial boundary conditions must be introduced.

7 Conclusions

We have presented three different particle approximations of a kinetic semiconductor equation, which are based on different mathematical concepts. For all of them we find satisfying numerical results in the literature. They run with much smaller particle numbers than the Monte-Carlo method and do not show noisy solutions. The reduction of the particle number decreases the computer costs, but a higher mathematical effort must be paid. This differs for each of the methods and is not really equilized by its advantages and disadvantages. Hence a decision which one is preferable requires additional mathematical investigations.

References

[1] R. Burkhard: *Ein finites Partikelverfahren für Elektron–Phonon–Streuung in Halbleitern*, Diplomarbeit 1992 FB Mathematik Universität Kaiserslautern.

[2] R. Burkhard, S. Motta, J. Wick: *The CRF-method on general coordinates*, to appear

[3] G.H. Cottet, P.A. Raviart: *Particle methods for the one-dimensional Vlasov-Possion equation*, J. Num. Anal. SIAM 21 (1984) 52-76

[4] T. Geyer, J. Wick: *A Deterministic Particle-Method Solving the Linearized Boltzmann Equation*, Computing 43 (1990) 199–207

[5] C. Jacobini, P. Lugli: *The Monte Carlo Method for Semiconductor Device Simulation*, Springer, Berlin etc. 1989

[6] S. Mas–Gallic: *A deterministic particle method for the linearized Boltzmann equation*, Transp. Th. St. Phys. 16 (1987) 855–887

[7] H. Moock: *Ein deterministisches Teilchenverfahren zur Simulation der Boltzmann-Vlasov-Gleichung für Halbleiter*, Dissertation Kaiserslautern 1992

[8] H. Moock, S. Motta, G. Russo, J. Wick: *Point approximation of a space-homogenous transport equation*, Num. Math. 56 (1990) 763–774

[9] S. Motta, J. Wick: *A New Numerical Method for Kinetic Equations in Several Dimensions*, Computing 46 (1991) 223–232

[10] H. Neunzert: *Neuere qualitative und numerische Methoden in der Plasmaphysik*, Paderborner Ferienkurs in angewandter Mathematik, 1975

[11] B. Niclot, P. Degond, F. Puopand: *Deterministic Particle Simulation of the Boltzmann Transport Equation of Semiconductors*, J. Comp. Phys. 78 (1988) 313–349

[12] J. Pfau: *Ein finites Teilchenverfahren für nichtlineare 1D Transportgleichung*, Diplomarbeit Kaiserslautern 1991

International Series of Numerical Mathematics, Vol. 117, © 1994 Birkhäuser Verlag Basel

THE NON–STATIONARY SEMICONDUCTOR MODEL WITH BOUNDED CONVECTIVE VELOCITY AND GENERATION/RECOMBINATION TERM.

Dariusz Wrzosek

Warsaw University, Warsaw

ABSTRACT. An evolutionary model of semiconductor devices accounting for saturated convective velocities and source terms is considered. The model includes a
generation/recombination term of Shockley-Read-Hall and Auger as well as current
dependent terms related to impact ionization.

For the one dimensional model with the standard avalanche term existence and
uniqueness of the global-in-time solutions are shown. In the case of several dimensions
the solvability of rather general reaction-diffusion-convection system coupled with the
Poisson equation is shown.

1. INTRODUCTION

We study the transient semiconductor device model with saturated convective
velocities and source terms. The model comprises Shockley-Read-Hall and Auger
generation/recombination terms and a term corresponding to the impact ionization.

Let Ω be a domain in $\mathbf{R}^n$, $n \geq 1$, with $\Gamma := \partial\Omega = \Gamma_D \cup \Gamma_N$, $\Gamma_D \cap \Gamma_N = \varnothing$.
The initial-boundary value problem, we are interested in, is given by

$$\operatorname{div}(a\nabla\psi) = \sum_{i=1}^{N} q_i u_i + f,$$

$$\dot{u}_i - \operatorname{div}J_i = -R_i(\underline{u})u_i + \hat{G}_i(\nabla\psi, \underline{u}, J_1, \ldots, J_N), \quad i = 1, \ldots, N, \tag{1.1}$$

in $\Omega \times (0, T) = \Omega_T$, where

$$J_i = D_i(\nabla\psi)\nabla u_i + \underline{v}_i(\nabla\psi)u_i,$$

$$\underline{u} = (u_1, u_2, \ldots, u_N), \quad q_i \in \{-1, 0, 1\} \quad i = 1, \ldots, N, \tag{1.2}$$

1980 *Mathematics Subject Classification* (1985 *Revision*). 35K60, 35Q20, 78A35.

Key words and phrases. Semiconductors, drift-diffusion model, impact ionization, nonlinear
parabolic systems, existence, uniqueness.

The research was partially supported by the grant 2 1168 91 01 of the Polish Scientific Research
Council(KBN) .

subject to boundary and initial conditions :

$$\underline{u} = \underline{U}, \quad \psi = \Psi \quad \text{on} \quad \Gamma_D \times (0,T),$$
$$J_\iota.\underline{n} = \nabla\psi.\underline{n} = 0 \quad \text{on} \quad \Gamma_N \times (0,T), \qquad (1.3)$$
$$\underline{u}(x,0) = \underline{U}(x,0) \quad \text{for} \quad x \in \Omega,$$

with $U_\iota \geq 0$, $\iota = 1,\ldots,N$, (in the sequel we shall write shortly $\underline{U} \geq \underline{0}$) and Ψ being given functions defined on Ω_T .

The system (1.1) along with the initial and boundary conditions (1.3) has the structure of a reaction-diffusion-convection model coupled with the Poisson equation determining the electrostatic potential. The boundary conditions imposed on Γ_N represent insulating boundary segments while Dirichlet boundary conditions correspond to Ohmic contacts (see, eg.[9]).Since only the nonnegative solutions represent concentrations and are thus physically relevant we shall impose appropriate conditions on the functions R_ι and G_ι to ensure nonnegativeness of u_ι provided initial conditions are nonnegative. The function $-R_\iota(\underline{u})u_\iota$, negative for $\underline{u} \geq \underline{0}$, may describe various recombination processes whereas $G_\iota(\nabla\psi,\underline{u},\nabla\underline{u})$ represents generation of charged particles (ions).

The system (1.1) is a generalization of the non-stationary semiconductor equations :

$$-\mathrm{div}(a\nabla\psi) = u_1 - u_2 + f,$$
$$\dot{u}_1 - \mathrm{div}J_1 = G_R, \qquad (1.4)$$
$$\dot{u}_2 - \mathrm{div}J_2 = G_R, \quad \text{in} \quad \Omega \times (0,T),$$

where

$$G_R = G_R^1 + G_R^2 + H = \frac{n_\iota^2 - u_1 u_2}{r_1 u_1 + r_2 u_2 + r_3} + (C_1 u_1 + C_2 u_2)(n_\iota^2 - u_1 u_2) + \\ + H(\nabla\psi, u_1, u_2, J_1, J_2). \qquad (1.5)$$

The vectors of current density J_ι $\iota = 1,2$ and initial-boundary conditions are defined as in (1.2) and (1.3) respectively. The functions u_1 and u_2 represent here densities of mobile electrons and holes respectively. ψ is the electrostatic potential, f is a net concentration of dopants and ionized impurities, a denotes dielectric permitivity of semiconductor material, $n_\iota, r_\iota, c_\iota$, $\iota = 1,2$, are some positive constants and $\underline{n}$ denotes the outward unit normal vector of $\partial\Omega$.

The current densities of electrons and holes J_1 , J_2 split into diffusive and convective components which depend on the electric field. In contrast to the classical model [2] we account for the velocity saturation effect, i.e.: $|\underline{\nu}_1|$, $|\underline{\nu}_2| \leq \gamma$.

The generation-recombination term G_R comprises the components of Shockley-Read-Hall (G_R^1) and Auger (G_R^2) as well as a current dependent term H describing a process of impact ionization. Observe that taking $N = 2$ and grouping separately negative and positive terms in G_R we can reduce formally the system (1.4) to (1.1). In (1.1) however, we allow for more general form of the function R_ι than that arising from (1.5)(see condition (Eiii) below).

In [2] Gajewski and Gröger considered the model with linear convective velocity and generation-recombination term which is independent of current flow. In [7] and [8] Seidman dealt with the existence and uniqueness of solutions for rather general system allowing for velocity saturation and gradient dependent source terms. However, conditions imposed on source term in this work preclude Auger generation term.

The Shockley-Read-Hall term describes two-particles process of generation and recombination of mobile curriers whereas Auger's generation and impact ionization are three-particles processes. The difference between them lies in the source of energy stimulating generation of pairs of electrons and holes. Auger generation may prevail in regions with a high concentration of carriers and negligible current flow, whereas impact ionization depends mainly on current flow. We refer to [9] for details.

Our results depend heavily on the structural assumptions imposed on the generation-recombination term G_R. In the case of one space dimension we assume in (1.4) the standard avalanche term

$$H(\nabla \psi, J_1, J_2) = \alpha_1(\nabla \psi)|J_1| + \alpha_2(\nabla \psi)|J_2|, \tag{1.6}$$

where $\alpha_i(\nabla \psi) = \alpha_i^\infty \exp\left(-\frac{\beta_i}{|\nabla \psi|}\right)$, with some positive constants α_i^∞, β_i, $i = 1, 2$. The boundary conditions are reduced then to the Dirichlet conditions on the ends of interval $\Omega = [0, l]$.

In the multidimensional situation a more restrictive condition on H, precluding (1.6) in particular, is assumed (see Remark 7). The main question here concerns the L^∞-estimates.

Notation:
From now on we assume that Ω is a bounded Lipschitzian domain in $\mathbf{R}^n$, $n \geq 1$, with $\partial \Omega = \Gamma_D \cup \Gamma_N$, where Γ_D is closed and mes$(\Gamma_D) > 0$.

The vector valued functions will be consistently denoted by underlined characters: $\underline{u}, \underline{v}$, etc. We use a shortened notation: $H^1(\Omega), L^p(\Omega), \dots$ for the spaces $H^1(\Omega; \mathbf{R}^N), L^p(\Omega; \mathbf{R}^N)$, etc. We apply standard notation $\|.\|_E$ for a norm in the Banach space E. For short, however, the norm in the space $L^p(\Omega), p \geq 1$, is denoted by $\|\cdot\|_p$. By $C(0, T; E)$ or $L^p(0, T, E)$ we denote the usual spaces of functions defined on an interval $[0, T]$, $T > 0$, with values in the Banach space E.

We also introduce the spaces

$$H_D = \{v \in H^1(\Omega); v|_{\Gamma_D} = 0\} \quad \text{and} \quad Y = L^2(0, T; H_D).$$

Note, that due to the Poincaré inequality an equivalent norm in the space Y can be introduced by $\|u\|_Y^2 = \int_0^T \|\nabla u\|_2^2 ds$.

The brackets $\langle \cdot, \cdot \rangle$ denote the scalar product in $L^2(\Omega)$ as well as the dual pairing between H_D and H_D'. Denote also $\Omega_t = \Omega \times (0, t)$.

In the next section (see Thm.2) we shall show the existence of **weak solutions** of (1.1) subject to condictions (E,i-vi) stated below. System (1.4) comprising the impact ionization term in standard form (1.6), will be treated separately in **Theorem 3**, in the case of one space dimension.

(Ei) The diffusion coefficients $D_\iota : \mathbf{R}^n \to \mathbf{R}$ are uniformly bounded continuous functions satisfying the condition

$$0 < D_0 \le D_\iota \le D^0, \quad \iota = 1, \dots, N,$$

where D_0, D^0 are generic constants.

(Eii) The convective velocities $\underline{\nu}_\iota : \mathbf{R}^n \to \mathbf{R}$ are continuous and uniformly bounded functions

$$|\underline{\nu}_\iota| < \gamma, \quad \iota = 1, \dots, N.$$

(Eiii) The functions $R_\iota = R_\iota(\eta)$, $R_\iota : \mathbf{R}^N \to \mathbf{R}$ are nonnegative for $\eta \in \mathbf{R}^N_+$ and satisfy locally Lipschitz condition. Thus, for every $\delta > 0$ there exists a positive constant L_δ such that for $\eta', \eta'' \in K$ where $K = \prod_{\iota=1}^N [0, \delta]$ we have

  a) $|R_\iota(\eta') - R_\iota(\eta'')| \le L_\delta |\eta' - \eta''|$,

  b) $|R_\iota(\eta)| \le r_\delta$ for $\eta \in K$ $\iota = 1, \dots, N$.

(Eiv) The function $\hat{G}_\iota$ can be expressed in terms of $\nabla \psi, \underline{u}, \nabla \underline{u}$ which enter $\hat{G}_\iota$ through the fluxes $J_\jmath$. Thus, it is convenient to define the functions

$$G_\iota(\nabla \psi, \underline{u}, \nabla \underline{u}) := \hat{G}_\iota(\nabla \psi, \underline{u}, J_1, \dots, J_N).$$

In what follows we shall be concerned with G_ι rather than $\hat{G}_\iota$.

We assume that $G_\iota : \mathbf{R}^n \times \mathbf{R}^N \times \mathbf{R}^{nN} \to \mathbf{R}$ are continuous functions for each $\iota = 1, \dots, N$ satisfying the conditions:

  a) $G_\iota(\underline{\zeta}, \underline{\eta}, \underline{\varrho}_1, \dots, \underline{\varrho}_N) \ge 0$ if $\underline{\eta} \ge 0$ for any, $\underline{\zeta} \in \mathbf{R}^n$, $\underline{\varrho}_\iota \in \mathbf{R}^n$,

  b) $G_\iota(\underline{\zeta}, \underline{\eta}, \underline{\varrho}_1, \dots, \underline{\varrho}_N) \le \alpha(1 + \sum_{\jmath=1}^N \eta_\jmath + |\underline{\varrho}_\iota|)$ for $\underline{\zeta}, \underline{\varrho}_\iota \in \mathbf{R}^n, \underline{\eta} \in \mathbf{R}^N_+$,

  c) $|G_\iota(\underline{\zeta}, \underline{\eta}', \underline{\varrho}') - G_\iota(\underline{\zeta}, \underline{\eta}'', \underline{\varrho}'')| \le \alpha(\sum_{\iota=1}^\Lambda |\eta'_\iota - \eta''_\iota| + \sum_{\iota=1}^N |\underline{\varrho}'_\iota - \underline{\varrho}''_\iota|)$,

    for any $\underline{\zeta} \in \mathbf{R}^n$; $\underline{\varrho}', \underline{\varrho}'' \in \mathbf{R}^{n\Lambda}$; $\underline{\eta}', \underline{\eta}'' \in \mathbf{R}^N$.

(Ev) The boundary and initial conditions are defined as traces of the functions $\underline{U} \in H_1(\Omega_T) \cap L^\infty(\Omega_T)$, and $\Psi \in H_1(\Omega_T)$ with $\underline{U} \ge \underline{0}$, $\|\underline{U}\|_\infty < M$.

(Evi) The net dopant concentration f belongs to the space $L^2(\Omega)$.

Note that the generation-recombination term in (1.4) along with (1.6) satisfies conditions (E,iii-iv) but (Eiv,b) which requires a domination of the i-th gradient (flux) in the i-th equation.

Now we can introduce a weak form of problem (1.1). We shall refer to it as to problem (P):

For an arbitrary but fixed $T > 0$ find a couple of functions $(\underline{u}, \psi)$ such that:

(Pi) $\underline{u} \geq \underline{0}$,

(Pii) $\underline{u}(x,0) = \underline{U}(x,0)$, $\underline{u} - \underline{U} \in L^2(0,T;H_D)$,

$(Piii)$ $\underline{u} \in C(0,T;L^2(\Omega)) \cap L^\infty(\Omega_T)$,

(Piv) $\underline{\dot{u}} \in L^2(0,T;H_D')$,

(Pv) $\psi - \Psi \in L^2(0,T;H_D)$,

(Pvi) $\langle a\nabla\psi(t), \nabla\eta \rangle = \langle \sum_{i=1}^{N} q_i u_i(t) + f , \eta \rangle$ for a.e. $t \in [0,T]$,

$$\int_0^T \langle \dot{u}_i, \varphi_i \rangle + \int_0^T \langle D_i(\nabla\psi)\nabla u_i, \nabla\varphi_i \rangle + \int_0^T \langle \underline{\nu}_i(\nabla\psi)u_i, \nabla\varphi_i \rangle +$$

$$+ \int_0^T \langle R_i(\underline{u})u_i, \varphi_i \rangle = \int_0^T \langle G_i(\nabla\psi, \underline{u}, \nabla\underline{u}), \varphi_i \rangle, \quad i = 1, \ldots, N,$$

for all $\eta \in H_D$ and $\underline{\varphi} \in Y$.

In the sequel, wherever not confusing, we shall skip arguments of functions.

2. EXISTENCE OF SOLUTION

In order to show the solvability of problem (P) we first prove that the auxiliary "truncated" problems (P^k) have solutions. Next for k sufficiently large we shall estimate solutions of problems (P^k) in the L^∞ - norm independently of k. An observation that for k large enough the solution of problem (P^k) is also a solution of problem (P) will complete the proof. This method was applied by Gajewski and Gröger in [2] to the basic semiconductor equations.

Let us define the truncated function $y^{(k)}$:

$$y^{(k)} = \begin{cases} k, & \text{for} \quad y > k \\ y, & \text{for} \quad y \leq k. \end{cases}$$

The same notation is used for vectors $\underline{y}^{(k)} = (y_1^{(k)}, \ldots, y_N^{(k)})$.

Recall that if $u \in W^{1,2}(\Omega)$ then also $u^{(k)} \in W^{1,2}(\Omega)$ (see [4]). Let $u_- = \min(u,0)$, and $u_+ = \max(u,0)$. Now we can formulate the auxiliary problem (P_u^k), $k > M$, of determining functions $\underline{u}$ and ψ which satisfy the conditions $(P, i - v)$ and

$(P_u^k):$ $\langle a\nabla\psi(t), \nabla\eta \rangle = \langle (\sum_{i=1}^{N} q_i u_i(t) + f), \eta \rangle$,   for a.e. $t \in [0,T]$,

$$(2.1)$$

$$\int\limits_0^T \langle \dot{u}_i, \varphi_i \rangle + \int\limits_0^T \langle D_i(\nabla\psi)\nabla u_i, \nabla\varphi \rangle + \int\limits_0^T \langle \underline{\nu}_i(\nabla\psi)u_i, \nabla\varphi_i \rangle +$$

$$+ \int\limits_0^T \langle R_i(\underline{u}^{(k)})u_i^{(k)}, \varphi_i \rangle = \int\limits_0^T \langle G_i(\nabla\psi, \underline{u}^{(k)}, \nabla(\underline{u}^k)), \varphi_i \rangle, \qquad (2.2)$$

$$i = 1, \ldots, N,$$

for all $\eta \in H_D$ and $\underline{\varphi} \in Y$.

Theorem 1. *Problem* (P_u^k) *has a solution.*

Proof. Let us introduce a mapping $T : L^2(\Omega_T) \to L^2(\Omega_T)$, where $T(\underline{z}) = \underline{v}$ is defined in the following way. For fixed $\underline{z} \in L^2(\Omega_T)$, let ψ_z be a weak solution of the Poisson equation

$$-a\triangle\psi_z(t) = \sum_{i=1}^N q_i z_i(t) + f \quad \text{for a.e. } t \in [0, T].$$

Then $\underline{v}$ is, by definition, the unique solution, guaranteed by Lemma 1 below, of parabolic problem (P_v^k) with ψ_z plunged into (2.2) in the place of ψ.

$$(P_v^k): \qquad \int\limits_0^T \langle \dot{v}_i, \varphi_i \rangle + \int\limits_0^T \langle D_i(\nabla\psi_z)\nabla v_i, \nabla\varphi_i \rangle + \int\limits_0^T \langle \underline{\nu}_i(\nabla\psi_z)v_i, \nabla\varphi_i \rangle +$$

$$+ \int\limits_0^T \langle R_i(\underline{v}^{(k)})v_i^{(k)}, \varphi_i \rangle = \int\limits_0^T \langle G_i(\nabla\psi_z, \underline{v}^{(k)}, \nabla(\underline{v}^{(k)})), \varphi_i \rangle,$$

$$i = 1, \ldots, N,$$

for all $\underline{\varphi} \in Y$, where $\underline{v} - \underline{U} \in Y$. It is clear that the solutions of problem (P^k) are fixed points of the map T.

Note that due to (E,i-ii) for any ψ_z the functions D_i and $\underline{\nu}_i$ are uniformly bounded measurable functions. In the sequel we shall skip the arguments of D_i, $\underline{\nu}_i$ and the first argument of G_i.

Lemma 1. *The system* (P_v^k) *has a unique nonnegative solution satisfying regularity properties* $(P, ii - iv)$.

Proof. Let us introduce a new variable $\underline{w} = e^{-\lambda t}\underline{v}$ where λ is a positive number which will be specified later. After a transformation we obtain the equalities:

$$(P_w^{k,\lambda}): \qquad \int\limits_0^T \langle \dot{w}_i, \varphi_i \rangle + \int\limits_0^T \langle D_i\nabla w_i, \nabla\varphi_i \rangle + \int\limits_0^T \langle \underline{\nu}_i w_i, \nabla\varphi_i \rangle + \lambda \int\limits_0^T \langle w_i, \varphi_i \rangle +$$

$$+ \int\limits_0^T \langle e^{-\lambda t} R_i([e^{\lambda t}\underline{w}]^{(k)})[e^{\lambda t}w_i]^{(k)}, \varphi_i \rangle = \int\limits_0^T \langle e^{-\lambda t} G_i([e^{\lambda t}\underline{w}]^{(k)}, \nabla[e^{\lambda t}\underline{w}]^{(k)}), \varphi_i \rangle$$

$$i = 1, \ldots, N,$$

for all $\underline{\varphi} \in Y$, where $\underline{w} - \underline{W} \in Y$ and $\underline{W} = e^{-\lambda t}\underline{U}$.

Let us define a bilinear form $b_\imath^\lambda : Y \times Y \to \mathbf{R}$ $\imath = 1, \dots, N$,

$$b_\imath^\lambda(v, \varphi) = \int_0^T \langle D_\imath \nabla v_\imath, \nabla \varphi_\imath \rangle + \int_0^T \langle \underline{v}_\imath v_\imath, \nabla \varphi_\imath \rangle + \lambda \int_0^T \langle v_\imath, \varphi_\imath \rangle,$$

for v, $\varphi \in Y$. In what follows we shall use the following properties of $b_\imath^\lambda$:

   bi) $b_\imath^\lambda(v, v_\pm) = b_\imath^\lambda(v_\pm, v_\pm)$ for $v \in Y$.

   bii) For $\lambda > \frac{\gamma^2}{2D_0}$, $b_\imath^\lambda$ is Y-coercive. Taking $\varepsilon = \frac{D_0}{2}$ we have

$$b_\imath^\lambda(v, v) \geq D_0 \int_0^T \|\nabla v\|_2^2 - \frac{\gamma^2}{4\varepsilon} \int_0^T \|v\|_2^2 - \varepsilon \int_0^T \|\nabla v\|_2^2 + \lambda \int_0^T \|v_2^2$$

$$\geq \frac{D_0}{2} \int_0^T \|\nabla v\|_2^2 = \frac{D_0}{2} \|v\|_Y.$$

Denote $Y_+ = \{\underline{v} : \underline{v} \in Y, \underline{v} \geq 0\}$. A weak solution of the problem $(P_w^{k,\lambda})$ will be constructed as a fixed point of the nonlinear mapping $T_\lambda : , Y_+ \to Y_+$ defined by $T_\lambda(\underline{\sigma}) = \underline{w}$, where $\underline{w}$ is the solution of

$(P_w^{k,\lambda})$:
$$\int_0^T \langle \dot{w}_\imath, \varphi_\imath \rangle + b_\imath^\lambda(w_\imath, \varphi_\imath) + \int_0^T \langle e^{-\lambda t} R_\imath([e^{\lambda t}\underline{\sigma}]^{(k)})[e^{\lambda t}w_\imath]^{(k)}, \varphi_\imath \rangle$$

$$= \int_0^T \langle e^{-\lambda t} G_\imath([e^{\lambda t}\underline{\sigma}]^{(k)}, \nabla[e^{\lambda t}\underline{\sigma}]^{(k)}), \varphi_\imath \rangle,$$

$$\imath = 1, \dots, N,$$

for all $\underline{\varphi} \in Y$. where $\underline{w} - \underline{W} \in Y$ and $\underline{W} = e^{-\lambda t}\underline{U}$.

The definition of T_λ requires an explanation. To solve the inhomogeneous boundary problem with respect to $\underline{w}$ let us introduce new variable $\underline{y} = \underline{w} - \underline{W}$. We can reformulate $(P_w^{k,\lambda})$ as the abstract Cauchy problem

$$\dot{\underline{y}} + A_\lambda \underline{y} = F, \quad \underline{y}(0) = \underline{0},$$

where $A_\lambda : Y \to Y'$ is given by

$$\langle A_\lambda \underline{y}, \underline{\varphi} \rangle_Y = \sum_{\imath=1}^N \left\{ b_\imath^\lambda(y_\imath, \varphi_\imath) + \int_0^T \langle h_\imath[e^{\lambda t}(y_\imath + W_\imath)]^{(k)}, \varphi_\imath \rangle \right\},$$

with $h_i = e^{-\lambda t} R_i([e^{\lambda t}\underline{\sigma}]^{(k)})$. Note that $h_i \in L^\infty(\Omega_T)$ and $h_i \geq 0$. For $\lambda > \frac{\gamma^2}{2D_0}$, A_λ is coercive, monotone (it results immediately from the monotonicity of the cut-off function) and radially continuous (see [3] for definitions). The function F depends on $\underline{\sigma}$ and $\underline{W}$ through G_i , hence due to (Eiv), $F \in L^2(\Omega_T)$. From the theory of evolutionary equations (see e.g. [3]) we conclude that problem $(P_{\underline{\omega}}^{k,\lambda})$ has a unique solution $\underline{\omega}$ such that $\underline{\omega} - \underline{W} \in Y$ and $\underline{\dot{\omega}} \in Y'$, whence $\underline{\omega} \in C([0,T]; L^2(\Omega))$. Testing $(P_{\underline{\omega}}^{k,\lambda})$ with $\underline{\omega}_-|_{[0,t]}$ and using the nonnegativeness of G_i and bi) we obtain for $i = 1, \ldots, N$

$$\frac{1}{2}\|\omega_{i-}(t)\|_2^2 + \int_0^t b_i^\lambda(\omega_{i-}, \omega_{i-}) + \int_0^t \langle h_i[e^{\lambda s}\omega_i(s)]^{(k)} \cdot \omega_{i-}(s)\rangle ds \leq 0$$

Since the third integral is nonnegative, we deduce that for each $t \in [0,T]$ $\omega(t) \geq 0$ a.e. in Ω, hence the range of T_λ is in Y_+. It remains to show that T_λ is a contraction for sufficiently large λ.

To this end let us test either $(P_{\underline{\omega}'}^{k,\lambda})$ or $(P_{\underline{\omega}''}^{k,\lambda})$ with $\underline{\tilde{\omega}} = \underline{\omega}' - \underline{\omega}''$ where $\underline{\omega}' = T(\underline{\sigma}')$, $\underline{\omega}'' = T(\underline{\sigma}'')$ and $\underline{\sigma}'$, $\underline{\sigma}'' \in Y_+$. Denote also $\underline{\sigma}' - \underline{\sigma}'' = \underline{\tilde{\sigma}}$. Subtracting the appropriate equations and summing them with respect to i from 1 to N we obtain from (Eiv,c) and Young's inequality

$$\frac{1}{2}\|\underline{\tilde{\omega}}\|_2^2 + \sum_{i=1}^{N} b_i^\lambda(\tilde{\omega}_i, \tilde{\omega}_i) \leq$$

$$\leq \sum_{i=1}^{N} \int_0^T \left\langle e^{-\lambda t}\left(R_i([e^{\lambda t}\underline{\sigma}'']^{(k)})[e^{\lambda t}\omega_i'']^{(k)} - R_i([e^{\lambda t}\underline{\sigma}']^{(k)})[e^{\lambda t}\omega_i']^{(k)}\right), \tilde{\omega}_i\right\rangle \quad (2.3)$$

$$+ 2\varepsilon \int_0^T \|\underline{\tilde{\sigma}}\|_2^2 + \varepsilon \int_0^T \|\nabla\underline{\tilde{\sigma}}\|_2^2 + \frac{2N^2 a^2}{4\varepsilon} \int_0^T \|\underline{\omega}\|_2^2 .$$

In view of (Eii,a) the first term on the R.H.S. of (2.3) can be estimated by

$$\sum_{i=1}^{N} \int_{\Omega_T} e^{-\lambda t} R_i([e^{\lambda t}\underline{\sigma}'']^{(k)}) \left|[e^{\lambda t}\omega_i'']^{(k)} - [e^{\lambda t}\omega_i']^{(k)}\right| |\omega_i| +$$

$$+ \sum_{i=1}^{N} \int_{\Omega_T} e^{-\lambda t}[e^{\lambda t}\omega_i']^{(k)} \left|R_i([e^{\lambda t}\underline{\sigma}'']^{(k)}) - R_i([e^{\lambda t}\underline{\sigma}']^{(k)})\right| |\tilde{\omega}_i| \leq$$

$$\leq r_k \int_0^T \|\underline{\tilde{\omega}}\|_2^2 + \varepsilon \int_0^T \|\underline{\tilde{\sigma}}\|_2^2 + \frac{k^2 L_k^2 N^2}{4\varepsilon} \int_0^T \|\underline{\tilde{\omega}}\|_2^2 .$$

Finally, from (2.3) we have

$$\frac{1}{2}\|\underline{\tilde{\omega}}\|_2^2 + \sum_{i=1}^{N} b_i^\lambda(\tilde{\omega}_i, \tilde{\omega}_i) \leq c_0 \int_0^T \|\underline{\tilde{\omega}}\|_2^2 + \varepsilon(3c_p + 1) \int_0^T \|\nabla\underline{\tilde{\sigma}}\|_2^2,$$

where c_p is the constant in the Poincaré inequality $\|u\|^2_{L^2(\Omega)} \le c_p \|u\|^2_{H_D}$, and $c_0 = \frac{N^2(2a^2+k^2L_k^2+r_k)}{4\epsilon}$. Using bii) and taking $\varepsilon = \frac{D_0}{4(3c_P+1)}$ and $\lambda > \max\{\frac{\gamma^2}{2D_0}, \frac{c_0}{4\epsilon}\}$ we conclude that T_λ is contractive,

$$\int\limits_0^T \|\nabla \tilde{\underline{\omega}}\|_2^2 \le \frac{1}{2}\int\limits_0^T \|\nabla \tilde{\underline{\sigma}}\|_2^2. \qquad \square$$

Lemma 2. $T : L^2(\Omega_T) \to L^2(\Omega_T)$ is a continuous mapping.

Proof. For $\underline{z}'$, $\underline{z}'' \in L^2(\Omega_T)$ let ψ_z' and ψ_z'' are the solutions of the Poisson equation (2.1) and $\underline{v}' = T(\underline{z}')$ and $\underline{v}'' = T(\underline{z}'')$. Denote $\tilde{\underline{z}} = \underline{z}' - \underline{z}''$, $\tilde{\underline{v}} = \underline{v}' - \underline{v}''$ and $E' = \nabla\psi_z'$, $E'' = \nabla\psi_z''$. Testing both equations with $\tilde{E} = E' - E''$, then subtracting and integrating over $[0, T]$ we find

$$\int\limits_0^T \|\tilde{E}\|_2^2 \le \frac{2Nc_p}{a^2}\int\limits_0^T \|\tilde{\underline{z}}\|_2^2, \tag{2.4}$$

where c_p is a constant in the Poincaré inequality. Taking $\tilde{\underline{v}}|_{[0,t]}$ as a test function in $(P_{v'}^k)$ and $(P_{v''}^k)$ and subtracting thus obtained equations side by side we obtain for $i = 1, \dots, N$

$$\int\limits_0^t \langle \dot{\tilde{v}}_\imath, \tilde{v}_\imath \rangle + \int\limits_0^t \langle D_\imath(E')\nabla v_\imath' - D_\imath(E'')\nabla v_\imath'', \nabla \tilde{v}_\imath \rangle +$$

$$+ \int\limits_0^t \langle \underline{\nu}_\imath(E')v_\imath' - \underline{\nu}_\imath(E'')v_\imath'', \nabla \tilde{v}_\imath \rangle = \int\limits_0^t \langle R_\imath(\underline{v}''^{(k)})v_\imath''^{(k)} - R_\imath(\underline{v}'^{(k)})v_\imath'^{(k)}, \tilde{v}_\imath \rangle \tag{2.5}$$

$$+ \int\limits_0^t \langle G_\imath(E', \underline{v}'^{(k)}, \nabla[\underline{v}'^{(k)}]) - G_\imath(E'', \underline{v}''^{(k)}, \nabla[\underline{v}''^{(k)}]), \tilde{v}_\imath \rangle.$$

Summing (2.5) over $\imath$ and using (E,i-iv) we obtain

$$\frac{1}{2}\|\tilde{\underline{v}}(t)\|_2^2 + \frac{D_0}{2}\int\limits_0^t \|\nabla \tilde{\underline{v}}\|_2^2 \le \int\limits_{\Omega_t} \sum_{\imath=1}^N |D_\imath(E') - D_\imath(E'')|\|\nabla v_\imath''\|\|\nabla \tilde{v}_\imath| +$$

$$+ \int\limits_{\Omega_t} \sum_{\imath=1}^N |\underline{\nu}_\imath(E') - \underline{\nu}_\imath(E'')|\|v_\imath''\|\|\nabla \tilde{v}_\imath| + \gamma \int\limits_{\Omega_t} \sum_{\imath=1}^N |\tilde{v}_\imath|\|\nabla \tilde{v}_\imath| + (r_k + k)\int\limits_0^t \|\tilde{\underline{v}}\|_2^2$$

$$+ \int\limits_{\Omega_t} \sum_{\imath=1}^N |G_\imath(E', \underline{v}'^{(k)}, \nabla[\underline{v}'^{(k)}]) - G_\imath(E', \underline{v}''^{(k)}. \nabla[\underline{v}''^{(k)}])|\|\tilde{v}_\imath|$$

$$+ \int\limits_{\Omega_t} \sum_{\imath=1}^N |G_\imath(E', \underline{v}''^{(k)}, \nabla[\underline{v}''^{(k)}]) - G_\imath(E'', \underline{v}''^{(k)}, \nabla[\underline{v}''^{(k)}])|\|\tilde{v}_\imath| \le$$

$$\leq \frac{1}{2}\Upsilon + 4\varepsilon \int\limits_0^t \|\nabla \tilde{\underline{v}}\|_2^2 + c_\varepsilon \int\limits_0^t \|\tilde{\underline{v}}\|_2^2 \,, \tag{2.6}$$

where $c_1 = c_1(\gamma, r_k, k, \alpha, N; \varepsilon)$ and

$$\Upsilon = 2\Big(\int\limits_{\Omega_T} |\underline{D}(E') - \underline{D}(E'')|^2 |\nabla \underline{v}''|^2 + \int\limits_{\Omega_T} |\underline{\nu}(E') - \underline{\nu}(E'')|^2 |\underline{v}''|^2 +$$

$$+ \int\limits_{\Omega_T} |\underline{G}(E'', \underline{v}''^{(k)}, \nabla[\underline{v}''^{(k)}]) - \underline{G}(E'', \underline{v}''^{(k)}, \nabla[\underline{v}''^{(k)}])|^2 \,.$$

With $\varepsilon = \frac{D_0}{16}$ in (2.6) we obtain

$$\|\tilde{\underline{v}}(t)\|_2^2 \leq \Upsilon + 2c_1 \int\limits_0^t \|\tilde{\underline{v}}\|_2^2 \,,$$

whence

$$\sup_{t \in [0,T]} \|\tilde{\underline{v}}(t)\|_2^2 \leq \Upsilon \exp(2c_1 T) \,. \tag{2.7}$$

To complete the proof let us take a sequence $\underline{z}_m \to \underline{z}$ in $L^2(\Omega_T)$. From (2.4) we conclude that $E_m \to E$ in $L^2(\Omega_T)$. Note that from this sequence one can choose a subsequence, indexed still by m, convergent almost everywhere in Ω_T to the same limit. Substituting in (2.7) $\underline{z}''$ by $\underline{z}$, $\underline{z}'$ by $\underline{z}_m$ and respectively v'' by v and v' by v_m and in consequence Υ by Υ_m we find

$$\sup_{t \in [0,T]} \|\underline{v}_m(t) - \underline{v}(t)\|_2^2 \leq \Upsilon_m \exp(2c_1 T) \,.$$

Since $E_m \to E$ almost everywhere in Ω_T, we conclude from the continuity of D_i, ν_i and G and an application of the Lebesgue majorated convergence theorem, that $\Upsilon_m \to 0$. To control the integrals in Υ_m observe that $|D_i(E_m) - D_i(E)|^2 |\nabla \underline{v}|^2 \leq 2D^2 |\nabla \underline{v}|^2$ and a similar estimate is true for $\underline{\nu}_i$. The uniform estimate of the third integral results from the assumption (Eiv,b).
Finally, we obtain that $\underline{v}_m \to \underline{v}$ in $L^2(\Omega_T)$. $\quad\square$

From the bounds assumed on $D_i, \underline{\nu}_i$ and G_i it follows that the range $V = \{T(\underline{z}) : \underline{z} \in L^2(\Omega_T)\}$ is bounded in $L^2(\Omega_T)$. In fact, testing (P_v^k) with the function $\hat{\underline{v}} = (\underline{v} - \underline{U})|_{[0,t]}$ we get

$$\frac{1}{2}\|\hat{\underline{v}}(t)\|_2^2 + \frac{D_0}{2} \int\limits_0^t \|\nabla \hat{\underline{v}}(s)\|_2^2 ds \leq \frac{c_1}{2\varepsilon} \int\limits_0^t \|\hat{\underline{v}}(s)\|_2^2 ds + \frac{\varepsilon}{2} \int\limits_0^t \|\nabla \hat{\underline{v}}\|_2^2 + c_2 \,, \tag{2.8}$$

where $c_1 = c_1(D^0, \gamma, r_k, \alpha, M, N)$ and $c_2 = c_2(\|\underline{U}\|_{H^1(\Omega_T)}, N)$. Choosing $\varepsilon = \frac{D_0}{2}$ we have

$$\|\hat{\underline{v}}(t)\|_2^2 + \frac{D_0}{2} \int\limits_0^t \|\nabla \hat{\underline{v}}(s)\|_2^2 ds \leq \frac{2c_1}{D_0} \int\limits_0^t \|\hat{\underline{v}}(s)\|_2^2 ds + 2c_2 \,.$$

The inequalities

$$\int\limits_0^T \|\hat{\underline{v}}(t)\|_2^2 dt \le 2c_2 T \exp\left(\frac{2c_1}{D_0}T\right) = c_3 \,,$$

$$\int\limits_0^T \|\nabla\hat{\underline{v}}(t)\|_2^2 dt \le \frac{2}{D_0}\left(\frac{2c_1}{D_0}c_3 + 2c_2\right) = c_4 \,,$$

follow then from Gronwall's lemma. Hence. we find immediately

$$\int\limits_0^T \|\underline{v}(t)\|_2^2 dt \le 2(c_3 + \|\underline{U}\|_{H^1(\Omega_T)}) = C_z \tag{2.9}$$

and

$$\int\limits_0^T \|\nabla\underline{v}(t)\|_2^2 dt \le 2(c_4 + \|\underline{U}\|_{H^1(\Omega_T)}) = C_z' . \tag{2.10}$$

Consider a set $V = \{\underline{v} : \int_0^T \|\underline{v}\|_2^2 \le C_z\}$, where c_z is given by (2.9). Estimating the terms of the parabolic system due to (2.9) and (2.10) one finds a bound on $\dot{\underline{v}}$ in the space $L^2(0,T;H'_D)$. From the Aubin compactness theorem we see that the closure of $T(V)$ is compact in $L^2(\Omega_T)$.

Now, we can conclude the proof of Theorem 1. The above properties of the mapping T allow us to apply the Schauder fixed point theorem. It ensures the existence of a function $\underline{u} \in L^2(\Omega_T)$ such that $\underline{u} = T(\underline{u})$. Thus, the problem $(P_{\underline{u}}^k)$ has at least one solution $(\underline{u}, \psi)$. The nonnegativeness of $\underline{u}$ and regularity properties $(P, iii - iv)$ result from the Lemma 1 while the condition (Pv) follows immediately from (2.4). $\square$

Remark 1. Without essential changes of the proofs of Lemmas 1 and 2 we can treat a more general situation in which hypothessis (Eiv,b) is replaced by

$$|G_i| \le \alpha(1 + \sum_{i=1}^{\Lambda} |\eta_i| + \sum_{i=1}^{N} |\underline{\varrho}_i|).$$

The condition (Eiv,b) is, however, essential in Lemmas 3 and 4 .

It remains to show that for $k > k_0$, where k_0 is a positive number the solution $\underline{u}$ of $(P_{\underline{u}}^k)$ is bounded in the space $L^\infty(\Omega_T)$ uniformly with respect to k. To this end we show at first that the function $(\underline{u} - \underline{U})_+$ is bounded in the space $C(0,T;L^2(\Omega))$ independently of k.

Lemma 3. *Any solution $\underline{u}$ of $(P_{\underline{u}}^k)$ satisfies the inequality*

$$\|(\underline{u} - \underline{U})_+\|_{C(0,T,L^2(\Omega)} \le c_\infty$$

with some constant c_∞ independent of k.

Proof. Introducing a new variable $\underline{w} = \underline{u} - \underline{U}$ we can reduce the parabolic system in $(P_{\underline{u}}^k)$ to

$$
\int_0^T \langle \dot{w}_\imath, \varphi_\imath \rangle + \int_0^T \langle (D_\imath(\nabla\psi)\nabla w_\imath, \nabla\varphi_\imath \rangle + \int_0^T \langle \underline{\nu}_\imath(\nabla\psi)w_\imath, \nabla\varphi_\imath \rangle +
$$

$$
+ \int_0^T \langle R_\imath([\underline{w} + \underline{U}]^{(k)})(w_\imath + U_\imath)^{(k)}, \varphi_\imath \rangle = - \int_0^T \langle D_\imath(\nabla\psi)\nabla U_\imath, \nabla\varphi_\imath \rangle - \tag{2.11}
$$

$$
- \int_0^T \langle \underline{\nu}_\imath(\nabla\psi)U_\imath, \nabla\varphi_\imath \rangle + \int_0^T \langle G_\imath(\nabla\psi, [\underline{w} + \underline{U}]^{(k)}, \nabla[(\underline{w} + \underline{U})^{(k)}]), \varphi_\imath \rangle,
$$

$$
\imath = 1, \ldots, N,
$$

for each $\underline{\varphi} \in Y$. Let us take $\underline{\varphi} = \underline{w}_+|_{[0,t]}$ as a test function in (2.11). We shall estimate separately the last term on the R.H.S. Three facts will be useful : a) $w_\jmath = w_{\jmath-} + w_{\jmath+}$, b) $|w_{\jmath-}| = |(u_\jmath - U_\jmath)_-| \le \|U_\jmath\|_\infty$ and c) $|\nabla w_\imath||w_{\imath+}| \le |\nabla w_{\imath+}||w_{\imath+}|$. From (Eiv.b) we obtain

$$
\int_{\Omega_t} |G_\imath(\nabla\psi, [\underline{w} + \underline{U}]^{(k)}, \nabla[(\underline{w} + \underline{U})^{(k)}])| \, |w_{\imath+}| \le
$$

$$
\le \alpha \int_{\Omega_t} \left(|w_{\imath+}| + \sum_{\jmath=1}^N [w_\jmath + U_\jmath]^{(k)} |w_{\imath+}| + |\nabla[w_\imath + U_\imath]^{(k)}||w_{\imath+}| \right) \le
$$

$$
\le \frac{\alpha}{2}|\Omega|T + \frac{\alpha}{2} \int_0^t \|w_{\imath+}\|^2 + \alpha \int_{\Omega_t} \left(\sum_{\jmath=1}^N (|w_{\jmath+}||w_{\imath+}| + |w_{\jmath-}||w_{\imath+}| + \right.
$$

$$
\left. + |U_\jmath||w_{\imath+}|) + |\nabla w_{\imath+}||w_{\imath+}| + |\nabla U_\imath||w_{\imath+}| \right) \le
$$

$$
\le \frac{\alpha}{2}|\Omega|T + \left(3\alpha + \frac{\alpha^2}{4\varepsilon} \right) \int_0^t \|\underline{w}_+\|_2^2 + \left(\frac{3}{2}\alpha \right) \|\underline{U}\|_{H^1(\Omega_T)} + \varepsilon \int_0^t \|\nabla\underline{w}_+\|_2^2.
$$

Summing (2.11) with respect to i and estimating as in (2.8) we have

$$
\frac{1}{2}\|\underline{w}_+(t)\|_2^2 + D_0 \int_0^t \|\nabla w_+\|_2^2 \le \frac{1}{2}\|\underline{w}_+(0)\|_2^2 + 4\varepsilon \int_0^t \|\nabla w_+\|_2^2 +
$$

$$
+ \frac{1}{2}c_1(\varepsilon) \int_0^t \|\underline{w}_+\|_2^2 + \frac{1}{2}c_2(\varepsilon), \tag{2.12}
$$

where $c_1(\varepsilon) = (3a + \frac{a^2}{4\varepsilon})N$ and $c_2(\varepsilon) = c_2(D^0, \gamma, a, \|\underline{U}\|_{H^1(\Omega_T)}, |\Omega|, T, N)$. Observe that the first term in (2.12) vanishes. Choosing $\varepsilon = \frac{D_0}{8}$ and applying Gronwall's inequality we obtain

$$\sup_{t \in [0,T]} \|\underline{w}_+\|_2^2 \leq c_2 \exp(c_1 T) = c_\infty^2. \qquad \square$$

Lemma 4. *If $\underline{u}$ is a solution of (P_u^k) then $\underline{u} \in L^\infty(\Omega_T)$. Moreover*

$$\|(\underline{u} - \underline{U}')_+\|_{L^\infty(\Omega_T)} \leq c^2(c_\infty^2 + 1), \tag{2.13}$$

where c is a constant independent of k.

Proof. The method of the proof is closely related to the iterative technique introduced by Moser in [6]. Different variants of this method were applied to a reaction-diffusion system in [1] and to the basic semiconductor equations in [2].

Let us introduce some auxiliary functions :

i) $\theta(\eta) = (\eta^{(\kappa)} - M)_+ \quad, \eta \in \mathbf{R}, \ M < k < \kappa.$
ii) $\Phi_p^{(\kappa)}(y) = p \int_0^y \theta^{p-1}(\eta) d\eta \quad, y \in \mathbf{R}, \ p \geq 2.$

It is easy to see that

$$\Phi_p^{(\kappa)}(y) \nearrow (y - M)_+^p \quad \text{if} \quad \kappa \to +\infty \quad \text{for} \quad y \in \mathbf{R}. \tag{2.14}$$

Let us recall (see [4]) that if $u \in H^1(\Omega)$ then also $\theta(u) \in H^1(\Omega)$ and

$$\nabla \theta(u) = \begin{cases} \nabla u & \text{for} \quad M < u < \kappa, \\ 0 & \text{for} \quad u \leq M \text{ or } u \geq \kappa, \end{cases}$$

whence $\nabla \Phi_p^{(\kappa)}(u) = p\theta^{p-1}(u)\nabla\theta(u)$. Let $q_\iota = \theta^{\frac{p}{2}}(u_\iota) = (u_\iota^{(\kappa)} - M)_+^{\frac{p}{2}}$. Testing the parabolic system in (P_u^k) with $\varphi_\iota = p\theta^{p-1}(u_\iota)|_{[0,t]}, \iota = 1, \ldots, N$, we obtain

$$\sum_{\iota=1}^N \int_\Omega \Phi_p^{(\kappa)}(u_\iota(t)) - \sum_{\iota=1}^N \int_\Omega \Phi_p^{(\kappa)}(u_\iota(0)) =$$

$$-p\sum_{\iota=1}^N \int_{\Omega_t} D_\iota \nabla u_\iota \nabla \theta^{p-1}(u_\iota) - p\sum_{\iota=1}^N \int_{\Omega_t} \nu_\iota u_\iota \nabla \theta^{p-1}(u_\iota)$$

$$-p\sum_{\iota=1}^N \int_{\Omega_t} R_\iota([\underline{u}]^{(k)})u_\iota^{(k)}\theta^{p-1}(u_\iota) + p\sum_{\iota=1}^N \int_{\Omega_t} G_\iota(\nabla\psi, [\underline{u}]^{(k)}, \nabla[\underline{u}^{(k)}])\theta^{p-1}(u_\iota) \tag{2.15}$$

$$= J_1 + J_2 + J_3 + J_4.$$

Note that $\sum_{i=1}^{N}\int_{\Omega}\Phi_p^{(\kappa)}(u_i(0))$ vanishes, since $|u(0)|_\infty \leq M$. We will estimate separately each component on the R.H.S. :

$$J_1 = -p(p-1)\sum_{i=1}^{N}\int_{\Omega_t} D_i \nabla u_i \nabla \theta(u_i)\theta^{p-2}(u_i)$$

$$= \frac{-4(p-1)}{p}\sum_{i=1}^{N}\int_{\Omega_t} D_i |\nabla(\theta^{\frac{p}{2}}(u_i))|^2 \leq -2D_0\int_0^t \|\nabla q_i\|_2^2\,,$$

$$|J_2| \leq p\sum_{i=1}^{N}(\int_{\Omega_t}|\underline{v}_i|\|\theta(u_i)\nabla\theta^{p-1}(u_i)| + \int_{\Omega_t}|\underline{v}_i|M|\nabla\theta^{p-1}(u_i)|) \leq$$

$$\leq 2(p-1)\gamma\sum_{i=1}^{N}(\int_{\Omega_t}|\frac{p}{2}\theta^{\frac{p}{2}-1}(u_i)\nabla\theta(u_i)|\,|\theta^{\frac{p}{2}}(u_i)|+$$

$$+ M\int_{\Omega_t}|\frac{p}{2}\theta^{\frac{p}{2}-1}(u_i)\nabla\theta(u_i)|\,|\theta^{\frac{p}{2}-1}(u_i)|) \leq p\varepsilon\int_0^t\|\nabla q\|_2^2 + \frac{pc_2}{\varepsilon}(\int_0^t\|q\|_2^2+1)$$

where $c_2 = c_2(|\Omega|, T, M, \gamma, N)$,

$J_3 \leq 0$,

$$|J_4| \leq p\alpha\sum_{i=1}^{N}\int_{\Omega_t}\left\{\theta^{p-1}(u_i) + \sum_{j=1}^{N}|u_j^{(k)}\theta^{p-1}(u_i)| + |\nabla[u_i^{(k)}]|\theta^{p-1}(u_i)\right\} \leq$$

$$\leq p\alpha\left\{\frac{|\Omega|TN}{4} + \int_0^t\|q\|_2^2+\right.$$

$$+ \sum_{i,j=1}^{N}\int_{\Omega_t}(|(u_j^{(k)}-M)_+| + |(u_j^{(k)}-M)_-| + M)\theta^{p-1}(u_i)+$$

$$\left.+ |\nabla[u_i^{(\kappa)}-M]\theta^{p-1}(u_i)|\right\} \leq pc_4(\varepsilon)(\int_0^t\|q\|_2^2+1) + p\varepsilon\int_0^t\|\nabla q\|_2^2\,,$$

where $c_4(\varepsilon) = c_4'(|\Omega|, T. M, \alpha, N; \varepsilon)$. Finally, from (2.15) we obtain

$$\sum_{i=1}^{N}\int_{\Omega}\Phi_p^{(\kappa)}(u_i(t)) \leq \int_0^t -2D_0\|\nabla q\|_2^2 + 2p\varepsilon\|\nabla q\|_2^2 + p(\frac{c_2}{\varepsilon}+c_4(\varepsilon))(\int_0^t\|q\|_2^2+1).$$

Hence, choosing $\varepsilon = \frac{D_0}{2p}$ we find

$$\sum_{i=1}^{N}\int_{\Omega}\Phi_p^{(\kappa)}(u_i(t)) \leq \int_0^t -D_0\|\nabla q\|_2^2 + p^2 c_5(\int_0^t\|q\|_2^2+1) \qquad (2.16)$$

with $c_5 = (\frac{2}{D_0}(c_2 + 1) + c_4')$. Now, we use the following inequality which is a consequence of the Gagliardo-Nirenberg interpolation inequality

$$\|z\|_2^2 \leq \sigma \|\nabla z\|_2^2 + K\sigma^{n/2}\|z\|_1^2,$$

for any $z \in H_D$ and $\sigma > 0$ where K is a constant depending on $\partial\Omega$. Applying this inequality to (2.16) with $\sigma = \frac{D_0}{c_5 p^2}$ we obtain

$$\sum_{i=1}^{N} \int_{\Omega} \Phi_p^{(\kappa)}(u_i(t)) \leq c_6 p^{n+2}(\int_0^t \|g\|_1^2 + 1) \tag{2.17}$$

where $c_6 = c_5 \cdot \max\{K(\frac{c_5}{D_0})^{n/2}, 1\}$. Putting $p = 2^r$, $r = 2, 3\ldots$ in (2.17) we get

$$\sum_{i=1}^{N} \int_{\Omega} \Phi_{2^r}^{(\kappa)}(u_i(t)) \leq c_0^r \left\{ \left(\sup_{s \in [0,T]} \sum_{i=1}^{N} \int_{\Omega} (u_i^{(\kappa)}(s) - M)_+^{2^{r-1}} \right)^2 + 1 \right\} \tag{2.18}$$

with $c_0 = \max\{1, c_6, 2^{n/2+1}, T\}$. Denote $d_r = \sup_{s \in [0,T]} \sum_{i=1}^{N} \int_{\Omega}(u_i - M)_+^{2^r}$. Lemma 3 gives $d_1 < \infty$. Plunging $r = 2$ into (2.18) we have

$$\sup_{s \in [0,T]} \sum_{i=1}^{N} \int_{\Omega} \Phi_{2^r}^{(\kappa)}(u_i(t)) \leq c_0^2(d_1^2 + 1).$$

Passing to the limit as $\kappa \to \infty$ and using (2.14) we obtain

$$d_2 \leq c_0^2(d_1^2 + 1).$$

Further, iterating with respect to r we get

$$d_r \leq c_0^r(d_{r-1}^2 + 1). \tag{2.19}$$

Denote $\delta_r = d_r + 1$. After suitable transformation we find from (2.19)

$$\delta_r \leq c^r \delta_{r-1}^2, \quad r = 2, 3 \ldots, c = c_0^2 \tag{2.20}$$

Applying successively (2.20) we obtain

$$\delta_r \leq c^r \delta_{r-1}^2 \leq c^{(r+2(r-1)+\ldots+2^{r-2}\cdot 2 + 2^{r-1})} \delta_1^{2^r} \leq c^{(r(\sum_{i=0}^{r-1} 2^i) - \sum_{i=1}^{r-1} i2^i)} \delta_1^{2^r}$$

$$= c^{(2^{r+1}-r-2)} \delta_1^{2^r},$$

whence

$$d_r^{2^{-r}} \leq \delta_r^{2^{-r}} \leq c^2 \delta_1. \tag{2.21}$$

Therefore,

$$(u_i - M)_+ \in L^\infty(0, T; L^s) \quad \text{for} \quad s \geq 1.$$

As a result, it follows from (2.21) that

$$\|(u_i - M)_+\|_{L^\infty(\Omega_T)} \leq c^2(\sup_{s \in [0,T]} \sum_{i=1}^{N} \|(u_i - M)_+\|_2^2 + 1)$$

for each $i = 1, \ldots, N$. Now, applying Lemma 3 we obtain (2.13). $\square$

Theorem 1 with Lemma 4 lead us to the following conclusion : if $(\underline{u}, \psi)$ is a solution of problem (P_u^k) with $k > M + c^2(c_\infty^2 + 1)$ then it is also a solution of problem (P). This proves the main theorem.

Theorem 2. *The problem (P) has a solution.* $\square$

Remark 2. If $f \in L^p(\Omega)$, $p > n/2$, $p \geq 1$ and in addition $\Psi \in L^\infty(\Omega_T)$ then from the elliptic regularity theory (see e.g. [4]) follows that $\psi \in L^\infty(\Omega_T)$.

3. THE EXISTENCE AND UNIQUENESS OF SOLUTIONS IN THE CASE OF ONE SPACE DIMENSION .

Now we shall be concerned with system (1.4) in the case of one space dimension, i.e., we account for the Shockley-Read-Hall and Auger terms as well as impact ionization component in the standard form (1.6).
Let $\Omega = [0, l]$. System (1.4) can be formulated in the following way :

$$
\begin{aligned}
&- \partial_x(a\partial_x\psi) = u_1 - u_2 + f \\
&\dot{u}_\imath - \partial_x(J_\imath) = -r(u_1, u_2)u_1 u_2 + G(\nabla\psi, u_1, u_2, \partial_x u_1, \partial_x u_2)
\end{aligned}
\tag{3.1}
$$

where $J_\imath = D_\imath(\partial_x\psi)\partial_x u_\imath + \nu_\imath(\partial_x\psi)u_\imath$, $\imath = 1, 2$, are subject to boundary and initial conditions :

$$
\begin{aligned}
u_\imath(0, t) &= U_\imath(0, t), \quad u_\imath(l, t) = U_\imath(l, t), \quad \imath = 1, 2; \\
\psi(0, t) &= \Psi(0, t), \quad \psi(l, t) = \Psi(l, t), \quad \text{for} \quad t \in (0, T), \\
u_\imath(x, 0) &= U_\imath(x, 0), \quad \imath = 1, 2, \quad \text{for} \quad x \in [0, l]
\end{aligned}
\tag{3.2}
$$

where $U_\imath \geq 0$ and Ψ are given functions defined on $\Omega_T = [0, l] \times [0, T]$.
To show the existence of solutions of (3.1). (3.2) we assume that the hypotheses (Ei), (Eii), (Ev) and (Evi) (with $n = 1$ and $N = 2$) still hold and (Eiii) and (Eiv) are replaced by :

(E'iii) The function $r : \mathbf{R}^2 \to \mathbf{R}_+$ such that $r \geq 0$ for $\underline{u} \geq \underline{0}$ satisfies uniformly the Lipschitz condition with a constant L_r and $|r(\underline{\eta})| \leq r_0 + L_r|\underline{\eta}|$ for $\underline{\eta} \in \mathbf{R}_+^2$.

(E'iv) The function G satisfies the conditions a) and c) of (Eiv) and b) in the more general form

$$
|G(\zeta, \underline{\eta}, \underline{\varrho})| \leq \alpha\left(1 + \sum_{\imath=1}^2 \eta_\imath + \sum_{\imath=1}^2 |\varrho_\imath|\right) \quad \text{for} \quad \underline{\eta}, \underline{\varrho} \in \mathbf{R}^2.
$$

A weak form of (3.1) is the same as in (P) (we shall refer to it as to problem (Q)) with an obvious modification $H_D = H_0^1(0\,l)$.
Theorem 1 is applicable here (see Remark 1), the L^∞-bound. however, has to be obtained in another way. Denote by (Q^k) the "truncated " problem corresponding to problem (Q) as in (2.1).

Lemma 6. *If $\underline{u}$ is a solution of problem (Q^k) then*

$$
\sup_{t \in [0,T]} \|\underline{u}\|_2^2 + \int_0^T \|\partial_x\underline{u}\|_2^2 \leq c_0
$$

where c_0 does not depend on k .

Proof. We proceed much in the same way as in the proof of Lemma 3. After the change of variables we choose, in contrast to (2.12) the test function $\underline{w} = (\underline{u} - \underline{U})|_{[0,t]}$ .We shall show how to estimate the term $r(u_1,u_2)u_1 u_2$. Using the inequality $0 \geq w_{i-} \geq -U_i$ we find

$$-\int_{\Omega_t} r([\underline{w} + \underline{U}]^{(k)})[w_1 + U_1]^{(k)}[w_2 + U_2]^{(k)}(w_{1+} + w_{1-}) \leq$$

$$\leq \int_{\Omega_t} r([\underline{w} + \underline{U}]^{(k)})[w_1 + U_1]^{(k)}[w_2 + U_2]^{(k)}|w_1-| \leq$$

$$\leq 2M^2 L_r \int_{\Omega_t}(1 + |w_1 + U_1| + |w_2 + U_2|)|w_2 + U_2| \leq c_1(\int_0^t \|\underline{w}\|_2^2 + 1)$$

with a positive constant $c_1 = c_1(M, L_r, \|U\|_{H^1(\Omega_T)}$ which is independent of k.

As far as G is concerned, the terms $\int_{\Omega_t}|\partial_x[(w_j + U_j)^{(k)}]||w_i|$, $i,j = 1,2$, enter. There admit the bounds

$$\int_{\Omega_t}|\partial_x w_j||w_i| + |\partial_x U_j||w_i| \leq \varepsilon\int_0^t \|\partial_x w_j\|_2^2 + \frac{1}{4\varepsilon}\int_0^t \|w_i\|_2^2 +$$

$$+ \frac{1}{2}\|\underline{U}\|_{H^1(\Omega_T)}^2 + \frac{1}{2}\int_0^t \|w_i\|_2^2$$

Choosing ε small enough to control the gradients terms we eventually obtain

$$\|\underline{w}(t)\|_2^2 + \int_0^t \|\partial_x \underline{w}\|_2^2 \leq c_2(\int_0^t \|\underline{w}\|_2^2 + 1),$$

where c_2 depends only on the generic constants. Hence, due to Gronwall's lemma we have

$$\sup_{t\in[0,T]} \|\underline{u}(t)\|_2^2 + \int_0^T \|\partial_x \underline{u}\|_2^2 \leq c_0,$$

with $c_0 = c_0(D^0, \gamma, \alpha, \|\underline{U}\|_{H^1(\Omega_T)}, l, T)$. $\square$

Now we are able to show the following existence theorem.

Theorem 3. *Problem (3.1) has a weak solution (u_1, u_2, ψ) satisfying the conditions $(P, i - v)$.*

Proof. Let us treat individually each parabolic equation in (3.1) . From the above lemma it follows that $G_R = -r(\underline{u}^{(k)})u_1^{(k)}u_2^{(k)} + G(\nabla\psi, \underline{u}^{(k)}, \partial_x[\underline{u}^{(k)}])$ is bounded in the space $L^2(\Omega_T)$ uniformly with respect to k. Therefore, we can apply Theorem 7.1, Chapter III of [5] on the linear parabolic L^∞-bounds. The arguments of this theorem require $G_R \in L^{q,r}(\Omega_T)$ to satisfy the conditions : $\frac{1}{r} + \frac{n}{2q} = 1 - \kappa$ with $q \in [1, \infty]$, $r \in [\frac{1}{1-\kappa}, \frac{2}{1-2\kappa}]$, $0 < \kappa < \frac{1}{2}$ for $n = 1$. It is easy to verify that in our case, i.e., for $n = 1$, $q = r = 2$, $\kappa = \frac{1}{4}$, these conditions are satisfied. $\square$

Next, we shall be concerned with the uniqueness of solutions of problem (3.1). To this end we assume somewhat stronger hypotheses i.e. :

(Ui)   In addition to (Ei) and (Eii) the functions D_i and ν_i $i = 1, 2$, satisfy (uniformly) the Lipschitz condition with constants L_D and L_ν respectively.

(Uiv) The generation term G satisfies the conditions a), b), c) of (E'iv) as well as

$$d)\quad |G(\zeta', \underline{\eta}, \underline{\varrho}) - G(\zeta'', \underline{\eta}, \underline{\varrho})| \le \alpha |\zeta' - \zeta''| \sum_{i=1}^{2} (|\eta_i| + |\varrho_i|), \quad \text{for } \underline{\eta}, \underline{\varrho} \in \mathbf{R}^2.$$

Since the coeficients α_1 and α_2 in the avalanche term (1.6) are uniformly Lipschitzian functions of $|\nabla \psi|$ it is clear that H in (1.6) satisfies (Uiv,d).

Theorem 4. *Suppose the hypotheses (Ui), (E'iii) and (Uiv) hold. Then system (3.1) has a unique weak solution.*

Proof. The potential ψ splits into two components $\psi = \psi_f + \psi_u$ where ψ_f is a weak solution of the inhomogeneous problem

$$-\partial_x(a\partial_x\psi_f(t)) = f, \quad \text{for a.e. } t \in [0, T], \tag{3.3}$$

along with Dirichlet boundary condition (3.2), while ψ_u is a weak solution of

$$-\partial_x(a\partial_x\psi_u(t)) = u_1(t) - u_2(t), \quad \text{for a.e. } t \in [0, T], \tag{3.4}$$

with homogeneous boundary conditions. Thus, only ψ_u gives an interactive contribution to the potential ψ whereas, ψ_f is known *a priori*.

Suppose that $(\psi', \underline{u}')$ and $(\psi'', \underline{u}'')$ are two different solutions of problem (3.1). Denote $\partial_x\psi'_u - \partial_x\psi''_u = \tilde{E}$ and $\tilde{u} = \underline{u}' - \underline{u}''$. Let $\|\underline{u}'\|$, $\|\underline{u}''\| \le \beta$.

Applying to (3.4) the L^p-elliptic theory and the Sobolev imbedding theorem we obtain

$$\|\tilde{E}(t)\|_\infty^2 \le c_l \|\underline{\tilde{u}}(t)\|_2^2 \tag{3.5}$$

for a.e. $t \in [0, T]$, with a positive constant c_l. To obtain a bound on $\underline{\tilde{u}}$ we use the test function $\underline{\varphi}(x, s) = 2e^{-rs}\underline{\tilde{u}}(x, s)|_{[0, t]}$. Note that

$$\int_0^t \langle \underline{\dot{\tilde{u}}}(s), 2e^{-rs}\underline{\tilde{u}}(s)\rangle ds = e^{-rt}\|\underline{\tilde{u}}(t)\|_2^2 - \|\underline{\tilde{u}}(0)\|_2^2 + r\int_0^t e^{-rs}\|\underline{\tilde{u}}(s)\|_2^2 ds.$$

Using (Ui), (E'iii) as well as (Uiv;c,d) and applying Young's inequality to the components with $\partial_x \tilde{\underline{u}}$ on the R.H.S. we obtain

$$e^{-rt}\|\tilde{\underline{u}}(t)\|_2^2 + D_0 \int_0^t e^{-rs}\|\partial_x \tilde{\underline{u}}(s)\|_2^2 ds + r \int_0^t e^{-rs}\|\tilde{\underline{u}}(s)\|_2^2 ds \leq$$

$$\leq \varepsilon \int_0^t e^{-rs}\|\partial_x \tilde{\underline{u}}(s)\|_2^2 ds + c_\varepsilon' \int_0^t e^{-rs}\|\tilde{\underline{u}}(s)\|_2^2 ds + \tag{3.6}$$

$$+ c_\varepsilon'' \int_0^t e^{-rs} \int_\Omega \{|\underline{u}''(s)|^2 + |\partial_x \underline{u}''(s)|^2\} \|\tilde{E}(s)\|_\infty^2 ds,$$

where $c_\varepsilon' = c'(\gamma, r_0, L_r, \alpha, \beta; \varepsilon)$ and $c_\varepsilon'' = c''(L_D, L_\nu, \alpha, \beta; \varepsilon)$ are positive constants. Denote $\vartheta(s) = \int_\Omega (|\underline{u}''(s)|^2 + |\partial_x \underline{u}''(s)|^2)$. First, let us choose in (3.6) $\varepsilon = D_0$ then $r = c_\varepsilon'$. Hence,

$$e^{-rt}\|\tilde{\underline{u}}(t)\|_2^2 \leq \int_0^t e^{-rs}\vartheta(s) \|\tilde{E}(s)\|_\infty^2 ds. \tag{3.7}$$

Multiplying each side of (3.7) by $e^{-\mu t}$ we find

$$e^{-(r+\mu)t}\|\tilde{\underline{u}}(t)\|_2^2 \leq c'' \int_0^t e^{-\mu(t-s)}\vartheta(s)e^{-(r+\mu)s}\|\tilde{E}(s)\|_\infty^2 ds \leq$$

$$\leq c'' \int_0^t e^{-\mu(t-s)}\vartheta(s) ds \sup_{s\in[0.T]} \{e^{-(r+\mu)s}\|\tilde{E}(s)\|_\infty^2\}. \tag{3.8}$$

Now we use a fact which follows from uniform continuity of the Lebesgue integral: For each $\sigma > 0$ and arbitrary $\vartheta \in L^1(0,T)$, $\vartheta > 0$ there exists μ_0 such that for $\mu > \mu_0$

$$\int_0^t e^{-\mu(t-s)}\vartheta(s) ds < \sigma. \tag{3.9}$$

uniformly with respect to $t \leq T$. Choosing μ large enough that $\sigma = \frac{1}{2c_1 c''}$ from (3.5) and (3.8) we conclude that

$$\sup_{t\in[0,T]} \{e^{-(r+\mu)t}\|\tilde{E}(t)\|_\infty^2\} \leq \frac{1}{2} \sup_{s\in[0\ T]} \{e^{-(r+\mu)s}\|\tilde{E}(s)\|_\infty^2\}.$$

Thus, $\psi' = \psi''$ and $\underline{u}' = \underline{u}''$. □

Remark 3. The existence and uniqueness results remain true if we assume that the functions r and G in (3.1) are different in each equation.

Remark 4. In [8] Seidman proved the uniqueness of solutions for a similar system of equations (in many space dimesions) in which the diffusion coefficients are independent of the electric field and additional regularity assumptions on the boundary geometry are imposed.

4. FINAL REMARKS.

Remark 5. It is possible to allow for more general assumptions on the diffusion co-effcients and convective velocity, i.e.: $D_i = D_i(x, t, \underline{u}, \nabla\psi)$ and $\underline{\nu}_i = \underline{\nu}_i(x, t, \underline{u}, \nabla\psi)$ and in addition to (Ei) and (Eii) D_i and $\underline{\nu}_i$ satisfy the Carathéodory conditions. The existence theorem (Thm.2) remains valid in this case. However, the construction of the fixed point mapping in the Theorem 1 requires a slight modification namely (assuming the notation from Thm.1) for fixed $\underline{z}$ we solve the parabolic system (see Lemma 2) with frozen arguments of $D_i(\cdot, \underline{z}(\cdot), \psi_z(\cdot))$ and $\underline{\nu}_i(\cdot, \underline{z}(\cdot), \psi_z(\cdot))$.

Remark 6. The effect of velocity saturation, appearing e.g. in *GaAs*, can be modelled by choosing appropriate asymptotic properties of the mobility $\mu = \mu(|\nabla\psi|)$ such that $|\mu(|\nabla\psi|)\nabla\psi| < \gamma$. We can assume that the diffusion coefficient is linked with the mobility, via the Einstein relation, for electric fields less then an arbitrary but fixed value E_0 i.e.:

$$D_i(|\nabla\psi|) = c_E\mu(|\nabla\psi|) \quad \text{for} \quad |\nabla\psi| < E_0\,,$$

where c_E is a constant dependent on temperature. For higher values of the electric field diffusion and convection are assumed to be independent.

Remark 7. Suppose that the functions $R_i(\underline{u})u_i$ in (1.1) are Lipschitzian, then we can show the existence of solutions of (1.1) even if we admit in the place of (Eiv,b) that G_i satisfies more general hypothesis:

$$G_i(\underline{\zeta}, \underline{\eta}, \underline{\varrho}_1, \dots, \underline{\varrho}_N) \leq \alpha(1 + \sum_{j=1}^{N} \eta_j + \sum_{j=1}^{N} |\underline{\varrho}_j|) \quad \text{for} \underline{\zeta}, \underline{\varrho}_i \in \mathbf{R}^n, \underline{\eta} \in \mathbf{R}_+^N\,.$$

The above assumptions permit us to account for the Shockley-Read-Hall generation/recombination term and avalanche term (1.6) in the same model. The solvability of such problem (for arbitrary dimension) one can obtain by means of the method used in the proof of Theorem 1.

However, assuming restriction on R_i we preclude the Auger term. On the other hand, this term can be allowed for if we take the impact ionization term in one of the following modified versions :

$$a) \qquad H^\kappa = a_1(\nabla\psi)h_1(|J_1|) + a_2(\nabla\psi)h_2(|J_2|),$$

where $h_1, h_2 : \mathbf{R}_+ \to \mathbf{R}_+$ are Lipschitzian functions bounded by an arbitrary positive number κ (e.g.:$h_1(\cdot) . h_2(\cdot) = (\cdot)^{(\kappa)}$, where $(\cdot)^{(\kappa)}$ is the cut-off function) or

$$b) \qquad H^{conv} = a_1(\nabla\psi)|J_1^{conv}| + a_2(\nabla\psi)|J_2^{conv}|,$$

with α_i, $i = 1, 2$, as in (1.6). In the first case H_κ is a certain approximation of the standard avalanche term (1.6) the more accurate the higher is κ. The essential point here is that G in (1.1) is bounded with respect to gradients of concentration thus, the assumption (Eiv,b) is satisfied. On the contrary, in the second case impact ionization term is assumed to be dependent only on convective component of current.

Acknowledgments : Several fruitful discussions with Professor T.I.Seidman during the IMA Summer Program on Semiconductors (held on July 1991 in Minneapolis) on the topics related to the paper are to be acknowledged. Also, I want to thank Dr. G.Lukaszewicz for correcting the first version of this paper.

REFERENCES

1. N.D. Alikakos, *An application of the invariance principle to reaction-diffusion equations*, J.Differantial Equations **33** (1979), 201–225.
2. H.Gajewski, K.Gröger, *On the basic equations for carrier transport in semiconductors*, J. Math. Anal. Appl. **113** (1986), 12–35.
3. H.Gajewski, K.Gröger, and K.Zacharias, *Nichtlineare Operatorgleichungen und Operatordifferentialgleichungen*, Akademie-Verlag, Berlin, 1974.
4. D.Gilbarg, N.S.Trudinger, *Elliptic Partial Differential Equations of Second Order*, Springer-Verlag, Berlin, 1977.
5. O.A. Ladyzhenskaya, V.A. Solonnikov, N.N. Ural'ceva, *Linear and Quasilinear Equations of Parabolic Type*, Amer.Math. Soc., Providence, 1968.
6. J.Moser, *A new proof of De Giorgi's theorem concerning the regularity problem for elliptic differential equations*, Comm. Pure Appl. Math. **13** (1960), 457–468.
7. T.I.Seidman, *The transient semiconductor problem with generation terms*, Computational Aspects of VLSI Design and Semiconductor Device Simulation, Amer. Math. Soc., Providence, 1988.
8. T.I.Seidman, *The transient semiconductor problem with generation terms, II*, (preprint).
9. S. Selberherr, *Analysis and Simulation of Semiconductor Devices*, Springer-Verlag, Wien, 1984.

† WARSAW UNIVERSITY, INSTITUTE OF APPLIED MATHEMATICS AND MECHANICS, UL. BANACHA 2, 02-097 WARSAW, POLAND

E-mail: darekw@mimuw.edu.pl

ISNM

A series with a long-standing reputation

Since its foundation in 1963 more than 100 volumes have been published by Birkhäuser Verlag in the **International Series of Numerical Mathematics.**

John Todd's *Introduction to the Constructive Theory of Functions,* published as Volume 1, was a remarkable start. Proceedings volumes and further monographs such as Fenyö/Frey, *Moderne mathematische Methoden in der Technik*, Ghizzetti/Ossicini, *Quadrature Formulae*, Todd, *Basic Numerical Mathematics* (two volumes) and Heinrich, *Finite Difference Methods on Irregular Networks* followed, always presenting the state of the art in exposition and research.

Originally the Editorial Board consisted of Ch. Blanc, A. Ghizzetti, A. Ostrowski, J. Todd, H. Unger, A. van Wijngaarden. Despite a number of changes, it has shown long years of continuity; Prof. Ostrowski and Prof. Henrici, for instance, were members of the Board all their lives.

At present the series is being edited by
Karl-Heinz Hoffmann, München, **Hans D. Mittelmann**, Tempe, **John Todd**, Pasadena.

As in the past, we do not intend to restrict the series a piori to certain subjects. The series is open to all aspects of numerical mathematics with emphasis on mathematical content. At the same time, we wish to include practical applications in science and engineering, with emphasis on mathematical content.

Some of the topics of particular interest to the series are:
Free boundary value problems for differential equations, phase transitions, problems of optimal control, other nonlinear phenomena in analysis; nonlinear partial differential equations, efficient solution methods, bifurcation problems; approximation theory.

If possible, the topic of each volume should be discussed from three different angles, namely that of Mathematical Modelling, Mathematical Analysis, Numerical Case Studies.

The editors particularly welcome research monographs; furthermore, the series is to contain advanced graduate texts, dealing with areas of current research interest, as well as selected and carefully refereed proceedings of major conferences or workshops sponsored by various research centers. Historical material in these areas would also be considered.

We encourage preparation of manuscripts in LaTeX or AMSTeX for delivery in camera-ready copy which enables a rapid publication, or in electronic form for interfacing with laser printers of typesetters.

Titles previously published in the series

INTERNATIONAL SERIES OF NUMERICAL MATHEMATICS
BIRKHÄUSER VERLAG

ISNM 100   **W. Desch, F. Kappel, K. Kunisch (Eds.):** Estimation and Control of Distributed Parameter Systems, 1991 (3-7643-2676-X)

ISNM 101   **G. Del Piero, F. Maceri (Eds.):** Unilateral Problems in Structural Analysis IV, 1991 (3-7643-2487-2)

ISNM 102   **U. Hornung, P. Kotelenez, G. Papanicolaou (Eds.):** Random Partial Differential Equations, 1991 (3-7643-2688-3)

ISNM 103   **W. Walter (Ed.):** General Inequalities 6, 1992 (3-7643-2737-5)

ISNM 104   **E. Allgower, K. Böhmer, M. Golubitsky (Eds.):** Bifurcation and Symmetry, 1992 (3-7643-2739-1)

ISNM 105   **D. Braess, L.L. Schumaker (Eds.):** Numerical Methods in Approximation Theory, Vol. 9, 1992 (3-7643-2746-4)

ISNM 106   **S.N. Antontsev, K.-H. Hoffmann, A.M. Khludnev (Eds.):** Free Boundary Problems in Continuum Mechanics, 1992 (3-7643-2784-7)

ISNM 107   **V. Barbu, F.J. Bonnans, D. Tiba (Eds.):** Optimization, Optimal Control and Partial Differential Equations, 1992 (3-7643-2788-X)

ISNM 108   **H. Antes, P.D. Panagiotopoulos:** The Boundary Integral Approach to Static and Dynamic Contact Problems. Equality and Inequality Methods, 1992 (3-7643-2592-5)

ISNM 109   **A.G. Kuz'min:** Non-Classical Equations of Mixed Type and their Applications in Gas Dynamics, 1992 (3-7643-2573-9)

ISNM 110   **H.R.E.M. Hörnlein, K. Schittkowski (Eds.):** Software Systems for Structural Optimization, 1992 (3-7643-2836-3)

ISNM 111   **R. Burlisch, A. Miele, J. Stoer, K.H. Well:** Optimal Control, 1993 (3-7643-2887-8)

ISNM 112   **H. Braess, G. Hämmerlin (Eds.):** Numerical Integration IV. Proceedings of the Conference at the Mathematical Research Institute at Oberwolfach, November 8-14, 1992, 1993 (3-7643-2922-X)

ISNM 113   **L. Quartapelle:** Numerical Solution of the Incompressible Navier-Stokes Equations, 1993 (3-7643-2935-1)

ISNM 114   **J. Douglas, U. Hornung (Eds.):** Flow in Porous Media. Proceedings of the Oberwolfach Conference, June 21-27, 1992 (ISBN 3-7643-2949-1)

ISNM 115   **R. Bulirsch, D. Kraft (Eds.):** Computational Optimal Control, 1994 (ISBN 3-7643-5015-6)

ISNM 116   **P.W. Hemker, P. Wesseling (Eds.):** Multigrid Methods IV. Proceedings of the Fourth European Multigrid Conference, Amsterdam, July 6-9, 1993 (ISBN 3-7643-5030-X)

ISNM 117   **R.E. Bank, R. Bulirsch, H. Gajewski, K. Merten (Eds.):** Mathematical Modelling and Simulation of Electrical Circuits and Semiconductor Devices, 1994 (ISBN 3-7643-5053-9)

Progress in Numerical Simulation for Microelectronics

R. Kircher / W. Bergner
SIEMENS AG, München,
Germany

Three-Dimensional Simulation of Semiconductor Devices

1991. 128 pages. Hardcover
ISBN 3-7643-2644-1

Please order through your bookseller or write to:
Birkhäuser Verlag AG
P.O. Box 133
CH-4010 Basel / Switzerland
FAX: ++41 / 61 / 271 76 66

For orders originating in the USA or Canada:
Birkhäuser
333 Meadowlands Parkway
Secaucus, NJ 07094-2491 / USA

Birkhäuser

Birkhäuser Verlag AG
Basel · Boston · Berlin

Prices are subject to change without notice. 5/94

In the last two decades the simulation of semiconducting device structures has become an increasingly important tool for developing and improving devices by predicting their electrical behaviour. Today the two-dimensional device simulation is regarded as a standard engineering tool, containing sophisticated models describing transport phenomena. The challenge arising from the increasing integration density is to consider the influence of a more and more three-dimensional geometry of the devices.

Insight is given here into modeling complex device structures in three dimensions. Starting from the basic differential equations and the physical models describing the transport in semiconductors, the work shows how to obtain the discretized equations, and discusses important numerical methods for solving these equations. How three-dimensional simulations can be utilized to analyze complex device structures and to optimize their electrical characteristics is illustrated by two specific examples.

Progress in Numerical Simulation for Microelectronics

C.R. Kleijn, TU Delft, The Netherlands / **C. Werner,** Siemens ZFE, München, Germany

Modeling of Chemical Vapor Deposition of Tungsten Films

1993. 138 pages. Hardcover
ISBN 3-7643-2858-4

Please order through your bookseller or write to:
Birkhäuser Verlag AG
P.O. Box 133
CH-4010 Basel / Switzerland
FAX: ++41 / 61 / 271 76 66

For orders originating in the USA or Canada:
Birkhäuser
333 Meadowlands Parkway
Secaucus, NJ 07094-2491 / USA

Birkhäuser

Birkhäuser Verlag AG
Basel · Boston · Berlin

Prices are subject to change without notice. 5/94

Numerical modeling of reactors for chemical vapor deposition has in recent years become a field of great interest, because it offers the potential to support development and optimization of manufacturing equipment and hence reduce the cost and improve the quality of the reactors.

This book is the result of two parallel lines of research dealing with the same subject – Modeling of Tungsten CVD Processes –, which were performed independently under very different boundary conditions. On the one side, Chris Kleijn, working in the adacemic research environment of Technical University Delft, was able to go deep enough into the subject to lay a solid foundation and prove the validity of all the assumptions made in his work. On the other side, Christoph Werner, working in the context of an industrial research lab at Siemens corporate research and development, was able to closely interact with manufacturing and development engineers in a modern submicron semiconductor processing line.